INTEGRATED PEST MANAGEMENT FOR
CITRUS

THIRD EDITION

INTEGRATED PEST MANAGEMENT FOR
CITRUS
THIRD EDITION

UNIVERSITY OF CALIFORNIA

STATEWIDE INTEGRATED PEST MANAGEMENT PROGRAM

AGRICULTURE AND NATURAL RESOURCES

PUBLICATION 3303

2012

PRECAUTIONS FOR USING PESTICIDES

Pesticides are poisonous and must be used with caution. READ THE LABEL BEFORE OPENING A PESTICIDE CONTAINER. Follow all label precautions and directions, including requirements for protective equipment. Use a pesticide only on the plants or site specified on the label or in published University of California recommendations. Apply pesticides at the rates specified on the label or at lower rates if suggested in University of California recommendations. In California, all agricultural uses of pesticides must be reported, including use in many nonfarm situations, such as cemeteries, golf courses, parks, roadsides, and commercial plant production including nurseries. Contact your county agricultural commissioner for further details. Laws, regulations, and information concerning pesticides change frequently, so be sure the publication you are using is up-to-date.

Legal Responsibility. The user is legally responsible for any damage due to misuse of pesticides. Responsibility extends to effects caused by drift, runoff, or residues.

Transportation. Do not ship or carry pesticides together with food or feed in a way that allows contamination of the edible items. Never transport pesticides in a closed passenger vehicle or in a closed cab.

Storage. Keep pesticides in original containers until used. Store them in a locked cabinet, building, or fenced area where they are not accessible to children, unauthorized persons, pets, or livestock. DO NOT store pesticides with foods, feed, fertilizers, or other materials that may become contaminated by the pesticides.

Container Disposal. Consult the pesticide label, the County Department of Agriculture, or the local waste disposal authorities for instructions on disposing of pesticide containers. Dispose of empty containers carefully. Never reuse them. Make sure empty containers are not accessible to children or animals. Never dispose of containers where they may contaminate water supplies or natural waterways. Offer empty containers for recycling if available. Home use pesticide containers can be thrown in the trash only if they are completely empty.

Protection of Nonpest Animals and Plants. Many pesticides are toxic to useful or desirable animals, including honey bees, natural enemies, fish, domestic animals, and birds. Certain rodenticides may pose a special hazard to animals that eat poisoned rodents. Plants may also be damaged by misapplied pesticides. Take precautions to protect nonpest species from direct exposure to pesticides and from contamination due to drift, runoff, or residues.

Permit Requirements. Certain pesticides require a permit from the county agricultural commissioner before possession or use. When such materials are mentioned in this publication, they are marked with an asterisk (*).

Plant Injury. Certain chemicals may cause injury to plants (phytotoxicity) under certain conditions. Always consult the label for limitations. Before applying any pesticide, take into account the stage of plant development, the soil type and condition, the temperature, moisture, and wind. Injury may also result from the use of incompatible materials.

Personal Safety. Follow label directions carefully. Avoid splashing, spilling, leaks, spray drift, and contamination of clothing. NEVER eat, smoke, drink, or chew while using pesticides. Provide for emergency medical care IN ADVANCE as required by regulation.

Worker Protection Standards. Federal Worker Protection Standards require pesticide safety training for all employees working in agricultural fields, greenhouses, and nurseries that have been treated with pesticides, including pesticide training for employees who do not work directly with pesticides.

ORDERING

For information about ordering this publication and/or a free catalog, contact

University of California
Agriculture and Natural Resources
Communication Services
1301 S. 46th Street
Building 478-MC 3580
Richmond, CA 94804-4600
Telephone 1-800-994-8849
(510) 642-2431 FAX (510) 643-5470
E-mail: anrcatalog@ucdavis.edu
Visit the ANR Communication Services Web site at: http://anrcatalog.ucdavis.edu

Publication 3303

Other books in this series include:
Integrated Pest Management for Alfalfa Hay, Publication 3312
Integrated Pest Management for Almonds, Second Edition, Publication 3308
Integrated Pest Management for Apples and Pears, Second Edition, Publication 3340
Integrated Pest Management for Avocados, Publication 3503
Integrated Pest Management for Cole Crops and Lettuce, Publication 3307
Integrated Pest Management for Cotton, Second Edition, Publication 3305
Integrated Pest Management for Floriculture and Nurseries, Publication 3402
Integrated Pest Management for Potatoes, Second Edition, Publication 3316
Integrated Pest Management for Rice, Third Edition, Publication 3280
Integrated Pest Management for Small Grains, Publication 3333
Integrated Pest Management for Stone Fruits, Publication 3389
Integrated Pest Management for Strawberries, Second Edition, Publication 3351
Integrated Pest Management for Tomatoes, Fourth Edition, Publication 3274
Integrated Pest Management for Walnuts, Third Edition, Publication 3270
Natural Enemies Handbook: The Illustrated Guide to Biological Pest Control, Publication 3386
Pests of Landscape Trees and Shrubs, Second Edition, Publication 3359
Pests of the Garden and Small Farm, Second Edition, Publication 3332
ISBN-13: 978-1-60107-696-0
Library of Congress Control Number: 2011943829
Third edition ©2012 by the Regents of the University of California
Division of Agriculture and Natural Resources
All rights reserved.

No part of this publication may be reproduced, stored in a retrieval system, or transmitted, in any form or by any means, electronic, mechanical, photocopying, recording, or otherwise, without the written permission of the publisher and the authors.

 Printed in Canada on recycled paper.

 The University of California Division of Agriculture & Natural Resources (ANR) prohibits discrimination against or harassment of any person participating in any of ANR's programs or activities on the basis of race, color, national origin, religion, sex, gender identity, pregnancy (which includes pregnancy, childbirth, and medical conditions related to pregnancy or childbirth), physical or mental disability, medical condition (cancer-related or genetic characteristics), genetic information (including family medical history), ancestry, marital status, age, sexual orientation, citizenship, or service in the uniformed services (as defined by the Uniformed Services Employment and Reemployment Rights Act of 1994: service in the uniformed services includes membership, application for membership, performance of service, application for service, or obligation for service in the uniformed services) or any person in any of its programs or activities. University policy also prohibits retaliation against any employee or person participating in any of ANR's programs or activities for bringing a complaint of discrimination or harassment pursuant to this policy. This policy is intended to be consistent with the provisions of applicable State and Federal laws. Inquiries regarding the University's equal employment opportunity policies may be directed to Linda Marie Manton, Affirmative Action Contact, University of California, Davis, Agriculture and Natural Resources, One Shields Avenue, Davis, CA 95616, (530) 752-0495. **For information about ordering this publication, telephone 1-800-994-8849.** To simplify information, trade names of products have been used. No endorsement of named or illustrated products is intended, nor is criticism implied of similar products that are not mentioned or illustrated.

4m-pr-2/12-SB/CR

Contributors and Acknowledgments

Prepared by the University of California Statewide IPM Program at Davis

Steve H. Dreistadt, Writer

Jack Kelly Clark and David Rosen, Principal Photographers

Tunyalee Martin, Content Supervisor

Kassim Al-Khatib, Joyce Strand, and Peter B. Goodell, Directors

James J. Stapleton, ANR Associate Editor

Technical Coordinators for the Third Edition

Elizabeth E. Grafton-Cardwell, Department of Entomology, University of California, Riverside

Neil V. O'Connell, University of California Cooperative Extension, Tulare

Phil A. Phillips, UC IPM Program and University of California Cooperative Extension, Ventura

Joseph G. Morse, Department of Entomology, University of California, Riverside

Ben Faber, University of California Cooperative Extension, Ventura

James E. Adaskaveg, Department of Plant Pathology, University of California, Riverside

Contributors to the Third Edition

Entomology: Joe Barcinas, Joseph H. Connell, Ben Faber, Peter B. Goodell, Elizabeth E. Grafton-Cardwell, David R. Haviland, David H. Headrick, Mark S. Hoddle, Craig E. Kallsen, John H. Klotz, Robert F. Luck, Joseph G. Morse, Neil V. O'Connell, Phil A. Phillips, Phillip S. Ward

Nematology: J. Ole Becker, Antoon T. Ploeg

Plant Pathology and Horticulture: James E. Adaskaveg, Akif Eskalen, Ben Faber, Elizabeth E. Grafton-Cardwell, John A. Menge, Neil V. O'Connell, Howard R. Ohr, Mikeal Roose, Georgios Vidalakis

Vertebrate Biology: Rex O. Baker, Roger A. Baldwin, Rex E. Marsh, Neil V. O'Connell, Terry P. Salmon

Weed Science: Neil V. O'Connell, Anil Shrestha, Scott Steinmaus

Contributors to the First and Second Editions

Entomology: E. Laurence Atkins, Jr., J. Blair Bailey, O. L. Brawner, Robert D. Brown, Paul DeBach, T. W. Fisher, Donald L. Flaherty, Peter B. Goodell, J. Daniel Hare, Charles E. Kennett, Robert F. Luck, James A. McMurtry, Daniel S. Moreno, Joseph G. Morse, Neil V. O'Connell, Phil A. Phillips, Louis A. Riehl, Mike Rose, Lynell Tanigoshi, Gregory P. Walker

Horticulture: Tom W. Embleton, David Goldhamer, Carol J. Lovatt, C. Dean McCarty, John E. Pehrson, Robert G. Platt

Nematology: Sahag Garabedian, Seymour van Gundy, Reinhold Mankau, John D. Radewald

Pesticide Application: Louis A. Riehl, John E. Pehrson, Glenn Carman

Plant Pathology: David J. Gumpf, John A. Menge, Howard D. Ohr, John E. Pehrson

Vertebrates: Rex E. Marsh, William Clark

Weed Science: Bill B. Fischer, Lowell S. Jordan

Special Thanks

The following have generously provided information, offered suggestions, reviewed draft manuscripts, identified pests, or helped obtain photographs: A. Agnelo, M. L. Arpaia, J. B. Bailey, T. Batkin, G. S. Bender, W. J. Bentley, M. L. Bianchi, T. Boswell, K. Chapman, D. M. Clarke, S. Cohen, C. C. Collahan, M. J. Costello, K. Daane, J. DeBenedictis, J. M. DiTomaso, F. Dlott, J. A. Downer, W. Dreyer, R. E. Duncan, R. Dunn, C. L. Elmore, A. Engilis, L. Ferguson, M. L. Flint, H. Forster, L. Forster, M. Freeman, M. Garnier, J. F. Germain, D. Ken Giles, R. J. Gill, J. Gorden, H. J. Griffiths, D. J. Gumpf, K. J. Hembree, J. M. Heraty, R. L. Hix, T. Kahn, A. Kapranas, D. Kellum, S. King, R. R. Krueger, C. J. Lovatt, R. Mankau, M. Martino, P. Mauk, J. McClain, C. W. McCoy, R. S. Melnicoe, G. Montez, V. Newman, F. J. Niederholzer, C. A. O'Donnell, B. L. P. Ohlendorf, K. Olsen, M. J. O'Neil, B. Pankey, V. S. Polito, C. Reynolds, F. Rinder, T. Roberts, M. E. Rogers, N. Sakovich, R. Savage, L. J. Schwankl, K. Severns, T. Shea, J. J. Stapleton, J. Stewart, L. L. Strand, T. V. Suslow, J. Sweet, B. Taylor, E. F. Thatcher, D. E. Ullman, A. A. Urena, S. Van Gundy, R. Walther, P. Washburn, J. West, B. B. Westerdahl, S. L. Wu

Photographs were obtained with financial support from the Citrus Research Board and through the cooperation of California citrus growers.

Previous Editions

Writers: Brunhilde Kobbe (first edition), Steve H. Dreistadt (second edition)

Technical Editor: Mary Louise Flint

Editing: Margaret Klein

Drawings: Pamela Fabry, Marvin Ehrlich, David Kidd

Production

Design: (this edition): Celeste Rusconi, ANR Communication Services

Design: (IPM manual series): Seventeenth Street Studios

Digital Photo Processing: Evett Kilmartin, Celeste Rusconi

Drawings: Celeste Rusconi, Robin Walton, Valerie Winemiller

Contents

Integrated Pest Management for Citrus 1
The Citrus Tree: Development and Growth Requirements ...5
    The Seasonal Cycle of Citrus 5
    Growth Requirements 9
Managing Pests in Citrus 13
    Pest Prevention 13
    Identification and Diagnosis 15
    Monitoring 16
        Monitoring Pests 16
        Monitoring Weather 17
        Accumulating Degree-Days 18
    Control Action Guidelines 19
    Year-Round IPM 20
    Management Methods 20
        Scion and Rootstock Cultivar Selection 22
        Sanitation 22
        Soil and Water Management 24
            Irrigation Methods 25
            Irrigation Efficiency and Scheduling 25
        Fertilizing 26
        Frost Protection 27
        Harvest 27
        Pruning 27
        Cover Crops 28
        Biological Control 28
            Augmentation 29
            Classical Biological Control 29
            Conservation 29
        Pesticides 30
            Oils 31
    Using Pesticides Effectively 31
        Application Timing 32
        Coverage 32
        Calibration 32
        Equipment 33
    Problems with Pesticide Use 33
        Pesticide Resistance 33
        Pest Resurgence 34
        Secondary Pest Outbreak 34
        Phytotoxicity 35
        Hazards to Honey Bees 35
        Hazards to People 35
        Hazards to Wildlife 36
        Hazards to Water Quality 36
        Air Pollution 36
Diseases 39
    Monitoring and Diagnosis 40
    Prevention and Management 41

ROOT ROTS 42
    Management 42
    Phytophthora Root Rot 43
    Dry Root Rot 44
    Armillaria Root Rot 46
    Rosellinia Root Rot 48
TRUNK DISEASES 48
    Management 49
    Phytophthora Gummosis 50
    Dothiorella Gummosis (Dothiorella Blight) 51
    Hendersonula Tree and Branch Wilt (Sooty Canker) 52
    Hyphoderma Gummosis 53
    Wood Decays 53
    Psorosis 54
    Exocortis 55
TRUNK DISORDERS 55
    Shell Bark and Dry Bark 55
    Bud Union Disorders 56
    Sunburn 56
    Frost and Freeze 57
FRUIT DISEASES 57
    Management 57
    Brown Rot 57
    Alternaria Rot 58
    Anthracnose 59
    Septoria Spot 59
    Bacterial Blast (Citrus Blast, or Black Pit) 60
    Blue and Green Mold (Clear Rot) 61
    Botrytis Rot (Gray Mold) 61
    Sooty Molds 61
FRUIT DISORDERS 62
    Chimeras 62
    Wind Injury 62
    Sunburn 63
    Hail 63
    Frost and Freeze 63
    Rind Stipple of Grapefruit 64
    Rind Disorder (Mandarin Rind Disorder) 64
    Peteca of Lemon 64
    Puff and Crease 65
    Oleocellosis (Oil Spotting) 65
    Split Fruit 65
    Spray Injury (Phytotoxicity) 66
LEAF AND TWIG DISEASES 66
    Bacterial Blast (Citrus Blast, or Black Pit) 66
    Botrytis Rot 67
    Anthracnose 68
    Vein Enation (Woody Gall) 68

LEAF AND TWIG DISORDERS 69
- Aeration Deficit or Root Asphyxiation 69
- Water Deficit or Stress 69
- Twig Dieback 70
- Frost and Freeze 70
- Wind Injury 71
- Hail 71
- Mesophyll Collapse 71
- Fire 71
- Sunburn 72
- Air Pollution 72
- Spray Injury (Phytotoxicity) 72
- Chimeras 74
- Mineral Deficiencies and Toxicities 74

EXOTIC DISEASES 77
- Citrus Bacterial Canker (Citrus Canker) 77
- Huanglongbing (HLB) or Citrus Greening 78
- Citrus Variegated Chlorosis 79
- Citrus Leprosis 80

DISEASES AFFECTING GROWTH HABIT AND YIELD 80
- Management 80
- Stubborn Disease 81
- Tristeza 82
- Cachexia (Xyloporosis) 83
- Lemon Sieve Tube Necrosis 84

Insects, Mites, and Snails 85
- Invertebrate Damage 86
- Identifying Insects and Mites 86
- Monitoring Insects and Mites 86
  - Monitoring Methods 87
    - Visual Search 87
    - Traps 87
    - Shake and Beat Sampling 88

GENERAL PREDATORS 88
- Green Lacewings 88
- Brown Lacewings 90
- Dustywings 91
- Minute Pirate Bugs 92
- Assassin Bugs 93
- Lady Beetles 93
- Syrphid Flies 94
- Predatory Mites 94
- Spiders 95

SCALES 96
- Scale Insect Management 96
- Armored Scales 97
  - California Red Scale and Yellow Scale 97
  - Purple Scale 106
- Soft Scales 107
  - Citricola Scale 107
  - Brown Soft Scale 110
  - Black Scale 112
- Cottony Cushion Scale 113

THRIPS 116
- Identifying Thrips Species 118
- Management 118
- Citrus Thrips 119
- Neohydatothrips 125
- Greenhouse Thrips 126
- Bean Thrips 128

CATERPILLARS (ORANGEWORMS) 130
- Management 131
- Cutworms 136
- Orange Tortrix 138
- Fruittree Leafroller 139
- Omnivorous Leafroller 141
- Amorbia 143
- Light Brown Apple Moth 144
- Western Tussock Moth 145
- Scavenger Caterpillars 146
- California Orangedog 147
- Giant Orangedog 148
- Citrus Looper 148
- Cabbage Looper 149
- Beet Armyworm 149

MINERS 150
- Citrus Peelminer 150
- Citrus Leafminer 152

HEMIPTERA (HOMOPTERA) 154
- Mealybugs 154
- Whiteflies 159
- Aphids 161
- Asian Citrus Psyllid 166
- Glassy-Winged Sharpshooter 167
- Potato Leafhopper 169

CHEWING INSECTS 170
- Forktailed Katydid 170
- Grasshoppers 171
- European Earwig 172
- Diaprepes Root Weevil 173
- Fuller Rose Beetle 174

ANTS 175
- Argentine Ant 180
- Native Gray Ant 180
- Southern Fire Ant 181
- Red Imported Fire Ant 181

Contents

- MITES 181
  - Spider Mites 183
    - Spider Mite Management 183
    - Citrus Red Mite 183
    - Twospotted Spider Mite 186
    - Texas Citrus Mite 187
    - Yuma Spider Mite 188
    - Sixspotted Mite 189
    - Lewis Spider Mite 189
    - Hydrangea Mite 190
  - Other Types of Mites 190
    - Broad Mite 190
    - Citrus Bud Mite 192
    - Citrus Rust Mite (Silver Mite) 193
    - Flat Mite 193
  - Mite Predators 194
    - *Euseius* Predatory Mites 194
    - Spider Mite Destroyer 195
    - Sixspotted Thrips 196
    - Predaceous Midges 196
    - Predatory Rove Beetle 197
  - OTHER INVERTEBRATE PESTS 198
    - Exotic Fruit Flies 198
    - Brown Garden Snail 199

Nematodes 201
- Description and Biology 202
- Symptoms and Damage 203
- Monitoring and Diagnosis 204
  - Sampling Nematode Larvae in Soil 205
  - Sampling Female Nematodes on Roots 205
  - Interpreting Sample Results 205
- Prevention and Management 205
  - Rootstock Selection 206
  - Sanitation and Cultural Practices 206
  - Natural Enemies 206
  - Chemical Control 207

Vertebrates 209
- Managing Pest Vertebrates 210
  - Observation and Identification 210
  - Monitoring 210
  - Control Actions 210
  - Habitat Modification 212
  - Biological Control 212
  - Exclusion 213
  - Frightening Devices 213
  - Shooting 213
  - Trapping 213
  - Baits and Fumigants 213
  - Endangered Species Guidelines 214
- Vertebrate Pests 215
  - Pocket Gophers 215
  - Ground Squirrel 216
  - Tree Squirrels 217
  - Rabbits 218
  - Voles (Meadow Mice) 219
  - Deer Mice 219
  - Roof Rat 220
  - Deer 221
  - Coyote 221
  - Wild Pig 223
  - European Starling 223

Weeds 225
- Life Cycles 226
- Weed Control Strategies 227
- Prevention 228
- Monitoring 228
- Control Methods 229
  - Irrigation 229
  - Cultivation 229
  - Ground Cover 229
  - Mulch 230
  - Biological Control 230
  - Herbicides 230
    - Preemergence Herbicides 232
    - Postemergence Herbicides 232
    - Herbicide-Resistant Weeds 232
    - Environmental Considerations 233
  - Identifying Major Weed Species 233
    - Identification Resources 234
  - Special Weed Problems in Citrus 236
    - Johnsongrass 236
    - Nutsedges 237

Harvesting and Pest Management 239
- Prevent Fruit Rot Diseases 240
- Pick a Good Harvest Time 240
- When Picking Fruit 241
- After Fruit Are Picked 242
- Good Agricultural Practices 242

World Wide Web Sites 245
Suggested Reading 247
Literature Cited 253
Tables and Figures 257
Glossary 259
Index 263

Integrated Pest Management for Citrus

This chapter summarizes the citrus cultivars and growing locations in California and the challenges growers face as economics and pest problems change. Subsequent chapters will help growers and pest control professionals apply the principles of integrated pest management (IPM) to California citrus production. IPM uses the most effective and sustainable biological, chemical, and cultural tactics. IPM decisions are based on established monitoring techniques and treatment thresholds. Pesticides are used only when necessary, and according to available treatment guidelines. Management relies on methods that are the least disruptive to natural enemies and that minimize adverse environmental effects.

Citrus is grown in four major areas of California: the San Joaquin Valley, the southern coast, the southern interior, and the southern desert valleys. Smaller growing areas include the northern Sacramento Valley (Figures 1 and 2). Small-scale production occurs in most counties of central and southern California to provide fruit grown for farmers' markets. Citrus is the most common fruit tree in California landscapes. It occurs on hundreds of thousands of properties, often as one to several citrus trees per lot. Exotic pests of citrus introduced into California are often first discovered on these noncommercial backyard trees.

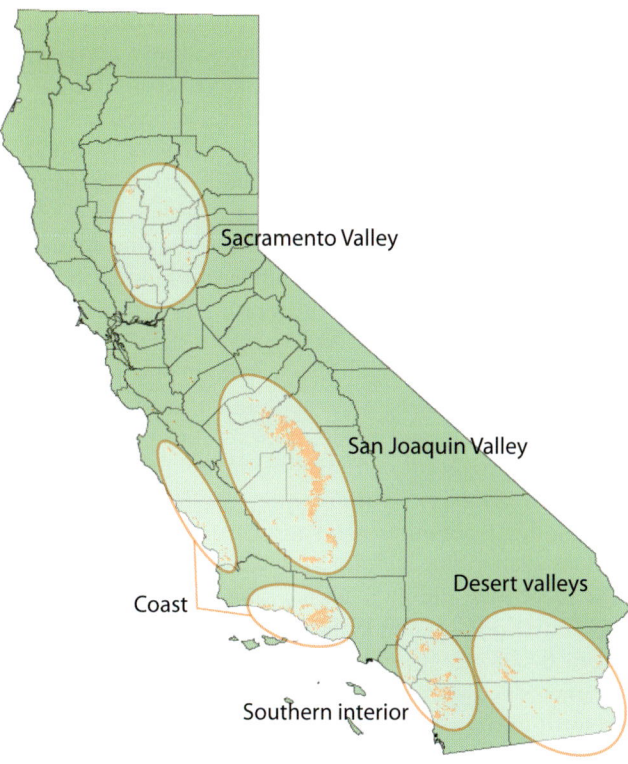

Figure 1. The citrus production areas of California. More than 70% of California's commercial citrus acreage is in the San Joaquin Valley, where summers are hot and dry, winters are cool and wet, and most of California's oranges (especially navels) and mandarins are grown. Other production areas in order of acreage are the coastal region, southern interior, desert valleys, and the northern Sacramento Valley. The Sacramento Valley and San Joaquin Valley combined are called the Central Valley.

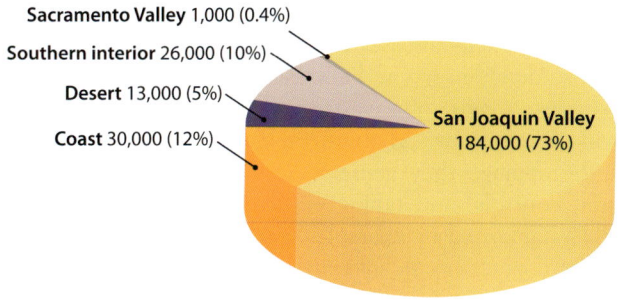

Figure 2. The acres of fruit-bearing citrus (and approximate percentage of total acreage) for the citrus-growing regions in California. See the Figure 1 map for these locations. *Source:* USDA 2008.

IPM Is...

Integrated pest management (IPM) is an ecosystem-based strategy that focuses on the long-term prevention of pests or their damage. Pests include fungi and other pathogens, insects, mites, vertebrates, and weeds that damage citrus. IPM combines techniques such as biological control, habitat modification, changes in cultural practices, and use of pest-resistant rootstock cultivars. Pesticides are applied only when monitoring indicates they are needed according to established guidelines, and applications are made with the goal of controlling only the target pest organism. Pesticides are selected and applied in ways that minimize risks to human health, beneficial and nontarget organisms, and the environment.

The major citrus-producing regions differ in climate. San Joaquin Valley summers are hot and dry, and winters are generally cool and wet. The coastal region has a milder climate because of temperature moderation from marine air. The interior district is separated from the coast by mountain ridges and is generally warmer and drier in the summer and colder in the winter in comparison with the coast. In the desert (including the Coachella and Imperial valleys), temperatures fluctuate widely between day and night, and the humidity is low most of the year.

Climate greatly influences choice of cultivar; however, all areas have some acreage of all major cultivars. About half of California's citrus are navel oranges (Figure 3), grown mostly in the San Joaquin Valley. The San Joaquin Valley contains more than 70% of California's 254,000 acres of bearing citrus trees. Lemon predominates on the coast. Grapefruit grows mostly in the desert valleys and interior areas of southern California. Valencia orange has historically been the major cultivar in coastal and southern interior areas, but its production has declined dramatically. The Valencia acreage statewide dropped from a peak of 150,000 acres in the 1940s to about 44,000 acres in 2008. Conversely, mandarins increased from about 1,000 acres in the 1950s to about 12,000 acres each of bearing and nonbearing (young) mandarins in 2006. Production, primarily in the San Joaquin Valley, then doubled to more than 25,000 bearing areas in 2008, spurred by consumers' desire for easy-to-peel citrus.

Citrus pests can vary based on the cultivar. For example, citrus thrips is a key pest of navels and mandarins but less of a pest in Valencias, and cottony cushion scale is more common in scion cultivars that grow densely, such as mandarins and grapefruit. Citrus pests can also vary based on rootstock cultivars. For example, certain rootstocks are highly susceptible to, or resistant to, citrus nematodes, Phytophthora root rot, or both. Using resistant rootstock cultivars is key to successfully managing these two problems.

Growing regions also influence the pests you might see. Citrus bud mite is a pest mostly on coastal lemons, where its damage is sometimes severe. Coastal lemons blossom and set fruit throughout the year, continuously providing this mite with its favored habitat of succulent new buds. Another example of localized pest differences is that although fall foliar damage caused by citrus leafminer does not affect most mature citrus varieties, the multiple flushes and multiple crops of coastal lemons make them more susceptible to citrus leafminer damage.

Management tactics also vary by cultivar and growing region. For example, biological control of insect and mite pests is generally less effective in the San Joaquin Valley in comparison with coastal and southern areas because the extremes of heat and cold in summer and winter synchronize pest populations and make it more difficult for natural enemies to maintain adequate levels necessary to control pest species effectively. Because of this, California red scale and citricola scale are difficult to control biologically in the San Joaquin Valley. As the majority of California citrus has shifted from Valencias grown in coastal and southern interior regions to navels and mandarins grown in the San Joaquin Valley, there has been a greater reliance on chemical control of key pests. For a thorough review of pest and natural enemy importance in California by citrus-growing region, see "Citrus IPM" (in preparation).

California produces most of the nation's fresh market citrus, especially sweet oranges and mandarins. Florida, traditionally the largest producer of citrus primarily for juice, has experienced decreased acreage due to the arrival of bacterial canker and Huanglongbing diseases. Arizona and Texas are also significant producers of citrus. Citrus is grown in most tropical, subtropical, and Mediterranean climate countries of the world (Figure 4), although some countries do not meet quality standards for fruit export. Fresh market citrus has especially high quality standards, particularly for export to markets such as Japan. Farm sales value of California citrus was about $1.2 billion in 2006. However, the percentage of world citrus produced by California and the United States has been declining. In 2004, China surpassed the United States in production. United States citrus production is currently third in the world, behind Brazil and China.

Increased world commerce and travel have facilitated the introduction of new pests, disrupting California's established IPM programs. More foreign production and fewer trade barriers have altered world citrus trade and economics. Integrated pest management can help growers resolve the challenges of new pests and foreign competition. IPM programs increase the grower's ability to produce high-quality fruit predictably and reliably.

Use this manual as one of several authoritative resources to help you plan an IPM strategy for your orchard. The manual's first few chapters summarize citrus growth and development, general crop management practices, and IPM principles. Subsequent chapters cover the biology and identification of citrus diseases (including abiotic or noninfectious disorders), insects and mites and their natural enemies, nematodes, weeds, and vertebrate pests. These chapters provide descriptions and photographs of the individual pests and the damage they cause. Pest monitoring, enhancing natural controls, and taking effective control actions are briefly summarized by examples. The final chapter reviews in-orchard practices to maximize postharvest fruit quality and food safety.

IPM is a flexible, evolving strategy that is periodically updated. IPM practices change due to the introduction of new citrus cultivars and exotic pests, changes in pesticide registrations, and the development of new information and tools. For more current and detailed monitoring and management guidelines, including regularly updated pesticide recommendations, consult the current UC IPM Pest Management Guidelines: Citrus and Citrus Year-Round IPM Program (online at www.ipm.ucdavis.edu). Also consult the Citrus Production Manual (Ferguson et al., in preparation), such as for details on crop cultural practices. Check regularly with your University of California Cooperative Extension farm advisor for new developments.

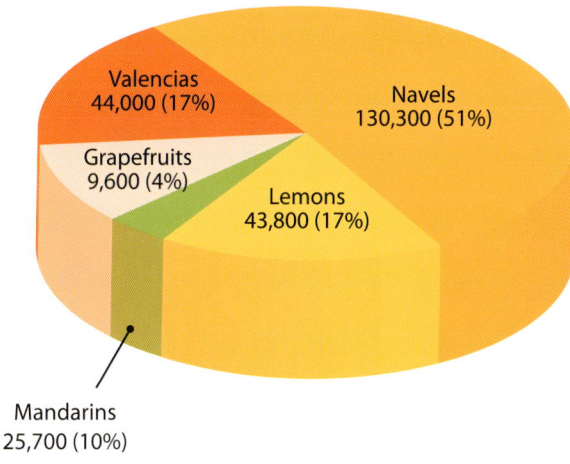

Figure 3. The fruit-bearing acreage (and percentage of total acres) for major citrus cultivars grown in California during 2008. Total bearing citrus was about 254,000 acres. Reported mandarin (tangerine) acreage includes mandarin hybrids, tangelos, and tangors. Grapefruits acreage includes pummelos and hybrids. Approximately 400 acres of limes (not shown) were also grown. *Source:* USDA 2008.

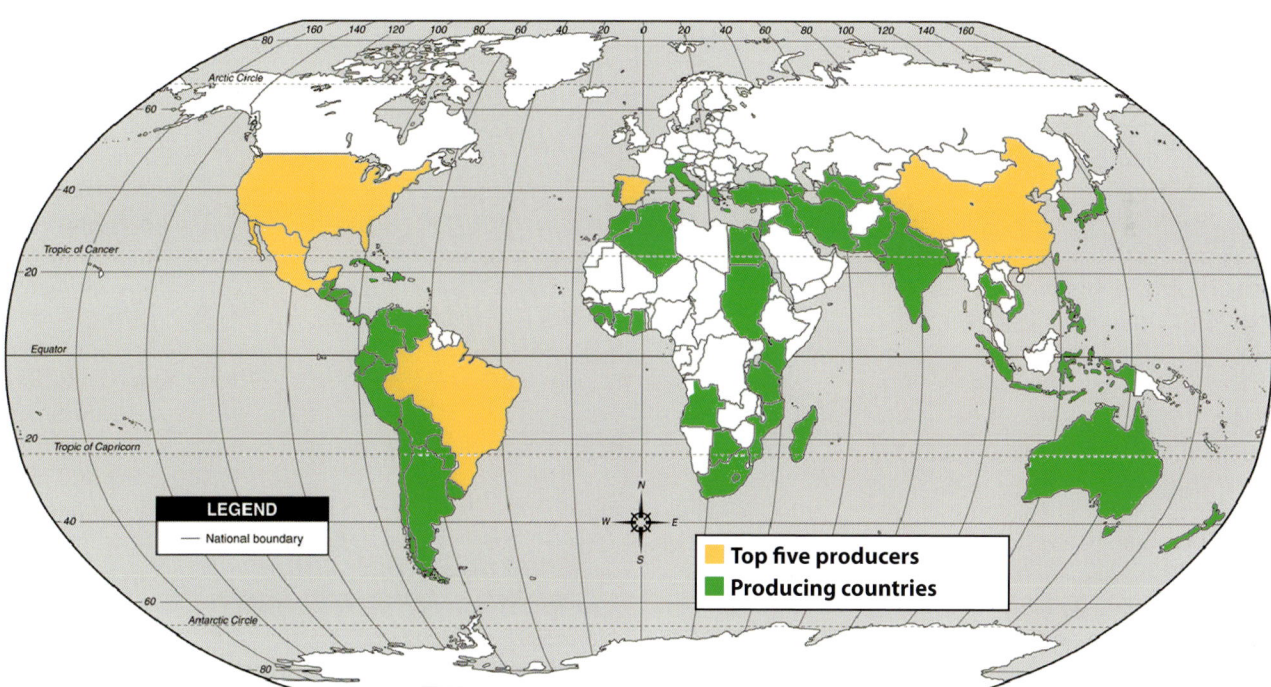

Figure 4. Citrus is grown in most countries with a tropical, subtropical, or Mediterranean climate. For most cultivars, the best-quality fruit develops in locations like California, which experience both prolonged warm weather and some chilling (a cold season). The top five citrus producers (6 to 20 million metric tons per year) are shown in orange. The U.S. production is third in the world, behind Brazil and China. Other citrus producers (shown in green) include some third-world countries that do not meet quality standards for fruit export. *Source:* FAO 2007, USDA 2007.

This flower in cross section reveals the main reproductive parts of citrus. In cultivars requiring fertilization, pollen (shed from anthers, top left and right) contacts the stigma (the tip of the pistil, top center), and each pollen grain forms a tube that grows down into the ovary. A sperm in each pollen tube fertilizes an egg cell (ovule), which develops into a seed. The ovary (center) develops into the fruit. See all major flower parts labeled in Figure 7.

The Citrus Tree: Development and Growth Requirements

Cultural care and pest problems vary by season and the stage of crop growth. Successful production requires that you understand the crop's seasonal development and its specific requirements for good growth. In order to apply the crop and pest management information in subsequent chapters of this manual, you must know the anatomy and developmental biology of citrus and the terms that describe them presented in this chapter.

The objective of citrus production is to harvest a large crop of high-quality fruit. High-quality fruit has firm, juicy flesh and a rich flavor resulting from a balance between sweetness and tartness. Fruit should have an aroma and size characteristic of the cultivar and, for fresh market, a smooth, deep-colored, blemish-free rind. Optimal fruit quality and yield, and less damage from pests, result when trees are well managed and healthy. A good IPM program includes evaluating and understanding how cultural practices and environmental conditions impact the total orchard system, as well as their importance in helping prevent and manage pest problems. Without adequate knowledge and regular monitoring, it is easy to overlook stress symptoms caused by cultural or environmental factors or to confuse those symptoms with pest damage. This chapter and those that follow will help growers meet these challenges.

The Seasonal Cycle of Citrus

In California's climate, trees of all citrus cultivars except coastal lemons stop growing during the winter. During this period, the tree maintains a base level of water transport and starch consumption. The main shoot growth and leaf flush appear in late February or March. Additional growth flushes typically occur in the summer and fall. Each leaf stays on the tree for 1 to 2 years. In healthy trees, older leaves shed gradually throughout the year, with the greatest leaf drop during spring flowering and early fruit set. Environmental factors such as excess soil salinity, high temperature, low relative humidity, nutrient disorders, too much or too little soil moisture, and pest problems can trigger premature leaf drop.

Root growth occurs during early summer. Additional root growth may occur during late summer and late fall. The occurrence of root growth and leaf flush roughly alternate, with two or three growth periods for each from late winter through late fall.

Most citrus cultivars flower in the spring. Although coastal lemons bloom throughout much of the year, they have three major bloom periods, with flowers produced most abundantly in the spring. With other cultivars, disease, rainfall, or dry soil conditions followed by irrigation may trigger irregular or "off" bloom. Fruit developing from "off" bloom are usually of inferior quality. Many cultivars, particularly Valencia and certain mandarins, have an alternate bearing habit, setting a heavy crop one year and a light crop the following year. The major fruit growth development stages in navel orange are illustrated in Figure 5. The typical seasonal occurrence of growth stages is shown in Figure 6.

Flowers develop in leaf axils on flushes of new shoot growth and on old shoots with few or no new leaves. Each inflorescence (flowering part) produces one to many flowers. A flower has male structures (multiple stamens, each consisting of a filament and anther) and a single female structure (pistil, composed primarily of the stigma, style, and ovary containing ovules), as illustrated in Figure 7. Most cultivars produce viable pollen, although some do not,

New shoots develop most abundantly during late winter to early spring in most cultivars.

Flower buds appear at leaf axils of growth flushes after a period of new shoot elongation. Flower buds may also develop on old shoots with few or no new leaves.

Navel oranges bloom during March and April in the San Joaquin Valley. For pest control purposes and to protect honey bees, in certain districts agricultural officials determine the official bloom period.

Young fruit after petal fall increase slowly in size for about 2 months. Internally, cell division is rapidly increasing the number of fruit cells.

Fruit rapidly increase in size beginning about 2 months after fruit set, in early summer for navels. Most fruit drop from the tree when they are small.

Navel oranges are ready for harvest by early winter. Fruit store well on the tree, and navels may not be picked until months later.

Figure 5. Citrus development stages.

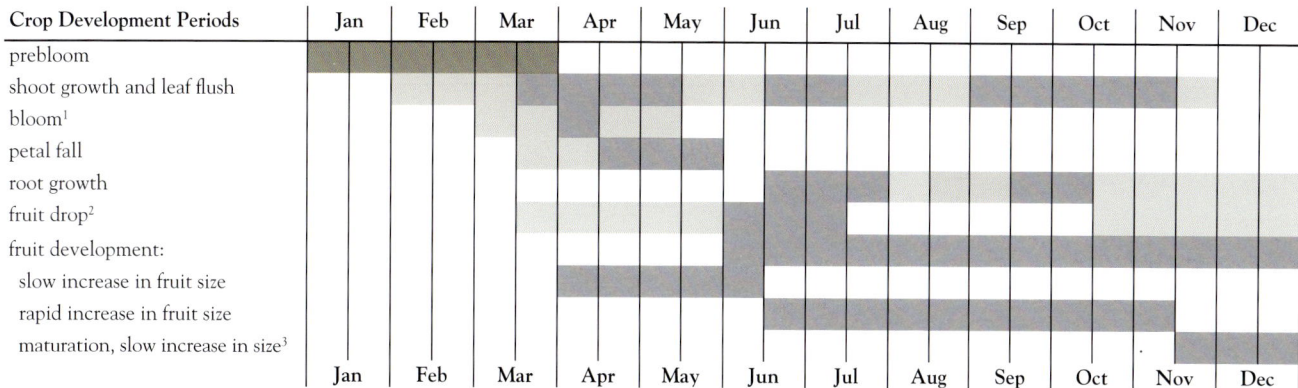

The average occurrence of each development stage is shown in dark gray. Periods when development is less vigorous or that stage is less prevalent are shown in light gray. There is overlap among stages, and actual development varies with cultivar, location, and weather.

1. For pest control purposes and to protect honey bees, in certain growing districts agricultural officials determine the official bloom period.
2. Includes "June drop" shown in dark gray and (in light gray) the preceding "early drop" and subsequent "preharvest drop."
3. Navel maturation can extend from early winter through May or longer if fruit are left on the tree.

Adapted from CCQC 2003; Lovatt 1999; Lovatt, in preparation.

Figure 6. Average seasonal development for navel orange grown in the San Joaquin Valley.

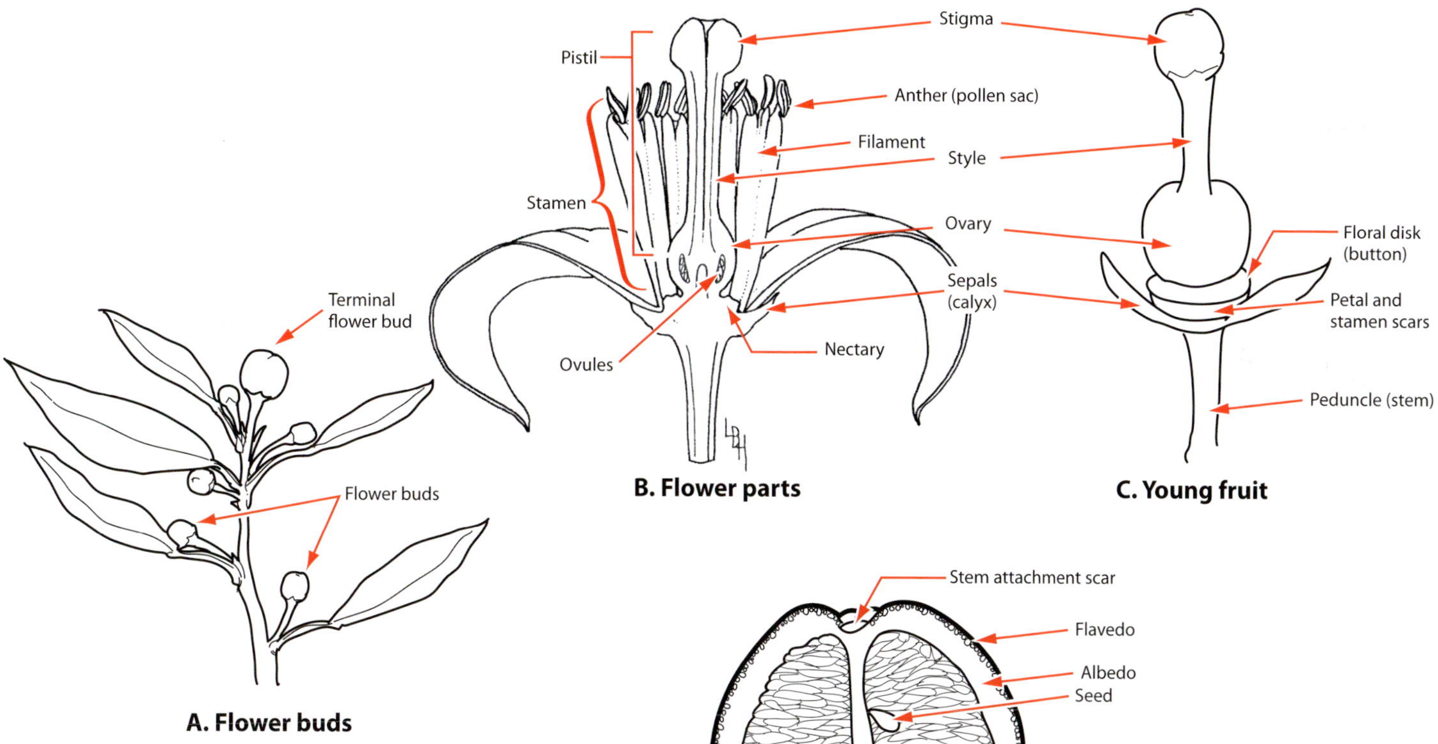

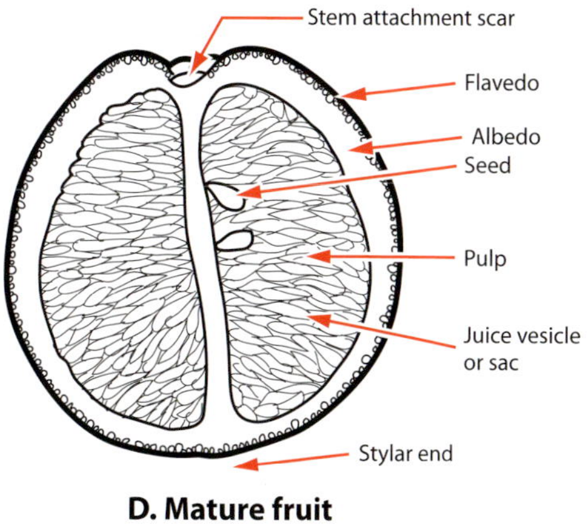

A. Flower buds

B. Flower parts

C. Young fruit

D. Mature fruit

Figure 7. Citrus flower and fruit parts and stages of their development. A. Flower buds develop in spring on succulent new shoots or on older stems with few or no new leaves. B. The parts of a citrus flower shown in cross section. In seedless cultivars (such as navel oranges and certain mandarins), the anthers (male part producing pollen) and ovules (egg cells) are sterile. C. After bloom, the petals, stamen, and style are shed as the ovary develops into a fruit. D. The thin outer layer (flavedo) of fruit contains pigments and oil cells. The rind's inner layer (albedo) is white and spongy. Adapted from Davenport 1990; McGregor 1976; Schneider 1968; Silva, Lovatt, and Arpaia 2002.

Table 1. Seedless and Seeded Cultivars of Mandarin, Tangelo, and Tangor Citrus.

Seeded[1]	Seedless[2]	Seedless to low-seeded[3]	Seedless only if...[4]
Clementine Monreal	Aoshima Satsuma	Shasta Gold	Clementine Algerian
Dancy	Armstrong Satsuma	Tahoe Gold	Clementine Caffin
Dweet	China S-9 Satsuma	Yosemite Gold	Clementine Carte Noir
Encore	Dobashi Beni Satsuma		Clementine Corsica #1
Fallglo	Frost Owari Satsuma		Clementine Corsica #2
Fortune	Gold Nugget		Clementine de Nules (Clemenules, Nules)
Freemont	Kawano Wase Satsuma		Clementine Fina
Honey (California Honey)	Kiyomi		Clementine Fina Sodea
Kara	Kuno Wase Satsuma		Clementine Marisol
Kinnow	Miho Wase Satsuma		Clementine Nour
Murcott tangor (Florida Honey)	Miyagawa Satsuma		Clementine Oroval
Pearl tangelo	Neopolitana Satsuma		Clementine Sidi Aissa
Ponkan	Okitsu Wase Satsuma		Clementine SRA 63
Sue Linda Temple	Pixie		Clementine SRA 92
Sunburst	Seedless Kishu		Ellendale
Temple	Silverhill Satsuma		Fairchild
Wekiwa	Tango		Koster
Wilking	Xie Shan Satsuma		Lee
Willowleaf			Minneola tangelo
			Nova
			Orlando tangelo
			Ortanique
			Page
			Robinson
			W. Murcott Afourer (W. Murcott, Afourer)

Varieties maintained at the California Citrus Clonal Protection Program, UC Riverside. Contact your commercial nursery for availability.

1. Seeded: Cultivars that will have seeds no matter where they are grown.
2. Seedless: Cultivars that will be "seedless" with only an occasional seed no matter where they are grown.
3. Seedless to low-seeded: Cultivars that commonly will be "seedless" or low-seeded but may have more seeds when cross-pollinated.
4. Seedless only if: Self-incompatible cultivars that will be seedless only if cross-pollination with certain other cultivars is prevented.

Source: Kahn 2007.

such as navel orange and certain mandarins. During pollination, pollen grains fall onto the stigma or are introduced there by insects (especially honey bees) attracted to flower nectar. Pollen germinates, producing pollen tubes that grow through the style into the ovary. Each pollen tube contains a sperm cell, which fertilizes an ovule (egg cell) in the ovary. Each fertilized ovule develops into a seed.

For many types of citrus, the number of seed produced depends primarily (or only) on the cultivar. Most citrus cultivars are self-compatible, that is, they can be fertilized by their own pollen. Most cultivars can produce fruit without fertilization and seed formation. However, pummelos and some mandarins and tangelos (mandarin hybrids) set a better crop when pollinated by certain other cultivars (cross-pollination). Navel orange has little or no viable pollen and few or no fertile ovules, so navels usually form no seed. Seed production in mandarins (sometimes called tangerines) is variable (see Table 1 and the section "Mandarins and Seediness"). Even in normally fertile (seed-producing) citrus cultivars, environmental conditions such as excessive heat or cold and low relative humidity can prevent seed formation or reduce the number of seeds by inhibiting the pollen's ability to germinate and produce the pollen tubes necessary for fertilization.

Citrus usually blooms abundantly, but most flowers drop (abscise) without forming (setting) fruit. How environmental and physiological factors influence which flowers develop into fruit that persist to harvest is not completely understood. In comparison with inflorescences with early-opening flowers, inflorescences whose flowers open late in the bloom period tend to occur on the more leafy shoots, and these later-maturing flowers produce more of the fruit that survive to harvest. In addition, faster-growing ovaries are more likely survive to harvest, whereas slower-growing fruit are more likely to drop at an early stage of development.

After petal fall, high temperatures during mid-May to mid-July coincide with the "June drop" of most young fruit. Some fruit also drop earlier (during March to April, called "early drop") or later just as they are maturing ("preharvest drop").

During early growth, young fruit undergo rapid cell division (replication) for about 9 weeks, depending on the temperature. Despite the great increase in cell numbers, young fruit increase slowly in size, with most of their size growth due to increasing peel thickness. Fruit next enter a period of rapid cell enlargement, during which fruit quickly increase in size. The peel, especially the inner, white albedo layer, reaches maximum thickness early in this period of rapid size increase. The peel then thins as the pulp continues to increase in size during the formation of the juice-containing acids, mainly citric acid. A maturation period follows, during which fruit can be picked. During maturation in grapefruit, mandarins, and oranges, the sugar concentration increases and the acid content decreases. Fruit continue to increase in size slowly during maturation, over a period of months if they are left on the tree. For more detailed discussion and illustration of citrus growth and development, consult "Citrus Physiology and Phenology" (in preparation).

Unlike deciduous fruit, citrus have no clearly identifiable point of maturity. Color is not a reliable indicator of ripeness. Rind color development depends largely on sufficiently low night temperatures and on the mineral nutrition of the tree. For oranges, mandarins, and grapefruit, the ratio of soluble solids (mainly sugars) to acids is important for determining legal maturity. Fruit on the tree continue to improve in flavor and taste for several weeks or even months after they first become suitable for harvest. Later, taste declines due to a breakdown of acid and flavor components. Lemons are considered mature when they have reached a certain percentage of juice per volume of fruit. Certain hormone sprays can affect fruit drop or prolong fruit quality on the tree; for recommended materials, consult the latest UC IPM *Pest Management Guidelines: Citrus* (online at www.ipm.ucdavis.edu).

Growth Requirements

Tree growth and fruit development require adequate light, water, and nutrients, and appropriate temperatures. Light is essential for photosynthesis, the process by which green plants manufacture sugars, their major food source (Figure 8). The

The young navel orange has a relatively thick rind, as shown in this fruit cross-section. The outer rind layer (flavedo) contains pigments and oil cells. The spongy, white inner layer (albedo) helps the fruit maintain moisture.

As citrus matures, the juice vesicles greatly enlarge as they fill with citric acid, sugars, and water. The ratio of acid to sugar varies by cultivar, fruit maturity, and growing conditions. When night temperatures have been sufficiently low, citrus skins change color, such as to bright orange in this navel.

Figure 8. Photosynthesis is a chemical process within green plant tissue. Energy in sunlight is used to convert carbon dioxide and water into oxygen and carbohydrates.

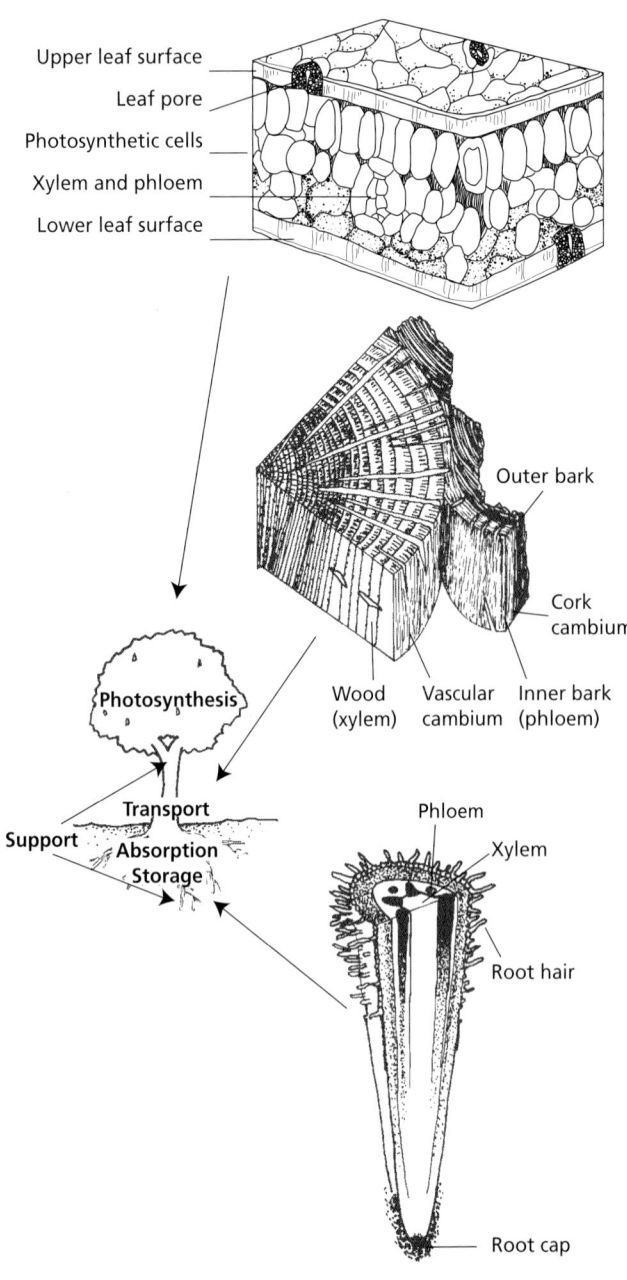

Figure 9. The basic anatomy and functions of key vegetative parts of a tree. The limbs, trunk, and roots support the plant and provide its vascular (transport) system. Small roots absorb water and nutrients, which are transported upward in the xylem. Leaves produce carbohydrates, which are used immediately for tissue growth and maintenance or transported down in phloem as sugars. Sugars are converted to other carbohydrates, such as cellulose and starch, and stored in roots, limbs, and the trunk. Cambium is vital tissue that grows to enlarge the plant and differentiate into other tissues.

plant may metabolize the sugars immediately, store them as starch for later use, or concentrate them in certain plant parts, such as the fruit. Sugars also serve as building blocks for thousands of other compounds, including pigments, flavor components, and structural molecules like cellulose.

Vital vegetative parts include the leaves, limbs, roots, and the trunk (Figure 9). All plant parts are closely interdependent. For example, leaf drop is greatest during the spring flowering period. The periods of new shoot growth and leaf flush (late winter and late summer) alternate with the times of root growth (early summer and fall), as illustrated in Figure 6. If leaves or roots fail to achieve sufficient healthy growth when environmental conditions and plant physiology are conducive to their development, trees may become stressed and fruit quantity and quality may decline.

All basic plant functions require water. Only a small fraction of the water taken up by the tree is retained; most of it passes through the plant and out leaf pores (stomata) during the process of transpiration. Transpiration is necessary to supply water to the leaves for photosynthesis, to carry nutrients to all parts of the tree, and to cool the plant. Climatic factors, such as the amount of heat, humidity, sunshine, and wind, affect the rate of transpiration and the trees' need for water.

Carbon, hydrogen, and oxygen (supplied by air and water) and 14 soil-derived nutrient elements are essential for good tree growth and fruit production. The nutrients needed in largest amounts (macronutrients) are (in approximate order of their abundance in dry-weight leaves) nitrogen (N), calcium (Ca), potassium (K), magnesium (Mg), sulfur (S), and phosphorus (P). The nutrients needed in small amounts (micronutrients) are chlorine (Cl), iron (Fe), boron (B), zinc (Zn), manganese (Mn), copper (Cu), molybdenum (Mo), and nickel (Ni).

California orchard soils contain sufficient amounts of most essential nutrients to meet the needs of the tree because the nutrients are already present in the soil, are returned periodically through the decay of organic matter, or may be present as impurities in air and irrigation water. Others, mainly nitrogen and potassium but also occasionally zinc and manganese, must be replenished through fertilization. The need to provide certain nutrients varies by location, as discussed in the "Mineral Deficiencies and Toxicities" section in the chapter "Diseases."

Beneficial fungi, called mycorrhizal fungi, greatly enhance citrus root uptake of nutrients and water. This symbiotic (mutually beneficial) association (mycorrhizae) increases citrus roots' ability to absorb water and many essential nutrients, such as phosphorus, zinc, and copper. Mycorrhizal fungi are efficient nutrient suppliers with threadlike hyphae that explore much larger areas than the root hairs. In return, the mycorrhizal fungi obtain carbohydrates from the plant roots.

Mandarins and Seediness

Some citrus cultivars naturally have few seeds. Other low-seeded cultivars have been developed through traditional selective breeding methods. Specially developed biotechnological methods and techniques such as irradiation are increasingly being used for citrus crop improvement.

Among the major citrus cultivars (grapefruit, lemon, lime, mandarin, and orange), seed production is the most variable in mandarins. Some cultivars of mandarin are always seeded; others always produce seedless fruit. Certain mandarins are seedless only when "isolated" from cross-pollination (see Table 1).

Satsuma mandarin have little or no viable pollen. Some Satsuma and other mandarin cultivars also have few or no fertile eggs, so these usually develop fruit with no seed. Clementine mandarin is self-incompatible. Traditional (older) Clementine varieties require pollination by another type of citrus to set fruit, and those Clementine fruit generally contain seed. Many more-recently developed Clementine cultivars do not require pollination to set fruit, and these Clementines can produce fruit with no seed. However, certain of these "seedless" mandarins (Clementine, W. Murcott, and others) develop seed if they are grown near, and cross-pollinated by, other citrus, such as tangelo, lemon, certain grapefruit and pummelo, and other mandarins. Because citrus have relatively heavy, sticky pollen that is not easily windblown, most cross-pollination of citrus results from honey bees moving between cultivars.

In potentially seedless cultivars that produce seed when cross-pollinated, the extent of undesirable seediness depends on the nearness of other cultivars and the distance from bee hives, as well as on the mandarin variety being grown. Because many consumers desire seedless fruit and often pay a higher price for it, mandarin "seediness" from cross-pollination in "seedless" varieties is commercially undesirable.

Mycorrhizae are sensitive to soil fumigants used against fungal pathogens. Mass-produced beneficial fungi for inoculating citrus seedlings are used successfully in some nurseries. Using mycorrhizae can substantially reduce the amount of fertilizer needed and improve the growth of nursery trees. In established orchards that are regularly fertilized, there is probably no benefit from adding amendments containing mycorrhizae.

Citrus produces best under a mild subtropical climate. Wide fluctuations between day and night temperatures promote acid formation, resulting in a rich flavor. Cool night temperatures trigger the development of the bright orange or yellow color of the rind. Total heat accumulation determines the amount of sugar produced and thus the time it takes for oranges and grapefruit to mature. Fruit grown in the desert reaches maturity first, followed by those grown in the Central Valley and the southern interior growing area; fruit on the coast ripens last. Citrus cultivars differ in their heat requirements. Grapefruit requires more heat than other cultivars, and in California grapefruit develop the best flavor in the desert regions. Acidic citrus, such as lemon and lime, do not need to sweeten, so they grow well in the mild-climate coastal areas. Citrus is sensitive to extreme temperatures, which can cause sunburn or frost damage and increase fruit and tree susceptibility to injury-related diseases. However, a season of cold weather (chilling) synchronizes plant development and flowering and induces better-quality fruit.

Managing Pests in Citrus

Integrated pest management (IPM) treats pests as part of a crop production system that includes not only the crop and its pests but also the physical and biological environment in which the crop is grown. To achieve economical and long-lasting solutions to pest problems, a good IPM program coordinates pest management activities with each other and with cultural operations.

By modifying cultural practices and orchard conditions, growers can prevent or control many pests and disorders (Table 2). The crop development stages when management actions may be needed, and when to expect and monitor for specific pest problems, are discussed later in the "Year-Round IPM" and "Management Methods" sections.

Using a good integrated pest management program will help you to

- anticipate and prevent problems whenever possible
- accurately diagnose the cause of problems and correctly identify pests and beneficials
- monitor regularly to gather information on which to base IPM decisions
- use control action guidelines to help you decide when pesticide applications are warranted
- manage pests by relying primarily on methods that are the least disruptive to natural enemies and minimize adverse environmental effects, as detailed in UC IPM *Pest Management Guidelines: Citrus* and the *Citrus Year-Round IPM Program* (both online at www.ipm.ucdavis.edu).

Pest Prevention

You can anticipate and avoid many crop problems. Prevention is often the least expensive and most effective method. For certain problems, prevention is the only control option. For example, most diseases can be effectively managed by action only during early stages of disease development. Once symptoms become obvious or severe, treatment options are more limited and the tree or fruit may have been seriously damaged. Key preventive management practices include choosing a good growing site, preparing the site well before planting, installing and maintaining appropriate irrigation and drainage systems, selecting disease-resistant cultivars and rootstocks certified as pathogen-free, using good sanitation, and providing appropriate cultural care (Table 2).

Table 2. Cultural Practices and Growing Conditions That Growers Can Manipulate (Y) in an IPM Program to Help Prevent and Control Citrus Pests and Disorders.

	Planting site selection	Scion cultivar	Rootstock cultivar	Pathogen-free nursery stock	Planting method (e.g., planting high)	Drainage	Irrigation amount, frequency, and method	Fertilization	Avoiding mechanical injury	Sanitation	Windbreaks	Weed and border vegetation control	Cultivation	Pruning extent and timing	Skirt pruning	Lesion excision	Root crown excavate & dry	Minimizing dust	Harvest timing
Diseases																			
Alternaria rot							Y												Y
Armillaria root rot	Y				Y	Y	Y			Y							Y		
bacterial blast							Y	Y		Y	Y			Y					
Botrytis rot									Y					Y	Y				
brown rot							Y								Y				
dry root rot					Y	Y	Y	Y	Y	Y							Y		
exocortis			Y	Y						Y				Y					
lemon sieve tube necrosis		Y																	
Phytophthora gummosis					Y	Y	Y									Y			
Phytophthora root rot			Y		Y	Y	Y												
psorosis			Y	Y						Y									
Septoria spot																			Y
stubborn disease					Y									Y					
tristeza virus		Y	Y	Y															
Disorders																			
frost tolerance	Y	Y	Y				Y	Y						Y					
nutrient disorders	Y				Y	Y	Y	Y											
salinity	Y	Y	Y		Y	Y	Y	Y											
sunburn	Y						Y	Y						Y					
Invertebrates																			
ants														Y	Y				
aphids							Y												
bean thrips	Y	Y										Y			Y				
brown garden snail							Y								Y				
California red scale and yellow scale																		Y	
citricola scale																		Y	
citrus leafminer							Y							Y					
citrus peelminer		Y										Y							
citrus red mite							Y	Y										Y	
citrus thrips											Y								
cottony cushion scale														Y	Y				
earwigs												Y			Y				
glassy-winged sharpshooter												Y	Y						
greenhouse thrips																			Y
mealybugs																		Y	
potato leafhopper														Y					
purple scale																		Y	
Texas citrus mite								Y										Y	
twospotted spider mite							Y	Y										Y	
whiteflies																		Y	
Yuma spider mite							Y											Y	
Nematodes	Y		Y	Y						Y									
Weeds	Y				Y	Y	Y	Y		Y		Y	Y						

Excluding Foreign Pests. Many of our worst citrus pests were introduced from other states such as Florida or other areas of the world including Asia, Australia, the Caribbean, Central America, and Mexico. For example, Diaprepes root weevil, which is native to the Caribbean, was discovered in California beginning in 2005 and likely arrived from Florida. Asian citrus psyllid, the vector of Huanglongbing (HLB, or citrus greening), was first found in California in 2008 and arrived from Mexico. Exotic (foreign) fruit flies, such as the Caribbean fruit fly and Mediterranean fruit fly, have arrived from many countries. Exotic fruit flies are discovered periodically in California and apparently have been eradicated many times.

Prevent new pest introductions during trade and travel:
- Do not bring uncertified fruit, plants, or soil into California.
- Buy only certified disease-free citrus trees from reputable, local nurseries.
- Report suspected fruit or plant smugglers to local authorities.
- Take any unfamiliar pests to your county agricultural commissioner or University of California Cooperative Extension office for identification or telephone 1-800-491-1899.

More recommendations on how you can stop the spread of exotic and invasive species are available online from the UC Statewide IPM Program (www.ipm.ucdavis.edu/EXOTIC/exoticabout.html). Other information sources include the California Department of Food and Agriculture (www.cdfa.ca.gov/invasive), California Invasive Plant Council (www.cal-ipc.org), UC Center for Invasive Species Research (www.cisr.ucr.edu), and USDA Agricultural Library National Invasive Species Information Center (www.invasivespeciesinfo.gov/unitedstates/ca.shtml).

Identification and Diagnosis

Correctly identify the cause of any damage or stress symptoms in the orchard and the species of pests and natural enemies so you can choose the appropriate management strategies. Most pest management tools, including pesticides, are effective only against a select group of species. Even closely related species may need to be managed differently. For example, citrus thrips and greenhouse thrips require different management methods, flower thrips are harmless and require no management, and sixspotted thrips are beneficial predators that should be conserved. In some cases, the symptoms caused by pest organisms closely resemble those caused by nutrient deficiencies or other abiotic (nonliving) causes, such as with fruit scarring (Table 3).

Diagnosing problems requires appropriate tools, such as a hand lens, hatchet, or shovel. The causes of some problems can be identified only when you send appropriate samples to a laboratory for testing. To obtain and submit adequate samples for diagnosis, you must know the proper collection procedures and use appropriate materials, such as containers and a cooler with ice.

Use the descriptions, photographs, and seasonal charts in this manual to help identify the important diseases, invertebrates, vertebrates, and weeds affecting citrus. Also consult the UC IPM *Pest Management Guidelines: Citrus* and *Citrus Year-Round IPM Program* for current, well-illustrated information.

Many citrus problems can be anticipated and avoided. These young citrus trees are planted on a slight berm, which improves drainage and keeps the root crown drier. Trunk guards help prevent damage from herbicide spray, sunburn, and vertebrate chewing.

Report suspected new pests to agricultural officials. These Asian citrus psyllids excrete copious wax and vector the pathogen that causes Huanglongbing (citrus greening). To help keep out exotic species, buy stock only from reputable, local sources. Do not bring uncertified fruit, plants, or soil into California.

Table 3. Citrus Fruit Scarring Causes.

Cause	Symptoms	Page
amorbia	Scar in ring or partial ring around sepals (calyx) is chewed deeper than with thrips. Webbing under calyx, around scar, or on leaves attached to fruit. Hole may be bored into fruit. Most prevalent in citrus near avocado.	143
anthracnose	Dull reddish green streaks, or "tearstaining," usually after foggy or wet weather.	59
Botrytis rot	Develops into soft decay at injury point. Preceded by prolonged wet periods, especially on coastal lemon and occasionally Valencia.	61
brown garden snail	Small brown to white chewed areas, usually in proximity to chewed or scraped leaves. Most prevalent on fruit nearest the ground and during warm and wet winters and cool and wet springs.	199
citrus cutworm	Elongate or meandering to circular scar, deeper than citrus thrips. Blossoms and spring leaf flush were previously chewed.	136
citrus peelminer	Long, winding mines beneath peel, especially on grapefruit and smooth-skinned navels. Leaves or twigs may also be mined.	150
citrus thrips	Yellowish scars with brown scabby center around the calyx in a thin to heavy ring or partial ring.	119
dust or sand	Shallow abrasions, often in relatively large and irregular patches, on fruit facing prevailing wind-carried particles.	—
equipment injury	Elongate or irregular scratches always on side of tree facing the equipment driving path.	—
frost	Brownish, watery specks or pits, mostly on the outside of fruit exposed to radiation frost. Most prevalent on more cold-sensitive cultivars, e.g., Eureka lemon and grapefruit.	63
fruittree leafroller	Fresh damage is to previous season's fruit still on the tree. Hole may be bored into fruit. Webbed foliage attached to fruit. Spring leaf flush was previously chewed.	139
hail	Small pits and irregular scars, including on fruit where equipment would not touch. May be leaf fragments on the ground. Causal weather event often observed or reported.	63
katydid	Roundish, relatively deep scar, usually in midsection of fruit.	170
phytotoxicity	Damage mostly on side of fruit facing away from the tree, usually appearing 3–4 days after application.	66
potato leafhopper	Small, uniformly yellowish or brownish scars in irregular-shaped patches any place on rind. Damage appears in late summer to fall.	169
rind stipple	Small brownish patches on grapefruit after prolonged wet periods. Pits may have a halo or concentric rings, which may coalesce into irregular patches. Mostly on fruit exposed on north side of the tree.	64
rust mite	Irregular, scattered scars on fruit surface. Develops on coastal lemon during summer.	193
Septoria	Small tan to reddish brown pits that may contain dark brown to black fruiting bodies. May be discolored streaks, or tearstaining. Mostly in San Joaquin Valley and interior southern California after cool, wet weather.	59
sunburn	Brownish leathery patches on side of fruit continually exposed to intense sunlight. Some fruit may be slightly lopsided.	63
wind	Scar usually superficial or very shallow. Shape and location often indicate source of wind abrasion against thorns or twigs.	62

These causes of rind injury can often be distinguished based on the scar pattern, growing region, location within the tree, recent environmental conditions, and cultivar, as described above. Also consult resources such as *A Photographic Guide to Citrus Fruit Scarring* (Grafton-Cardwell et al. 2003) for more information.

To identify the causes of some problems, you will need the assistance of experienced professionals. Do not hesitate to seek their help if you are not sure what is causing a problem. University of California Cooperative Extension (UCCE) farm advisors, the county agricultural commissioner, or a professional pest control adviser (PCA) can help you or direct you to other specialists when necessary.

Monitoring

Look regularly for pest problems, such as when you are in the orchard conducting routine cultural practices. Use established pest monitoring techniques to identify what pest and beneficial species are present and (when appropriate) determine their population levels to predict and evaluate potential problems. Also monitor the maturity and health of the crop, weather, and plant environment (including soil conditions). Because conditions vary even between neighboring orchards, individual orchards must be monitored.

Keep records of monitoring results and management activities. During the season, these records show whether pests or natural enemy populations are increasing or decreasing. They help forecast possible outbreaks or the next generation of a pest and provide the information on which IPM decisions are based, such as whether control actions will be needed. Simple tables and graphs of data help define patterns. Maps and the use of GPS (global positioning systems) can identify localized problems and pest distribution. Over the years, these records provide valuable historical data for long-term orchard management.

Monitoring Pests

When and how often to monitor pests varies with the individual pest species, crop development stage or season,

Diagnosing the cause of aboveground symptoms may include digging up and examining root crowns or small feeder roots. Correctly identifying causes of damage such as nematodes or Phytophthora root rot may require submitting properly collected samples to a diagnostic laboratory for testing.

Pest population counts, together with records of control measures, cultural practices, and weather conditions, provide the information on which IPM decisions are based. Insect monitoring methods include checking a certain number of fruit or leaves, shaking foliage into a sweep net, or using traps.

Monitoring methods vary depending on the pest and the purpose for monitoring. A species-specific attractant is placed in this reusable bucket trap to monitor certain beetles and moths, such as adults of cutworms.

growing location, and the age of trees and current weather conditions. Most problems need to be monitored only at certain times of the year. Start monitoring when pests and their natural enemies are likely to be present, but begin well before populations begin to build. Continue monitoring through the pests' damaging stages. The sampling frequency varies, but generally, weekly sampling is recommended with more frequent checking when monitoring indicates an outbreak.

Currently, no formal monitoring programs have been developed to assess most diseases, nematodes, vertebrates, and weeds in citrus. It is important, however, to note their presence and watch for changes in their status. Prepare a record at least twice per year for each orchard, noting the dominant diseases, vertebrates, weeds, and any other significant stressors. Sample for nematodes when you suspect they might be involved in poor tree vigor and might be causing losses or when the site history indicates they may become a problem.

Monitoring methods for insects include visually inspecting a certain number of fruit or leaves and using special traps. Monitoring for diseases may include cutting beneath bark and digging up root crowns and feeder roots to look for unhealthy plant tissue. Monitoring methods are summarized near the beginning of each pest chapter in this manual and, for invertebrate species with formal methods, are detailed in the individual pest sections. Current monitoring methods and sample record-keeping forms are available in the UC IPM *Pest Management Guidelines: Citrus* and the *Citrus Year-Round IPM Program*.

Monitoring Weather

Weather greatly influences the development of the citrus tree and its pests. Moisture from irrigation and rainfall promotes development of diseases such as brown rot of fruit. Because temperature in particular controls the rate at which plants, insects, and mites develop, monitoring temperatures helps to time management actions for certain pests. Temperature-based monitoring is explained below in "Accumulating Degree-Days."

Significant local variations in current weather conditions are common, especially variations in rainfall and temperature. A reliable source of local weather information is critical for crop management, helping to decide whether to protect trees from cold injury or how much irrigation water to apply, and for pest and pesticide management, such as when using degree-days, as discussed below. Many public news outlets report local weather. The National

Weather Service (www.nws.noaa.gov) broadcasts local and regional weather on NOAA Weather Radio: VHF channels 162.40 to 162.55 megahertz. Rainfall and evapotranspiration information (for irrigation management) is available from the California Irrigation Management Information System, CIMIS program (online at www.cimis.water.ca.gov).

For the most accurate data, set up a weather station in or near your orchard. If this is not feasible, seek reliable weather data from a location representative of your orchard, such as the sites accessible online at www.ipm.ucdavis.edu. Instruments available for monitoring in your orchard vary from simple maximum-minimum thermometers to electronic devices that continuously monitor and record weather information and can calculate degree-days or frost units. More sophisticated stations transmit the data to a remote computer. Set up and maintain the weather instruments according to the manufacturers' instructions. Calibrate the instruments regularly to ensure accuracy. Compare weather information to your other monitoring records to help plan your crop and pest management programs.

Accumulating Degree-Days

The growth and development rate of plants and invertebrates speeds up or slows down depending on the temperature of their environment. Plants and invertebrates require a certain amount of heat to develop from one point in their life cycle to another. This measure of accumulated heat is known as physiological time. Physiological time is approximated in units called degree-days (°D, or DD).

Degree-days are used to predict citrus bloom and to time controls for insect pests such as California red scale. Examples of using degree-days are presented in the chapter "Insects, Mites, and Snails." To calculate degree-days, you must know the organism's lower and upper developmental threshold temperatures and how much heat (how many °D) they require to develop from one stage to another. The lower developmental threshold is the temperature below which the organism's development stops. An upper threshold, the temperature above which the organism's development slows or stops, may also be needed when using degree-days.

The total heat requirement and the lower and upper thresholds are determined through laboratory and field

Significant local variations in rainfall and temperature are common, such as between this citrus high on a slope and that growing below in the valley. A reliable source of local weather information is critical to deciding when to protect trees from cold, how much irrigation water to apply, and when to monitor and manage certain pests.

A simple rain gauge for measuring precipitation within the grove helps in scheduling irrigation if it is properly maintained. Wait for the rain event to be completely over, then record the amount and empty the gauge. This stand includes a minimum-maximum thermometer (in the white cover), which should be reset following each time it is read.

experiments. A biofix date, the date to begin accumulating degree-days, is also needed, and it varies with the species. The biofix date is usually based on a biological event, such as the first pheromone trap catch of males for California red scale. Once the biofix is known, development can be estimated by accumulating degree-days. Degree-days can be measured using either the Fahrenheit or Celsius temperature scale; all the °D figures in this manual are presented in Fahrenheit units.

One degree-day is one day (24 hours) with the average daily temperature (maximum + minimum ÷ 2) 1 degree above the lower developmental threshold. For example, citrus bloom is predicted using a lower temperature threshold of 49°F (9.4°C). If the temperature averages 50°F (or 1° above the lower developmental threshold) for 24 hours, 1°D is accumulated. An upper development threshold is not used when calculating citrus bloom, because citrus normally continues to develop in warm temperatures.

Agricultural officials use this °D model to help them predict when citrus will blossom. For example, in navel oranges, the model predicts that 10% of the total citrus blossoms will be open when the orchard has accumulated 536.4°D, and that 75% of the blossom petals will have fallen at 766.8°D. This 10% anthesis (when flowers are open) to 75% petal fall is a time when pesticide use may be specially restricted to protect honey bees.

You can roughly estimate degree-days from the daily minimum and maximum temperatures as outlined in Table 4. However, this "manual" method of adding and dividing to estimate degree-days can be inaccurate, especially when temperatures are near the upper or lower thresholds. Computerized "sine wave" estimates are more accurate and are highly recommended instead of this manual calculation. A more thorough and well-illustrated explanation of degree-days, easy-to-use point-and-click software for calculating degree-days, and temperature records for numerous growing locations throughout the state are available online from the Statewide IPM Program. For pest-specific recommendations on using degree-days, consult the UC IPM *Pest Management Guidelines: Citrus*.

Control Action Guidelines

Growers and pest control advisers use control action guidelines to help decide whether management actions are necessary to prevent tree damage or fruit loss due to pests. Action guidelines are meaningful only when used in combination with careful field monitoring and accurate identification of pests and beneficials. For certain pests, such as California red scale, citricola scale, citrus red mite, and citrus thrips, numerical thresholds have been developed that indicate economically damaging population levels, and these are used to decide when to take control action. Some monitoring guidelines include assessing the population levels of beneficial species that provide biological control. Theoretically, a threshold level for an insect or mite pest is reached when the economic damage caused by that pest exceeds the cost of control.

There are currently no numerical guidelines for diseases,

Table 4. Approximating Degree-Days (°D) Manually.

Add the daily minimum and maximum temperature (from March 14 below) and divide by 2 to get the average daily temperature.

$$\frac{45°F + 65°F}{2} = 55°F$$

Subtract the lower threshold temperature (for example, 53°F for California red scale) from the average daily temperature. The result is the approximate number of degree-days accumulated that day.

$$55°F - 53°F = 2°D$$

Add up the degree-days accumulated for each day until you reach the sum when specific actions are recommended. For example, California red scale degree-day monitoring begins each year on the date of the first trapped male scale. Starting then (March 14 in this example), degree-days are calculated each day and added to the degree-days from previous days. Because the red scale crawler stage peaks at about 555°D after the peak in each male flight, 555°D is the best time to spray, assuming that scales are abundant enough that treatment is warranted.

Day	Temperature °F		Degree-days	
	Min	Max	Daily	Accumulated
Mar 14	45	65	2	2
Mar 15	49	64	3.5	5.5
Mar 16	49	63	3	8.5
—	—	—	—	—
—	—	—	—	—
—	—	—	—	—
May 13	56	84	17	533
May 14	52	89	17.5	550.5
May 15	58	96	24	574.5

CUMULATIVE TOTAL HAS EXCEEDED 555°D

Note: This "manual" method of adding and dividing to estimate degree-days can be inaccurate, especially when temperatures are near the threshold. Accurate, easy-to-use computerized sine wave calculations are recommended and are available at www.ipm.ucdavis.edu.

vertebrates, weeds, and most invertebrate species, so treatment decisions for these are based on qualitative assessments. When deciding whether and when to act, consider the extent of the pest infestation, potential for effective and economical control, and any delay between the time you take a management action and the time that control becomes effective. Also consider the history of an orchard or a region, the stage of crop development, and weather. Control action guidelines may change as new cultivars, cultural practices, and pests are introduced and as new information becomes available.

Year-Round IPM

Crop and pest management practices vary according to citrus growth development stage and time of year. Several pests (such as Phytophthora diseases, citrus red mite, and California red scale) often must be monitored and managed at the same time (Figures 10, 11). This pest management must be coordinated with cultural practices such as fertilization, irrigation, and harvest. Each management practice can help prevent, or may inadvertently cause, several different problems (see Table 2).

This manual organizes information by practice, disorder, and pest species. This topical organization makes information easy to find in some ways, but it can make it difficult for advisors and growers to determine the timing and choices of the different actions that need to be taken.

The online *Citrus Year-Round IPM Program* presents the major activities you might need to conduct and integrate during each crop growth development stage. The *Citrus Year-Round IPM Program* guides users through a full-year cycle of monitoring pests, making management decisions, and planning for the following season. In addition to helping advisors and growers improve crop yield and tree health, and better manage pests, this program shows users how to reduce water quality risks and other environmental problems.

Use this manual in combination with the online *UC IPM Pest Management Guidelines: Citrus* and *Citrus Year-Round IPM Program*. These online publications provide information that is not included in this manual, such as

- current monitoring instructions that include decision thresholds
- monitoring checklists and example forms to print and use for recordkeeping
- photo galleries to help you identify the pests and beneficials seen while monitoring
- pesticide recommendations
- a pesticide checklist to identify ways to prevent or minimize negative impacts of pesticide applications
- biological and cultural management alternatives

Management Methods

To promote crop health and minimize tree susceptibility to pests, choose a good growing site and prepare the site well before planting. Select disease-tolerant cultivars and certified, pathogen-free stock. Plant citrus trees properly, such as by placing the root crown high enough so it does not remain continually wet and spacing the trees to allow growth. Conduct cultural practices and monitor environmental conditions to avoid problems, as summarized in Table 2. Install and maintain appropriate irrigation and drainage systems. Use recommended procedures to determine the required amount and frequency of irrigation and fertilization based on tree size and current weather conditions. Prune trees to maintain their health and facilitate their management. Conserve natural enemies, in part by using pesticides only when careful field monitoring indicates that an application is needed to prevent economic loss. Rely on selective pesti-

This electronic unit continuously monitors and records weather in the orchard, such as humidity, precipitation, temperatures, and wind, and automatically transmits this information to a conveniently located remote computer. Weather data from this CIMIS station at Lindcove, and from hundreds of other locations in California, is available online at www.ipm.ucdavis.edu.

CROP DEVELOPMENT PERIOD	Jan	Feb	Mar	Apr	May	Jun	Jul	Aug	Sep	Oct	Nov	Dec
prebloom	• •	• •	• •									
bloom				•								
petal fall				•	• •							
fruit development						• •	• •	• •	• •	•		
fall										• •	• •	• •

PEST	Page in text						TIME OF YEAR						
DISEASES		Jan	Feb	Mar	Apr	May	Jun	Jul	Aug	Sep	Oct	Nov	Dec
Armillaria root rot mushrooms[1]	46	• •										• •	• •
brown rot, Septoria spot	57, 59	• •	• •	• •	• •							• •	• •
dry root rot	44	• •	• •	• •	• •	• •	• •	• •	• •	• •	• •	• •	• •
Phytophthora gummosis	50	• •	• •	• •	• •	• •	• •	• •	• •	• •	• •	• •	• •
Phytophthora root rot	43	• •	• •	• •	• •	• •	• •	• •	• •	• •	• •	• •	• •
stubborn disease	81						•	•	•	•			
tristeza	82				•	•	•				•	•	
NEMATODES	201		•	•	•	•	•	•	•	•	•	•	
VERTEBRATES	209	• •	• •	• •	• •	• •	• •	• •	• •	• •	• •	• •	• •
WEEDS	225		•	•	•	•	•	•	•	•	•		

1. Look for Armillaria root rot disease throughout the year; short-lived *Armillaria* mushrooms are commonly apparent only after fall/winter rains.

Figure 10. Approximate monitoring times (•) for diseases, nematodes, vertebrates, and weeds in citrus.

CROP DEVELOPMENT PERIOD	Jan	Feb	Mar	Apr	May	Jun	Jul	Aug	Sep	Oct	Nov	Dec
prebloom	• •	• •	• •									
bloom				•								
petal fall				•	• •							
fruit development						• •	• •	• •	• •	•		
fall										• •	• •	• •

LOCATION								TIME OF YEAR						
Central Valley	South Coast	INVERTEBRATES	Page in text	Jan	Feb	Mar	Apr	May	Jun	Jul	Aug	Sep	Oct	Nov	Dec
+	+	ants[1]	175		• •	• •	• •	• •	• •	• •	• •	• •	• •	•	
+	+	bean thrips[2]	128							•	• •	• •	• •	•	
−	+	black scale	112			• •	• •	• •			• •	• •	•		
−	+	broad mite	190							• •	• •	• •			
+	+	brown garden snail	199	• •	• •	• •	•		•	•					
−	+	brown soft scale	110							• •	• •	• •	•		
+	+	caterpillars	130	• •	• •	• •	• •	• •	• •	• •	• •				
+	+	California red scale	97				•	• •	•	•	•	•	•		
+	−	citricola scale	107						•			•			
+	+	citrus peelminer	150						•	•	• •	• •	• •	• •	•
+	+	citrus leafminer	152						•	•	• •	• •	• •	•	
+	−[3]	citrus red mite[3,4]	183		• •	• •	• •	• •	• •	• •	• •	• •	•		
+	+	citrus thrips	119				•	• •	•						
+	+	cottony cushion scale	113		• •	• •	• •	• •	• •	• •	• •	• •	• •	•	
+	+	earwigs	172				•	• •	•						
+	+	glassy-winged sharpshooter	167	• •	• •	• •	• •	• •	• •	• •	• •	• •	• •	• •	• •
+	−	grasshoppers	171					•	•						
−	+	greenhouse thrips	126			•	•	•	•	•	•	•			
+	+	katydids	170				• •	• •	•						
−	+	mealybugs	154						•	•	•	•			
+	+	potato leafhopper	169									•	• •	• •	•
+	−	Texas citrus mite[4]	187										•	• •	•
+	−	twospotted mite[4]	186			• •	• •	• •	• •	• •	• •	• •			
−	+	whiteflies	159							•	•	•			
+	−	Yuma spider mite[4]	188							•	•	•			

NATURAL ENEMIES														
		Aphytis melinus, *Comperiella bifasciata* red scale parasitism	100								•	• •	• •	•		
		Euseius tularensis and other predatory mites[3]	194			• •	• •	• •	• •	• •	• •	• •			
		decollate snail	200	• •	• •	• •	• •	•						• •	• •
		Metaphycus spp. soft scale parasitism	111						•	•	•	•			
		sixspotted thrips[3]	196				• •	• •	•						
		Stethorus spider mite destroyer beetle	195						•	• •	• •	• •			
		vedalia beetle	114			• •	• •	• •	• •	• •	• •				

+ Monitor routinely, commonly a pest.
− Routine monitoring probably warranted only in special circumstances, such as in orchards with a history of this problem.

1. When using sweet liquid ant baits, begin monitoring early (about February along the coast, somewhat later in the Central Valley) before Homopteran honeydew becomes abundant. Competition from abundant honeydew reduces ant attraction to sweet baits.
2. Monitor only on navels that may be exported to where bean thrips is a quarantined pest (e.g., Australia).
3. In the South Coast and interior southern California, monitoring for citrus red mite is generally warranted only in the late summer and fall. Monitoring for predatory mites and sixspotted thrips also may be warranted in the late summer and fall to help decide whether pest mites may be a problem then.
4. See "Pest Mites in California Citrus" (Table 10) in the chapter "Insects, Mites, and Snails" for more information on the seasonal importance of mite species by growing region.

Figure 11. Approximate monitoring times (•) for invertebrate pests and key natural enemies in citrus. For the most up-to-date recommendations, consult the online *UC IPM Pest Management Guidelines: Citrus*.

cides when feasible. Anticipate at what stages of crop development various cultural and pest management actions are most likely to be needed by consulting Figures 10, 11, and 12. Consult the *Citrus Production Manual* (Ferguson et al., in preparation) for details on these recommended practices.

Scion and Rootstock Cultivar Selection

The selection of scion and rootstock cultivars is a critical planning decision. The cultivar affects management practices, pest problems, and fruit yield and profits throughout the life of the orchard. First decide what scion cultivar will be used. Scion cultivars include navel, Valencia and pigmented or blood oranges, Eureka and Lisbon lemons, limes and limettas, grapefruit and pummelos, various mandarins (tangerines) and their hybrids (Clementines, Satsumas, tangelos, tangors), and different specialty or niche cultivars (kumquats and their hybrids, citrons, calamondins, microcitrus). Scion choice depends mostly on climatic conditions and the desired market qualities.

Next identify the rootstocks that are compatible with the scion. Rootstock selection influences scion quality. Not all rootstock-scion combinations are compatible, and no combination is resistant to all important diseases and environmental stresses. The best choice depends on climate, fruit quality, and the prevailing pest problems and soil conditions, as summarized in Table 5. For more information, see "Rootstocks for Citrus in California" (in preparation) and the Citrus Clonal Protection Program Web site (http://ccpp.ucr.edu), or consult your University of California Cooperative Extension farm advisor or nursery stock producer.

Sanitation

Practicing good sanitation is a key strategy to prevent problems and to reduce the buildup and spread of insects, pathogens, vertebrates, and weeds. Diseased nursery stock can be a major source of fungal pathogens, nematodes, and viruses. Obtain certified, pathogen-free stock at all times.

TIME OF YEAR

CROP DEVELOPMENT PERIOD	Jan	Feb	Mar	Apr	May	Jun	Jul	Aug	Sep	Oct	Nov	Dec
prebloom	• •	• •	• •									
bloom period				•								
petal fall					• •	•						
fruit development						• •	• •	• •	• •			
fall[1]										• •	• •	•

PRACTICE	Jan	Feb	Mar	Apr	May	Jun	Jul	Aug	Sep	Oct	Nov	Dec
monitor pests regularly[2]	• •	• •	• •	• •	• •	• •	• •	• •	• •	• •	• •	• •
harvest	• •	• •	• •	• •	• •	• •					• •	• •
survey weeds[3]	• •	• •	• •			• •	• •					
frost protection	• •	• •	• •								• •	• •
monitor soil moisture, irrigate if it is dry	• •	• •	• •									• •
irrigate regularly based on monitoring trees' varying need for water				• •	• •	• •	• •	• •	• •	• •	• •	
manage ants[4]			• •	• •	• •	• •	• •	• •	• •	• •		
fertilize				• •	• •	• •	• •	• •				
prune				• •	• •	• •	• •	• •				
sample leaves for nutrients								•	• •	• •		
protect fruit and manage fruit drop and size									• •	• •	• •	• •
whitewash trunks									• •	• •	• •	

1. Fruit continue to develop in some cultivars.
2. Most pests do not need to be monitored as often during the cooler periods (Oct. to Jan.) as during the rest of the year. Specific monitoring times vary according to species and location as listed in Figures 10 and 11.
3. Survey weeds four times each year. The most important times are at least once each during late winter and again in summer.
4. When using sweet liquid ant baits, begin monitoring early (about February along the coast, somewhat later in the Central Valley) before any Homoptera honeydew becomes abundant. Competition from abundant honeydew reduces ants' attraction to sweet baits.

Figure 12. Approximate times (•) for citrus-growing practices in California's Central Valley. Actual times vary according to location, weather, crop history, and crop cultivar and development stage. Testing foliage and whitewashing trees generally are done once during the indicated period. Practices such as frost protection, irrigation, and pest monitoring are ongoing or are repeated at appropriate intervals.

The selection of scion and rootstock cultivars affects management practices, pest problems, fruit yield, and profits throughout the life of the orchard. When planting, chose pest-resistant cultivars and certified, pathogen-free nursery stock.

For example, virus-free budwood is available to nurseries and growers through the Citrus Clonal Protection Program (http://ccpp.ucr.edu).

Other examples of good sanitation include

- staying out of orchards when the soil is wet and preventing soil and water movement from contaminated areas to clean areas of the orchard to stop the spread of soil- and water-borne pathogens and weed seeds
- cleaning all equipment and bins brought in from other areas to avoid introducing insects, pathogens, and weeds
- eradicating small infestations of a perennial weed in or around the orchard before it spreads
- managing vegetation in and near orchards and removing abandoned citrus trees to reduce the likelihood that invertebrates and the diseases they vector will move from there to nearby productive citrus
- reducing debris and weeds to minimize vertebrate cover and habitat

Table 5. Effects of Citrus Rootstocks on Tree Performance, Disease Tolerance, and Fruit Quality.

Rootstock	Yield/tree[1]	Tree size[2]	Tristeza declines tolerance	*Phytophthora* tolerance[3]	Citrus nematode tolerance[4]	Calcareous soils tolerance[5]	Salinity tolerance	Poorly drained soils tolerance	Freeze tolerance	Incompatible scions[6]	Fruit quality
Carrizo/Troyer citrange	I–H	I–Lg	T	T[3]	S–T	P–I	P	I	G	EL, FN?	G
C35 citrange	I–H	I	T	T	R	P	P	I	I	EL?, FN?	G
Swingle citrumelo	I	I	T	T	R	P	P	P	G	EL, O?	G
Trifoliate orange	I–H	I–Lg	T	T	R	P	P	G	G	EL, FNN	G
Cleopatra mandarin	L–I	Lg	T	S	S	I	G	P	I	EL, LL	I
Sour orange	I–H	I–Lg	S	T	S	G	G	G	G	—	G
Rough lemon	I–H	I–Lg	T	S	S	G	I	I	P	—	P
Volkameriana	H	Lg	T	S	S	G	I	I–G	P	—	P
Macrophylla (Alemow)	H	I	S–T	T	S	I	I	?	P	EL, LL[7]	P

Key to effects and relative rating symbols:
G = good
H = high
I = intermediate
L = low
Lg = large

P = poor
R = resistant
S = susceptible
T = tolerant
? = Information uncertain
— = No common incompatibilities

1. Relative comparisons are for 10- to 20-year-old trees.
2. When planted at a density optimized for tree size.
3. Relative tolerance to both Phytophthora gummosis and Phytophthora root rot. There is a range of susceptibility to *Phytophthora* within the "Tolerant" class: C35, Swingle, and Trifoliate are more tolerant to *Phytophthora* than Carrizo.
4. Based on greenhouse seedling tests, which have correlated well with field observations. C35, Swingle, and Trifoliate are resistant to the dominant citrus nematode strain ("Citrus") in California. The "Poncirus" strain of citrus nematode has been reported in California, and it reproduces well on trifoliate and its hybrids; the distribution of "Poncirus" in California is unknown.
5. Soil ratings include anecdotal observations due to insufficient comparative research and difficulties in reliably correlating laboratory soil tests with field observations of tree performance.
6. EL = Eureka lemon, FN = Fukumoto navel, FNN = Frost nucellar navel, LL = Lisbon lemon, O = oranges (all).
7. Eureka lemons often decline by about 10 years, Lisbon lemons by 15 to 20 years.

Adapted from Roose, in preparation.

Soil and Water Management

Proper irrigation and soil moisture management are critical for preventing and managing pest problems. The objective is to supply the tree with the right amount of quality water to produce the optimal yield of high-quality fruit and a healthy tree that is resistant to pests. Most root and trunk diseases, and many disorders and pathogens affecting fruit and shoots, are caused or aggravated by improper moisture conditions. Irrigation affects the location and abundance of weeds and damage by invertebrates such as mites and snails.

Proper irrigation and soil moisture management take into account soil texture, root depth, crop water requirements, water quality and availability, and irrigation method. Keep the soil well aerated by avoiding compaction and waterlogging. Adjust your irrigation practices to the intake

Direct wetting of the trunk by sprinkler irrigation promotes Phytophthora gummosis disease.

Water puddling beneath trees, slow drainage, and poor irrigation management contribute to problems such as root asphyxiation, root rot, and weed growth.

rate of the soil and the needs of the trees. Keep irrigation water off of trunks and prevent waterlogging around root crowns. Sometimes the health of trees already damaged by root diseases can be improved by increasing irrigation intervals, reducing run time, or switching to alternate middle row irrigation or a different irrigation system (such as mini-sprinklers).

See the chapter "Diseases" for discussion and photographs of disorders and pathogens related to soil and water problems. For site selection and planting recommendations to avoid these problems, consult "Establishing the Citrus Grove" (in preparation). For information on soil and water analysis and corrective measures, consult "Soil/Water Analysis and Amendment Strategies" (in preparation). For details on citrus irrigation system design, installation, maintenance, and operation, consult "Irrigation" (in preparation). Online sources of information include the California Irrigation Management Information System, CIMIS program (www.cimis.water.ca.gov) and California State University's Center for Irrigation Technology (www.wateright.org).

Irrigation Methods. Citrus is most commonly irrigated by low-volume (low-flow) systems. Furrows, high-pressure sprinklers, and (least often) basin and flood irrigation are also used. Each system can supply water adequately if managed properly, but irrigation methods differ in their affects on certain pest problems. All irrigation systems, but especially low-volume systems, must be carefully designed and well managed to meet trees' varying daily water requirements and to avoid problems.

Low-volume irrigation generally reduces the percentage of orchard soil surface between trees that is wet and exposed to sunlight. This reduces weed growth, especially when trees are young and most susceptible to weed competition and lack a large canopy to shade out weeds. However, under low-volume irrigations, weed control has to be adjusted because the continuously moist zone around the emitters favors weed growth and the rapid breakdown of herbicides.

Where soil or water is alkaline or salty, and on heavy soils with a low infiltration rate, well-managed low-volume irrigation can be the best irrigation method for avoiding mineral- and water-related disorders and diseases, such as root rots. With low-volume systems, fertilizers and pesticides can be injected into the irrigation water when needed. Disadvantages of low-volume systems include higher maintenance and a smaller amount of reserve moisture in the soil, requiring better system design and operation to avoid tree stress, as well as diseases and mite outbreaks due to water stress.

Basin or furrow irrigation can spread pathogens and weed seeds through the orchard as the water moves. Basin irrigation promotes the development of crown diseases if tree trunks remain wet too long. Root and crown disease can be favored by level furrows because they provide poor drainage for heavy winter rains. Graded furrows improve drainage but usually require a tailwater return system to minimize runoff of potentially contaminated water and to improve irrigation efficiency.

High-pressure sprinkler irrigation is suitable for all soil types and terrains, particularly for sloping or rolling land. The amount of water discharged can be adjusted to the intake rate of the soil, which helps to avoid diseases and disorders related to waterlogging and minimizes runoff. Sprinklers are convenient for short, unscheduled irrigations, for example, to reduce frost damage, which in turn helps to avoid injury-related maladies such as fruit disorders and rots.

Overhead sprinklers can be advantageous for pest control, such as by removing dust from foliage. This reduces mite populations and increases the effectiveness of natural enemies. However, overhead sprinklers are rarely used in citrus because they require a more expensive pumping system and high-quality water to avoid phytotoxicity.

Irrigation Efficiency and Scheduling. Avoid over- or underirrigation by improving your system's irrigation efficiency and water distribution uniformity. Schedule the frequency and amount of irrigation by developing a water budget and by monitoring the weather (local evapotranspiration, or ET), measuring soil moisture, or using a combination of reliable methods.

For example, tensiometers, gypsum blocks, and neutron probes are accurate, reliable devices preferred for monitoring soil moisture to schedule irrigation. These soil moisture monitoring devices are particularly useful for low-volume systems because they can indicate whether the system is maintaining appropriate moisture in the root zone.

The irrigation method and operation affect various pest problems. With low-volume irrigations, the continuously moist zone around the emitters favors the rapid breakdown of herbicides. Irrigation amounts need to be increased in weedy orchards to ensure that citrus roots obtain sufficient water.

With furrow irrigation, two to several furrows are made between each row of mature trees. Soilborne pathogens and weed seeds spread through the orchard as the water moves, so these problems may need to be more carefully managed when using surface irrigation.

Avoid over- or underirrigation by scheduling the frequency and amount of irrigation using reliable methods. This tensiometer with an analog vacuum pressure gauge is one of several accurate devices for measuring soil moisture and scheduling irrigation.

Accuracy, cost, ease of use, portability, reliability, and the extent to which salinity affects readings are primary considerations when choosing among soil moisture monitoring devices. Soil moisture monitoring devices and techniques are constantly being improved.

For more information, see publications such as *Determining Daily Reference Evapotranspiration ETo* (Snyder, Pruitt, and Shaw 1987), *Evapotranspiration and Irrigation Water Requirements* (Jensen, Burman, and Allen 1990), and "Irrigation" (in preparation). For current information, consult experts at your University of California Cooperative Extension office, local resource conservation district, or the Natural Resources Conservation Service.

Fertilizing

Proper nutrition prevents many disorder and pest problems, increases the trees' ability to withstand pest damage, and improves fruit quality and yield. Poor nutrition or inappropriate fertilization causes various disorders and increases damage from certain insects, mites, and pathogens. For example, excess nitrogen application increases populations of aphids, leafminers, and pest mites. Poor application placement or using excess amounts causes injury that promotes dry root rot and Hendersonula tree and branch wilt and increases weed growth. Inappropriate timing of applications (such as application during fall) increases susceptibility to bacterial blast and frost and increases the likelihood of nitrogen leaching from winter rain.

Usually the first indication of nutrient disorders is leaf symptoms, as pictured in the "Mineral Deficiencies and Toxicities" section in the chapter "Diseases." However, the cause of these symptoms is often difficult to diagnose. Symptoms of certain disorders resemble each other or damage caused by certain pathogens. Nutrient disorder symptoms are often secondary, developing as a result of other problems occurring at the same time, so symptoms of one disorder may be masked by those from other disorders or biotic diseases.

Mineral deficiencies occur because of low amounts in the soil or conditions that reduce the tree's ability to absorb nutrients. Excess soil moisture, low temperatures, too much of other minerals in water or soil, and root diseases or injuries can inhibit the plants' nutrient absorption. Use leaf and soil analyses and other knowledge of the local situation (such as soil conditions) to verify your field diagnosis of nutritional disorders. Correctly sample leaves between mid-August through October and have them analyzed by a reputable laboratory. If nutrients are deficient, remedy adverse conditions that prevent their absorption, apply nutrients, or both as summarized in the chapter "Diseases" and detailed in

Low-volume irrigation systems allow chemicals to be applied with the irrigation water. When fertilizing this way, use proper equipment (such as this metered injection pump) and good technique. For example, inject nitrogen only during the last half hour of a several-hour irrigation cycle, then flush the system with plain water for about an additional 30 minutes. This way, the nutrients will not remain in the system and will be retained near the soil surface where most roots occur. There also will be less risk of contaminating groundwater or surface water.

"Fertigation" (in preparation) and "Nutrient Deficiency and Correction" (in preparation).

Frost Protection

Cold damages citrus fruit, limbs, leaves, and twigs and sometimes kills trees, especially when they are young. Frost injury causes several rind disorders and increases the incidence of injury-related fruit rots, gummosis diseases of limbs and trunks, and leaf and twig diseases such as anthracnose and bacterial blast. See the chapter "Diseases" for information on these injury-related disorders and diseases and for cold damage photographs in the "Frost and Freeze" section. Consult "Citrus Frost Protection" (in preparation) for recommended cold-protection methods and when to use them.

Harvest

The harvest season may extend from 4 to 9 months, depending on the cultivar and the market conditions. Navel oranges and mandarins are harvested from November through May or June, Valencia oranges from February through November. Coastal lemons are picked three to six times per year, mainly from March through June. Lemons in the Central Valley are harvested during the winter and spring. Southern interior grapefruit are picked during July and August, after the harvest of the desert grapefruit is completed.

Pest management activities must be scheduled around the extended harvest periods. For the times of year when actions are recommended, consult the *Citrus Year-Round IPM Program*. When treatment is necessary, observe restricted entry intervals and preharvest intervals.

Mature fruit of many cultivars keep on the tree for a long time, but this can increase their exposure to damage by pests and adverse weather. Older fruit may drop or rinds may develop stains, turn puffy or sticky, and become susceptible to decay. Consult the *UC IPM Pest Management Guidelines: Citrus* for recommended plant growth regulators that can help alleviate some of these problems.

Fruit is especially susceptible to injury, rot diseases, and potential contamination with human pathogens during harvest and handling. Good harvest and handling practices can minimize fruit contamination and prevent rind wounds that promote the development of postharvest fruit rots. See the chapter "Harvesting and Pest Management" for more discussion.

Pruning

Citrus is pruned to facilitate access for cultural practices and harvest, to train trees (especially young trees), and to improve penetration of foliar sprays. Pruning off infected limbs can stop the spread of Dothiorella gummosis and Hyphoderma gummosis. Thinning canopies to improve air circulation and selectively removing dead limbs (sources of pathogen inoculum) reduce fruit rots. Pruning inner branches may reduce populations of scale insects by increasing their exposure to summer heat and allowing better access by their natural enemies. Thinning canopies also reduces cover for roof rats. However, depending on how and when it is done, pruning can increase tree susceptibility to sunburn and increase or decrease susceptibility to frost, which affect many injury-related diseases and disorders. Consult "Pruning" (in preparation) for details on how, when, and why to prune citrus.

Citrus are topped to facilitate harvest and spraying. Mechanical pruning machines are used to save labor costs. However, selective (hand) pruning to remove dead limbs and open canopies helps control certain diseases, insects, and vertebrates.

Skirt pruning helps control brown rot of fruit and flightless invertebrates. Pruning lower branches and applying sticky material on top of a trunk wrap to exclude ants can dramatically improve the biological control of many invertebrate pests.

Skirt Pruning. Skirt pruning is especially beneficial for pest control. Skirt pruning is the removal of branches within 24 to 30 inches (30–45 cm) of the ground. Pruning skirts reduces brown rot fruit disease and, when combined with trunk treatments, controls Fuller rose beetles, brown garden snails, earwigs, and ants. Controlling ants increases the biological control of many invertebrate pests. Begin skirt pruning when trees are young, thereby minimizing the temporary yield loss that may occur if skirt pruning is initiated when trees are mature. Raise the height of scaffold branches by removing lower branches and canopy from young trees. Wait until young trunks become corky and a crown has developed (expanded growth at the basal trunk), reducing the likelihood of sunburn to exposed bark. Whitewash exposed wood.

Cover Crops

Maintaining a cover crop in citrus is common only among organic growers and in orchards with soil problems or sloping land. The extra costs of water, nutrients, and management and the potential frost hazard due to the cover crop may outweigh the benefits in most cases. Pests such as pocket gophers, snails, and voles can be more damaging when using cover crops.

Benefits include improving water penetration and aeration of fine-textured or compacted soil, which affect water-related diseases. Cover crops minimize soil erosion (an important consideration on sloping land) and reduce the off-site movement of agricultural chemicals. A ground cover also reduces dust, which contributes to mite outbreaks and interferes with parasitoid activity. Cover crops may provide food and shelter for predators and parasites of citrus pests, but more research is necessary to determine whether this is a significant benefit for pest management.

Cover crops can consist of resident vegetation, one or more planted species (including prostrate winter legumes, such as vetch or clovers), or a blend of resident and seeded vegetation. The best cover crop depends on the age of the orchard, the type of irrigation system, grove location, soil conditions, and weather. The vegetation can be managed by mowing alone or by mowing the row middles and using herbicides around tree trunks. Cultivating cover crops is practical only in young orchards. See the "Weeds" chapter and especially "Integrated Weed Management in Citrus" (in preparation) for more discussion. Other resources include the UC Sustainable Agriculture Research and Education Program's Cover Crops Database (online at www.sarep.ucdavis.edu), *Covercrops for California Agriculture* (Miller et al. 1989), *Cover Cropping in Vineyards* (Ingels et al. 1998), "Ground cover height affects pre-dawn orchard floor temperature" (Snyder and Connell 1993), and *Managing Cover Crops Profitably* (Sustainable Agriculture Network 1998).

Biological Control

Biological control is the action of competitors, parasites, pathogens, and predators to control pests and reduce damage. Augmentation, classical biological control, and con-

Cover crops are beneficial in organic production and orchards on sloping land or problem soils. Cover crops can also increase or reduce certain pest problems, depending on the local situation. The extra need for water and a greater risk of cold damage are reasons why cover crops are not used in most citrus orchards.

servation are three tactics for using natural enemies. IPM programs in citrus depend on the many natural enemies that provide partial or complete control of various invertebrate pests. Some of the most effective natural enemies in citrus today were introduced from other citrus-producing countries. Native species, including parasitic wasps, lacewings, lady beetles, syrphid flies, predaceous mites, thrips, and various fungi and viruses, provide control in some cases or supplement the action of introduced beneficials.

Biological control in citrus is most important for insects, mites, and snails. Table 7 in the chapter "Insects, Mites, and Snails" lists some invertebrate pests, their specialized natural enemies, and the degree of biological control in citrus. The "General Predators" section near the beginning of the chapter "Insects, Mites, and Snails" reviews common natural enemies that prey on a variety of citrus pests. Consult "Citrus IPM" (in preparation) for discussion of biological control's importance by pest species and growing region. Natural enemies apparently play a relatively minor role in controlling plant pathogens, vertebrates, and weeds in commercial citrus. However, for problems such as soil-dwelling pathogens, the role of biological control is not well known. Consult the *Natural Enemies Handbook* for more discussion of these other types of biological control. For pest-specific recommendations on natural enemy conservation and releases, consult the UC IPM *Pest Management Guidelines: Citrus*.

Augmentation. Augmentation is the release of natural enemies when resident natural enemies are insufficient. Some species are commercially available, such as *Aphytis melinus* to control California red scale, the mealybug destroyer (*Cryptolaemus montrouzieri*) for citrus mealybug, and *Trichogramma* spp. for amorbia. Other natural enemies can be collected in the field and relocated and released, like vedalia beetle for cottony cushion scale, *Thripobius* for greenhouse thrips, and *Metaphycus* parasites of soft scales.

Sometimes only one, well-timed "inoculative" release each year is needed, such as when introducing the mealybug destroyer or vedalia beetles. In other cases, to control California red scale in the San Joaquin Valley for example, repeated "inundative" releases of relatively large numbers of *Aphytis* may be needed from spring through fall.

Classical Biological Control. Sometimes called introduction or importation, classical biological control involves the foreign collection and local release and establishment of exotic natural enemies. Many organisms that are not pests in their native habitats become unusually abundant when they are unintentionally introduced into new areas where their native natural enemies do not occur. By law, introduction can only be done by qualified university or government scientists using strict quarantine procedures. Before they make any releases in the new country, researchers use field research in the native habitat and controlled tests in quarantine facilities to confirm that the natural enemies are beneficial and will have little or no negative impact.

Over the past hundred years, many important citrus pests have been brought under biological control by introduced natural enemies. One of the earliest and most successful examples was the establishment of the predaceous vedalia beetle (*Rodolia cardinalis*) and a parasitic fly (*Cryptochaetum iceryae*) against the cottony cushion scale. Subsequently, other major pests were partially or completely controlled biologically, including black scale, brown soft scale, purple scale, yellow scale, and several species each of mealybugs and whiteflies, as listed in Table 7 near the beginning of the chapter "Insects, Mites, and Snails."

Conservation. Preservation of resident natural enemies is the most important biological control method. Help preserve natural enemies in citrus orchards by minimizing pesticides and dust and by selectively controlling ants.

Many naturally occurring parasites and predators are easily disrupted by pesticide applications, and this can result in outbreaks of pests such as caterpillars, mealybugs, mites, scales, and whiteflies. To improve your overall IPM program, carefully select the location, method, rate, and timing of any pesticide applications. Use the most selective and shortest residual (least persistent) insecticide or miticide when possible, as summarized in Table 6. Make spot applications if feasible to minimize harm to nontarget species and provide unsprayed reservoirs from which beneficials can recolonize trees. Minimize numbers and/or rates of applications. An occasional single application of broad-spectrum pesticides

This tiny *Aphytis melinus* wasp is the most important of several species of parasites that attack California red scale. Parasitism can sometimes be recognized by changes in host color or a hole in its cover. Or, it may be necessary to remove the scale cover and look for immature stages of the parasite or their signs, such as fecal pellets.

is less harmful to natural enemies than multiple broad-spectrum treatments.

Minimize dust in the orchard. Dusty conditions reduce the reproduction and host-finding ability of many natural enemies, and this in turn causes outbreaks of pests such as mites and scales. Drive through the orchard slowly and only when necessary. Periodically wet dirt roads during periods of high use. Plant windbreaks along orchard borders. Be careful about the timing of lime or kaolin treatments, as these materials will coat leaves like dust and limit natural enemy activity.

Control ants, which feed on honeydew and protect the honeydew-producing insects from their natural enemies. Because ants attack almost any parasite or predatory insect they encounter, they also interfere with the biological control of pests that do not produce honeydew, such as California red scale and citrus red mite. Prune skirts to eliminate canopy bridges and apply a polybutene-based sticky material on top of a trunk wrap as an ant barrier. Alternatively, apply an effective insecticide ant bait as recommended in the online *UC IPM Pest Management Guidelines: Citrus*.

If biological control has been disrupted, remedy the problem (such as by controlling ants and switching to selective pesticides), then consider purchasing or field-collecting and releasing additional parasites or predators of the problem pests. Consult the Association of Natural Biocontrol Producers (www.anbp.org) for commercial sources of natural enemies.

Pesticides

Within an IPM program, treat for specific pests only when monitoring indicates that economically damaging populations are present or anticipated. Pesticides are valuable tools because they can drastically reduce pest populations or prevent further damage within a short time after application. Where pests are already causing economic damage or are expected to soon, pesticide application may be the only effective action.

Pesticides include fungicides, herbicides, insecticides, miticides (acaricides), nematicides, and rodenticides. They belong to a variety of chemical groups with very different properties. To control one pest, you can often choose among several active ingredients or several different formulations that contain the same active ingredient. You may also be able to select one chemical to control several pests at the same time, or mix compatible materials together. The choice

Table 6. Examples of the Relative Selectivity of Selected Pesticides Used in Citrus.

	Selective or soft	Intermediate	Broad-spectrum or toxic
Botanicals	azadirachtin pyrethrins sabadilla		
Microbials or their by-products	abamectin[1] *Bacillus thuringiensis*		
Oil	oil ≤ 0.5%	oil > 0.5%	
Inorganics	cryolite		sulfur
Miticides	fenbutatin oxide hexythiazox	acequinocyl fenpyroximate propargite pyridaben spirodiclofen	dicofol
Fermentation products	spinetoram spinosad spirotetramat		
Insect growth regulators		buprofezin diflubenzuron pyriproxyfen[1]	
Neonicotinoids		acetamiprid imidacloprid (soil applied)[1]	imidacloprid (foliar spray)
Organophosphates		chlorpyrifos (low rates) phosmet	chlorpyrifos (higher rates) dimethoate malathion methidathion naled
Pyrethroids		cyfluthrin (low rates)	cyfluthrin (higher rates) fenpropathrin
Carbamates			carbaryl formetanate hydrochloride methomyl

For a comprehensive list of pesticide selectivity and problems such as pesticide resistance, consult the online *UC IPM Pest Management Guidelines: Citrus*.

1. When used in ant baits, these materials are "Selective or soft."

will depend on the degree of control you need, the effect on beneficial species and other pests, and the various economic and legal restrictions. Where feasible, choose the effective material that is least disruptive to biological control.

The microbial insecticide *Bacillus thuringiensis* (Bt), for example, is desirable in IPM programs because it selectively kills caterpillars that eat treated leaves, and it is nontoxic to natural enemies and other nontarget species. Certain other insecticides have modest adverse impacts on beneficials because they are chemically selective for a narrow group of invertebrates (sabadilla for thrips), have only short-term residual toxicity (oil and pyrethrins), or are applied in a manner that prevents exposure to most nontarget species (ant baits).

When choosing pesticides for your pest management program, keep in mind that insects, pathogens, and weeds can develop resistance if you repeatedly use the same pesticide or a group of pesticides from the same chemical class. Learn the chemical classes and rotate to different classes within and between seasons. See "Pesticide Resistance" below and consult "Managing Pesticide Resistance in Insects, Mites, Weeds, and Fungi Infesting California Citrus" (in preparation) for more information.

For discussion on integrating pesticide use with biological control and other IPM tools, consult "Citrus IPM" (in preparation). For details on pesticide selectivity and problems such as pesticide resistance, consult the *UC IPM Pest Management Guidelines: Citrus*.

Oils. Oils are uniquely important in citrus IPM because they have little residual toxicity for beneficial species and effectively control many invertebrates. Rates and coverage vary depending on the amount of oil you want to apply, the crop, time of year, temperature, the target pest, whether oils are used alone or combined with other insecticides, and which oil is used. You can choose from several grades of oil classified according to the size of their hydrocarbon molecules. The molecule size, which determines volatility and thus persistence, increases from narrow-range (NR) 415 to NR 440. The heavier the oil (NR 440 is heavier than NR 415), the better its insecticidal properties will be but also the greater the potential for phytotoxicity. Persistence should be long enough to kill the pest, yet short enough not to damage fruit or trees.

Extensive research on the use of oil sprays against various mite and scale insects has resulted in recommendations for specific rates and timing of treatments that vary according to the cultivar of citrus and region of California. Consult the *UC IPM Pest Management Guidelines: Citrus* for pest-specific recommendations and *Managing Insects & Mites with Spray Oils* (Davidson et al. 1991) for more details on oils.

Using Pesticides Effectively

The success of many IPM programs depends on the correct application of pesticides when they are needed. Careful use makes pesticides more effective and reduces the likelihood of pesticide resistance, pest resurgence, secondary outbreaks, crop injury, and hazards to humans and the environment. The safe and effective use of pesticides requires correct calibration and operation of equipment and the proper coverage, rate, and timing of application.

Consult the online *UC IPM Pest Management Guidelines: Citrus* for pest-specific application recommendations. For more information on pesticide mixing and applica-

Ants attack almost any beneficial parasite or predatory insect they encounter. Several approved bait stations can be purchased (such as this blue plastic one) or constructed to dispense sweet liquid bait to control honeydew-feeding Argentine ants and gray ants.

Air blast (air carrier) sprayers can provide good spray coverage using a relatively small volume of material. Because it uses a relatively high concentration of active ingredient in the mix and produces small droplets, extra care may be necessary to minimize worker exposure and drift.

Adjustable air tower sections are installed on this air blast sprayer to improve pesticide coverage on the tree tops.

tion, including equipment calibration, consult "Pesticide Principles" (in preparation), *Pesticide Safety: A Reference Manual for Private Applicators* (O'Connor-Marer and Cohen 2006), and *The Safe and Effective Use of Pesticides* (O'Connor-Marer 2000).

Application Timing. Try to apply a pesticide when the target pest is at its most vulnerable state, while also using treatment times and methods least likely to affect natural enemies. For example, use bait stations for vertebrates to exclude nontarget species and make sure it is a time of year when pest vertebrates are active and likely to feed on bait. Ground squirrels prefer to feed on green vegetation during late winter and spring; baiting then is ineffective because the squirrels will not eat bait at that time of year. (See Figures 51 and 55 in the chapter "Vertebrates.")

When treating weeds, make sure they are at the growth stage recommended on the pesticide label. An herbicide label may state "Apply to actively growing weeds" because the material will not kill weeds unless they are growing vigorously enough to absorb or systemically move the herbicide (see Figure 59 in the chapter "Weeds"). Herbicide rates may need to be adjusted for climatic conditions, irrigation method, and soil type.

Coverage. For many invertebrates, the most effective treatment time and what part of the canopy to treat vary according to pest species. Although certain systemic materials are moved throughout the plant, with most pesticides, obtaining the proper distribution of a spray on or within the tree (coverage) is essential. For example, outside coverage (OC) spray is best for aphids, caterpillars, citrus thrips, and mites because they occur mostly on the periphery of the tree. Armored scales need more thorough coverage (TC) because they occur deep in the interior of the tree. Thorough coverage is especially needed with petroleum spray oils because oils act only on contact to smother the pest or render the plant surface unusable by pests. Oil spray must completely cover infested plant parts to be effective.

Coverage is affected by the amount of material applied, the size of spray droplets, and the speed of the spray rig. The amount of spray mixture applied per acre or per tree (gallonage) varies according to tree height, density of foliage, application equipment and method, required coverage, and whatever instructions are provided on product labels. Pesticides are applied either as concentrate sprays (low-volume or no-drip sprays) or as dilute sprays (high-volume or runoff sprays). High-volume (dilute) sprays apply up to about 2,000 gallons per acre using larger droplets. Larger droplets are less likely to drift, so off-site movement in air is reduced. Dilute sprays generally pose less risk of worker hazard and phytotoxicity because of their lower concentration of active ingredient. Dilute sprays must be applied using very slow tractor speed to avoid causing the citrus foliage to form a sheet that prevents penetration of the spray.

Some pesticides are most effective when applied as a high-gallonage dilute spray, such as spray oils. Other materials are most effective when applied as a low-volume concentrate spray, such as Bt. For some pesticides, concentrate sprays can be as effective as dilute sprays as long as they are applied from a properly calibrated sprayer.

Calibration. Calibration is the correct adjustment of the sprayer so that it delivers the recommended gallonage. Good coverage and effective pest control depend on accurate calibration. The calibration steps differ according to the equipment and whether trees or acres are the basis for determining gallonage. First determine how many gallons per minute (gpm) must be discharged to deliver the proper amount of material to each tree or acre. Then equip your machinery with nozzle sizes and arrangements that will emit the gallons per minute of spray at the appropriate pressures and moving at the required ground speed. Once you have calibrated your sprayer, then calculate how much pesticide you need per tank to apply it at the rate directed on the product label and in the online *UC IPM Pest Management Guidelines: Citrus*.

Equipment. Citrus is usually sprayed with mechanized ground equipment. Alternatively, hand-spraying is used for spot treatments, dense plantings, and trees next to buildings, borders, or on hilly terrain. Aircraft are sometimes used for certain outside coverage applications, such as on small trees and when the terrain or ground conditions do not allow ground application.

Application equipment operates on various principles to form the spray droplets and propel them into trees. The oscillating boom sprayer uses relatively high hydraulic pressure and no forced air to move liquid pesticide mixture through guns oscillating on a vertical boom. Air blast (air carrier) sprayers pump a pesticide-water mixture to a series of atomizing nozzles on a manifold that directs the pesticide mixture into a blast of air from a fan that carries the spray into the tree.

Unlike conventional air blast sprayers, multi-fan tower sprayers release pesticide into adjustable fans that are evenly spaced on adjustable towers. These sprayers can reduce the volume of water used per acre and can improve spray deposit on the crop while reducing drift.

Low drift venturi nozzles use a flow-metering preorifice (an entrance-side flow restrictor) and air inlet to mix pesticide solution and air into relatively large droplets. Using venturi nozzles on air blast sprayers minimizes the formation of small droplets prone to drift.

Most types of sprayers can be used to effectively control many different pests, assuming correct calibration, coverage, spray volume, and application timing and that equipment is properly operated. For all equipment, proper operation and maintenance are critical for achieving effective pest control. Maintenance includes checking nozzles regularly for plugging and wear and replacing those that are worn. Correct operation includes using the appropriate equipment pressure settings and travel speed when you run the sprayer. For example, if ground speed is too fast, much of the spray will not reach the target (coverage will be poor) and the pest population will not be controlled. If you drive too slowly, the spray will overlap and be wasted and the risk of phytotoxicity and off-site movement may increase.

Problems with Pesticide Use

The attributes that make pesticides valuable control tools can sometimes cause problems that impede the success of an IPM program. Problems include the development of pesticide resistance, pest resurgence, and secondary pest outbreaks. Pesticide applications may adversely affect honey bees, nearby crops, wildlife, and human health. Minimize pesticide use and rely on alternate methods where feasible to help avoid these problems.

Pesticide Resistance. Repeated use of the same pesticide or pesticide class promotes resistance by selecting for (favoring the survival and reproduction of) pest individuals that are less susceptible to that pesticide (Figure 13). Resistance renders ineffective any subsequent treatments of that pesticide or of other pesticides that have the same mode of action

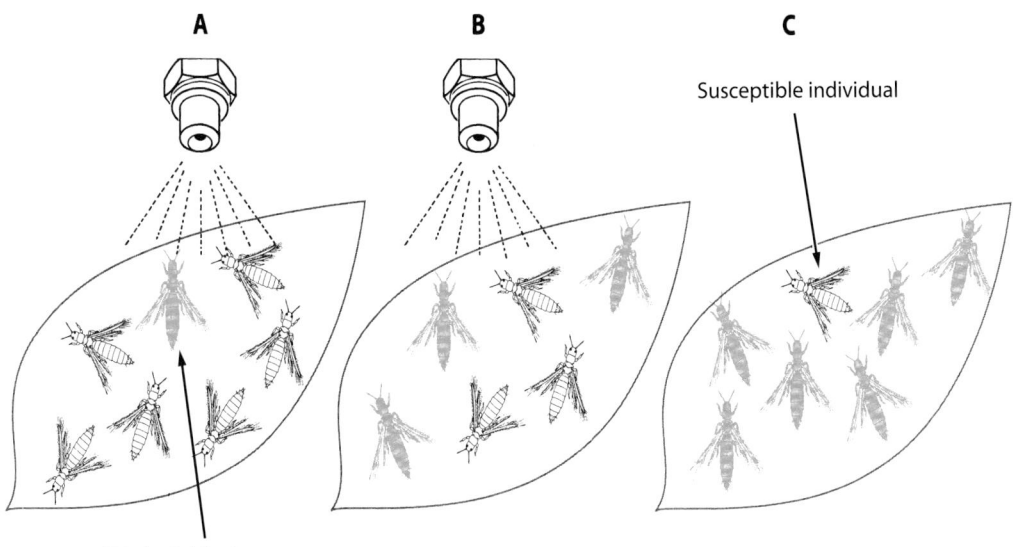

Figure 13. Resistance to pesticides develops through genetic selection in populations of pests, including insects, mites, pathogens, and weeds. A. Certain individuals in a pest population are naturally less susceptible to a pesticide than other individuals. B. These less-susceptible pest biotypes are more likely to survive an application and to produce offspring that are also less susceptible. C. After repeated applications over several generations, the population consists primarily of less-susceptible individuals, so that pest population is resistant to pesticides. Applying the same pesticide, or other chemicals with the same mode of action, is no longer effective.

(the same way of killing pests). Resistance is especially likely to develop in populations of pests that reproduce rapidly, including insects, mites, and pathogens. Although less common, resistance also occurs in weed species.

Historically, the development of resistance has required growers to periodically change how they manage citrus pests. At least some populations of California red scale, citricola scale, citrus red mite, citrus thrips, and twospotted mite are resistant to certain pesticides. Some populations of the natural enemies *Aphytis melinus*, *Euseius tularensis*, and vedalia beetle have also developed resistance to certain pesticides.

To slow the development of resistant pests, use control methods other than pesticides where feasible. Make applications only when needed and keep the area treated to a minimum to allow the survival of natural enemies. Leaving some areas unsprayed also allows populations of untreated and treated pests to interbreed, reducing the overall rate at which pesticide resistance develops. When you do apply pesticides, alternate (rotate) applications among several pesticides with different modes of action as listed in the *UC IPM Pest Management Guidelines: Citrus*. For more information, consult "Managing Pesticide Resistance in Insects, Mites, Weeds, and Fungi Infesting California Citrus" (in preparation) and the current versions of *Fungicides Sorted by Modes of Action* (online at www.frac.info), *Classification of Herbicides According to Mode of Action* (www.hracglobal.com), and *IRAC Mode of Action Classification* (www.irac-online.org).

Pest Resurgence. Pest resurgence is the rapid rebuilding of a pest population that was recently controlled by a pesticide application. Insecticides and miticides that kill or inhibit the effectiveness of natural enemies sometimes cause pest resurgence. Natural enemies depend on the pest for food, and so they take much longer than the pest to build to their former numbers. Meanwhile, pest individuals that survive the treatment reproduce without the restraint of natural enemies, sometimes building to greater numbers than existed before the treatment. Sometimes pests show hormoligosis, that is, they actually increase their reproduction when treated with certain pesticides. If faced with resurgent populations, you may have to spray repeatedly, increasing production costs and the chances for secondary pest outbreaks and promoting the development of pesticide-resistant pests. Minimizing applications and choosing materials least toxic to natural enemies whenever feasible help avoid pest resurgence.

Secondary Pest Outbreak. Secondary pest outbreaks can occur when a pesticide applied to control one pest unintentionally causes the buildup of other pests that were previously controlled by natural enemies or by competition with the primary pest. For example, broad-spectrum insecticides applied for citrus thrips or caterpillars are highly toxic to mite natural enemies and can cause outbreaks of the citrus red mite (Figure 14). To reduce the chances for pest resurgence and secondary pest outbreaks, choose selective materials when possible (Table 6). Time applications to be

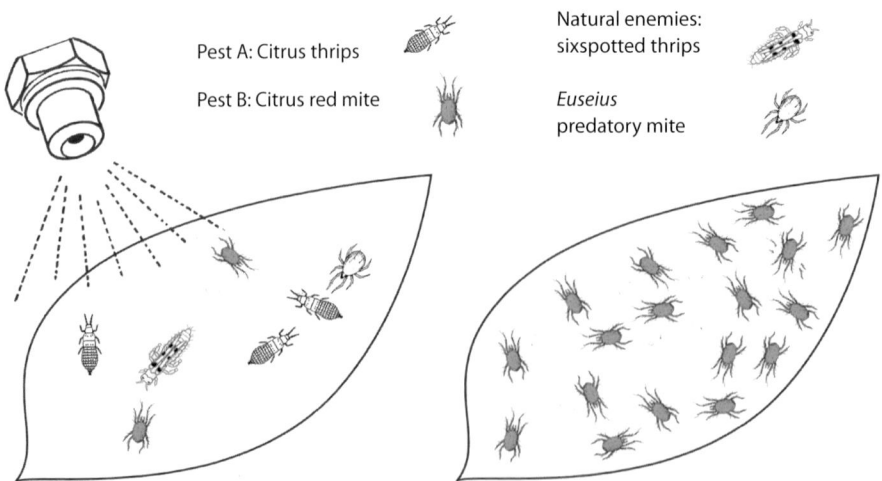

Figure 14. Disrupting biological control often results in secondary outbreaks of insects and mites. For example, many potential pests such as citrus red mite are present at nondamaging densities but become abundant and cause damage only when ants, dust, or broad-spectrum pesticides prevent natural enemies from being effective. Here, a nonselective persistent pesticide applied for citrus thrips (Pest A) also killed predaceous sixspotted thrips and *Euseius tularensis* predatory mites, leading to a secondary outbreak of citrus red mite (Pest B).

Citrus blooms are an important food source for honey bees. Healthy bee colonies are vital to many California crops. Learn and follow all regulations to avoid pesticide poisoning of honey bees.

most destructive to the pests and least destructive to the beneficial species.

Phytotoxicity. Many pesticides used in citrus can directly damage plants (cause phytotoxicity) if used incorrectly. Growth regulators (such as 2,4-D*), herbicides (which are designed to kill plants), and spray oils present the most risk of phytotoxicity. Almost any pesticide may damage foliage and fruit when applied under hot, dry conditions. Follow all precautions on product labels. Consult the UC IPM Pest Management Guidelines: Citrus for situation-specific recommendations on avoiding phytotoxicity. For more details and photos, see the "Phytotoxicity" sections of the chapters "Diseases" and "Weeds."

Hazards to Honey Bees. Many insecticides are toxic to honey bees. Insecticide and miticide toxicity to honey bees ranges from highly toxic to relatively nontoxic. Insecticides can kill bees directly while they forage in the orchard or indirectly through contaminated food brought back to the hive. Most citrus cultivars, except tangerine and tangelo, do not require pollination, but honey bees forage extensively in the orchards during bloom. As an early-blooming crop, citrus provides a valuable food source for honey bees as they build their colonies in the spring. Strong colonies are vital to California agriculture to pollinate summer crops, such as melon, squash, and seed alfalfa, and the following year's almond and fruit trees.

Check with your county agricultural commissioner to find out whether there are beekeepers in your area who-should be informed of your scheduled spraying and to learn about application restrictions to protect honey bees. For example, in the San Joaquin Valley Citrus/Bee Protection Area, certain applications may be entirely prohibited in citrus during the bloom period as declared by the agricultural commissioners. Consult the latest UC IPM Pest Management Guidelines: Citrus or How to Reduce Bee Poisoning From Pesticides (Mayer et al. 1999) for the honey bee toxicity rating of each pesticide.

Hazards to People. Pesticides are potentially hazardous to people. Applicators and handlers are most heavily exposed to pesticides and may be at risk of acute poisoning or chronic ailments if all safety regulations are not followed. Harvesting crews and other orchard workers also may be exposed to toxic residues that may remain in the orchard after application.

Make good choices on when and how to use pesticides to avoid human exposure. Take steps to prevent pesticides from drifting off the application site, such as by ensuring that wind speed and direction and spray droplet size are appropriate for that application. Prevent pesticide movement into sensitive areas, such as when making applications near an agricultural-urban interface.

Minimize health problems by following label directions, wearing appropriate work clothing and personal protective equipment, using closed mixing and loading systems where possible, observing state and local regulations, and confirming the availability of emergency medical care. Properly train and supervise employees, including field workers who do not work directly with pesticides. Ensure that restricted entry

Adopting IPM methods can reduce potential conflicts with neighbors concerned about pesticide drift and runoff. When using pesticides, be sure to follow all safety procedures and regulations.

*Restricted-use material. Permit required for purchase or use.

Consider the surrounding environment when planning and taking pest management actions. Take precautions to prevent pesticide runoff or spray drift from contaminating water bodies or sensitive wildlife habitat areas.

intervals and preharvest intervals are followed. Always post sprayed orchards as required by law. Consult *Pesticide Safety: A Reference Manual for Private Applicators* (O'Connor-Marer and Cohen 2006) and *The Safe and Effective Use of Pesticides* (O'Connor-Marer 2000) for more details.

Hazards to Wildlife. In areas where endangered or protected wildlife species occur, special guidelines apply to the types of pesticides you can use and the ways you can apply them. A list of the endangered species, their locations, and specific guidelines for using pesticides near endangered species habitat are available from the California Department of Pesticide Regulation (DPR) Web site (www.cdpr.ca.gov/docs/es/index.htm) and from your county agricultural commissioner.

Hazards to Water Quality. Fertilizers and pesticides may contaminate water, such as through leaching, runoff, or spray drift. The federal Clean Water Act requires a written plan (called a total maximum daily load, or TMDL) for every water body that is impaired and does not meet water quality standards. Many California rivers and creeks are impaired. Managing the off-site movement of soil and water is recommended and may even be required, especially for orchards in watersheds with impaired waterways (Figure 15). Take special precautions, select less-harmful materials, or avoid using pesticides where runoff or spray drift is likely to contaminate nearby bodies of water or sensitive areas. Stay informed about regulations that are designed to protect water supplies. Sources of information include the California State Water Resources Control Board, your regional water quality control board, and publications such as *Tillage and Crop Management Effects on Air, Water, and Soil Quality in California* (Horwath, Mitchell, and Six 2008). Advice or funding to help you reduce the off-site movement of materials may be available from the federal Natural Resources Conservation Service (NRCS) or your local resource conservation district.

Air Pollution. Air quality is below state and federal standards in much of California. Agricultural practices can contribute to air pollution. For instance, cultural operations can generate very small particulates that cause lung damage if inhaled. Certain pesticides (fumigants and emulsifiable concentrate [EC] formulations), vehicles, and livestock waste emit volatile organic compounds (VOCs). VOCs and nitrogen oxide combine with sunlight to form ozone. Ozone present at ground level can reduce crop yields and harm human health, causing respiratory irritation and sicknesses.

The federal Clean Air Act requires states to work toward meeting air quality standards, including limits on ozone in the lower atmosphere. The DPR and the California Air Resources Board are reducing pesticide VOC emissions through research, regulations, and education to promote alternatives to VOC-generating practices.

Employ IPM techniques to minimize pesticide use and the release of VOCs. To the extent that it is compatible with your other IPM goals, minimize the size of the treatment area, reduce application rates, and use methods that minimize volatilization—for instance, apply materials through drip or other low-volume irrigation systems instead of using aerial application or foliar sprays. If pests can be effectively controlled at various times of the year, consider avoiding applications when air quality is poor.

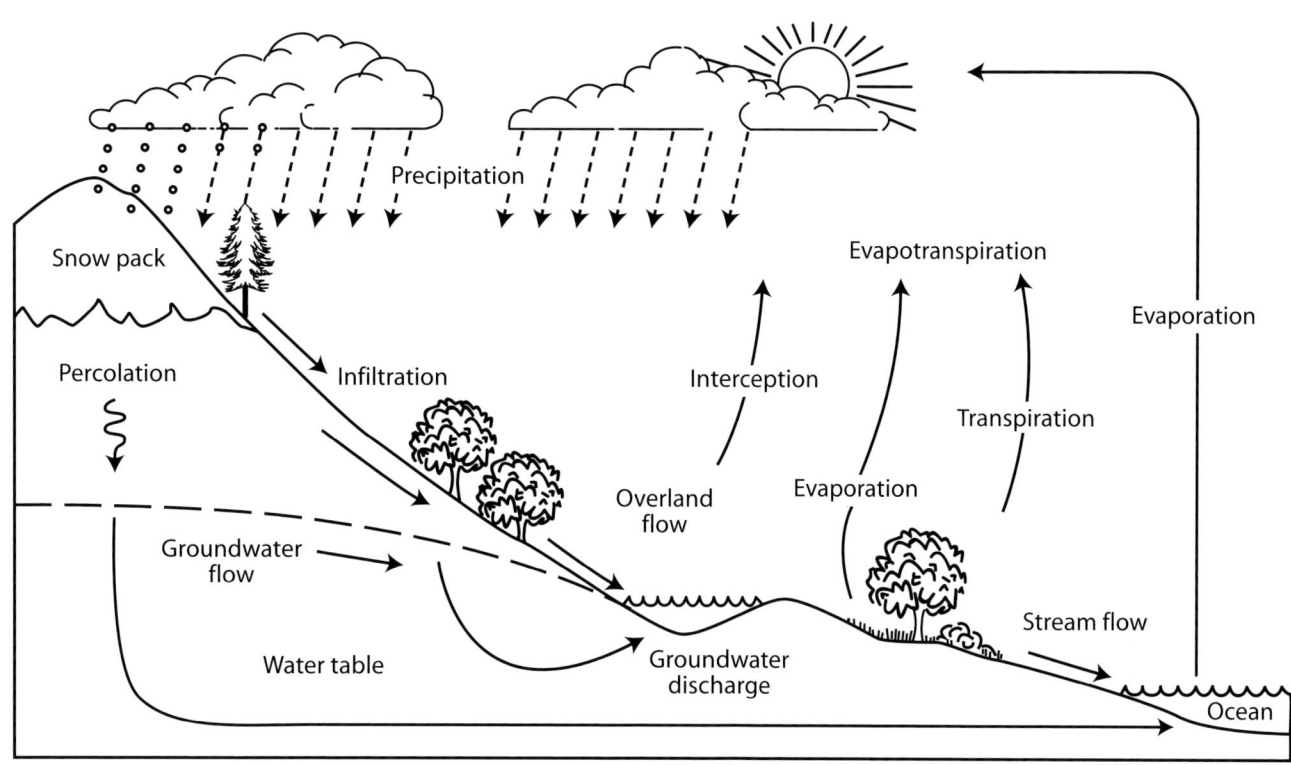

Figure 15. Pesticides and fertilizers applied in orchards can contaminate groundwater or surface water by moving as shown here by the hydrologic cycle. Avoid contamination by making good application decisions, carefully managing irrigation, and using methods such as untreated buffers and vegetative filter strips to minimize off-site movement. Adapted from Padgett-Johnson and Bedell 2002.

Diseases

Disease is an abnormal condition that damages plants. Diseases are caused by pathogenic microorganisms and environmental stresses. Microorganisms include bacteria, fungi, nematodes, phytoplasmas, viruses, and viroids that produce biotic or infectious diseases. Noninfectious or abiotic diseases (commonly called disorders) include those caused by adverse soil and weather conditions, genetic mutations, and nutrient deficiencies. These disorders can create conditions favoring the development of biotic diseases or produce symptoms of their own. Abiotic and pathogenic causes of diseases are discussed below, except that nematodes are discussed later, in their own chapter.

In California, several viruses, fungi, and funguslike Oomycetes, and one bacterial species cause the major biotic diseases in citrus. Among the fungi and Oomycetes, those infecting the roots and trunk are of greatest concern. Most diseases of fruit are less important in the semiarid climate of California. However, brown rot caused by *Phytophthora* species and postharvest decays of fruit caused by *Penicillium* species can be significant problems requiring regular preventive management.

The several viruses, viroids (subviral plant pathogens lacking a protein coat around their genetic material), and a spiroplasma (a phytoplasma- or bacteriumlike organism lacking a cell wall) can infect all parts of the tree. Many viruses and viroids infect citrus as complex populations of closely related but different pathogens interacting in manners that are not completely understood. Fortunately, careful propagation and planting of certified disease-free stock can prevent many infections.

The severity of a pathogenic disease is influenced by the virulence of the pathogen, the genetic susceptibility and growth stage of the host, and environmental conditions such as moisture and temperature. The interaction of these factors is referred to as the disease triangle (Figure 16). The effect of a pathogen on a tree also depends on the site of infection. If the feeder roots, which are important in the uptake of water and nutrients, are diseased, the tree will suffer more readily than when some leaves and twigs are infected.

The importance of some diseases varies by growing region. Botrytis rot is mostly a problem in coastal areas, which frequently experience mild temperatures and higher relative humidity. Stubborn disease is a problem in the hot climate of the San Joaquin and Coachella valleys. Citrus

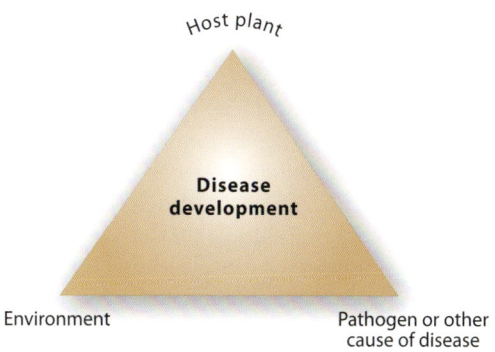

Figure 16. The disease triangle. Disease results from interactions among a susceptible host plant, an environment favorable for disease development, and virulent pathogens. Time (such as season or duration of disease-favoring conditions) is also a factor, and this disease-development interaction is sometimes illustrated as a pyramid with time as a fourth axis.

blast occurs almost exclusively in northern California, where long, wet and windy periods in winter and spring promote bacterial disease. However, many diseases occur in most growing areas. Their damage potential depends more on soil conditions, water management practices, and on the planted rootstock and scion cultivars than on climate or other regional differences.

California citrus is also threatened by exotic (foreign) diseases. Asia, Central and Latin America, Mexico, and southern states (from Florida to Texas) have many serious citrus diseases that have not yet established in California. Exotic diseases to look out for include citrus bacterial canker, Huanglongbing (citrus greening), citrus variegated chlorosis, and citrus leprosis. Help keep exotic pests out of California by obeying all quarantine regulations. Purchase plants only from reputable, local sources. Many citrus pests are harbored in ornamental or noncommercial citrus rela-tives. Do not carry any kind of plant or fruit across state lines or, most importantly, across country borders. Promptly report any suspected new problems to agricultural officials.

Many diseases have ultimately the same effect: a declining tree with light green foliage, poor growth flushes, premature leaf drop, and twig dieback. Primary symptoms can appear on roots and trunks, fruit, or leaves and twigs. Symptoms can also affect the overall growth habit and size of the tree and crop yield. Because primary symptoms help in identifying a disease, this chapter is organized according to where the major symptoms of a disease occur on the tree. Keep in mind that other pests, such as nematodes and gophers, or poor water management practices can also cause decline.

This chapter will help you to diagnose the cause(s) of diseases, understand the biology of their development, and learn the key preventive and management methods to keep trees healthy. For more details, consult the *Citrus Production Manual* (Ferguson et al., in preparation), especially the chapters "Bacterial Diseases of Citrus"; "Environmental, Physiological, and Cultural Injuries and Genetic Disorders"; "Fungal Diseases of Citrus"; and "Virus and Viroid Diseases of Citrus." Other good resources include the Citrus Clonal Protection Program (online at http://ccpp.ucr.edu) and publications listed in "Suggested Reading". For new problems and disease-specific treatment recommendations, consult the latest UC IPM *Pest Management Guidelines: Citrus* (online at www.ipm.ucdavis.edu).

Monitoring and Diagnosis

While walking through the orchard, look for weak trees and irregular growth patterns. Check problem areas such as low-lying sites and areas with fine-textured (high in clay content) or shallow soil. Monitor more frequently when adverse weather persists or where trees are susceptible to diseases prevalent in the growing district. Examine all parts of the plant to diagnose a problem. For example, look at the trunk for gumming and scaling or shelling (peeling off) of the bark. To reveal injured tissue, you may have to scrape off weak or discolored bark, dig up some feeder roots, or remove soil from the crown or the main lateral roots extending outward to the edge of the canopy. A chisel, hand lens, hatchet, knife, and shovel are useful tools for diagnosing many problems. Some diseases, such as viral diseases like tristeza, are difficult to identify by field symptoms; positive identification requires laboratory analysis.

Time of year influences what symptoms you may see: bacterial and fungal diseases of fruit and leaves cause the most damage during wet periods occurring in the winter months. Diseases affecting the roots and trunks can become more apparent when trees are likely to be stressed during the hot, dry summer months. Which problems to be especially alert for based on the time of year are summarized in Figures 10 and 11 in the chapter "Managing Pests in Citrus." For

The yellow leaves and dead leafless branches on this navel orange are the result of Phytophthora gummosis due to infection by *Phytophthora citrophthora*. Because other problems can produce this same damage, laboratory analysis of properly collected samples may be needed to confidently diagnose the cause of these symptoms.

To help diagnose the cause of disease, you often need to examine beneath the surface of bark or soil. Digging up feeder roots (shown here) to see whether or not they look healthy may be necessary to identify the cause of unhealthy fruit and leaves.

well-illustrated monitoring recommendations based on crop development stage, consult the latest *Citrus Year-Round IPM Program* (online at www.ipm.ucdavis.edu).

Certain rootstock-scion combinations as well as individual rootstocks or scions are prone to incompatibilities or genetic disorders. Some of these disorders can be diagnosed by field symptoms, such as shelling or necrosis; others can be identified only by microscopic examination of tissues. Some disorders do not become evident until long after the trees are in production, often 10 to 15 years after planting. Genetic mutations sometimes cause abnormal tissue growth, affecting fruit shape or color as well as the color or growth habit of leaves and twigs.

Compare your field observations with the descriptions and photographs in this manual. A combination of symptoms usually will identify a disease. If you find symptoms not described here, see the reference books on diseases listed in "Suggested Reading" or consult your University of California farm advisor or a PCA.

Keep accurate records of your observations, including the locations of infected trees or those suspected of being diseased, such as on a map of your grove or by using a GPS (global positioning system). Positive diagnosis, especially of slowly developing viral diseases, is often not possible at first inspection. Repeated monitoring and mapping will help you confirm your diagnosis, follow the development of the disease, make proper management decisions, and revisit problems after taking action to evaluate the effectiveness of your control efforts.

Prevention and Management

Keep trees healthy and vigorous to make them more resistant to infection and disease development and to help prevent numerous pest problems, as summarized in the chapter "Managing Pests in Citrus." Prevention is the most economical way to manage diseases in citrus. Moreover, for most diseases, prevention is the only effective strategy. Preventive measures include good site selection and preplant soil preparation, choosing disease-free and disease-resistant cultivars, proper planting, careful soil and water management, appropriate cultural care, and good sanitation. For certain diseases, fungicide application is an important tool, as recommended in the online *UC IPM Pest Management Guidelines: Citrus*.

Site Selection. Disease management starts before planting. Select a suitable site with deep, level soil and good drainage. Where needed, modify the soil and planting methods to improve soil aeration and drainage, as discussed below under "Root Rots."

Disease-Resistant and Pathogen-Free Stock. Select disease-resistant rootstock and scion cultivars. Use only budwood that is certified to be free of known pathogens. The Citrus Clonal Protection Program (CCPP, http://ccpp.ucr.edu/budwood/budwood.php) provides California nurseries and growers with true-to-type budwood of commercially important scion and rootstock varieties. This budwood is regularly tested according to the registration program of the California Department of Food and Agriculture to ensure that stock is free of graft-transmissible diseases. Consult "The California Citrus Clonal Protection Program" (in preparation) for more details.

Because no rootstock-scion combination is resistant to all important diseases and environmental stresses, choose cultivars based on the prevailing disease problems, soil conditions, and climate at that site. Regardless of cultivar, purchase only healthy stock that is certified as pathogen-free. For diseases caused by viruses, using resistant cultivars and pathogen-free stock are the only effective management methods. See Table 5 in the chapter "Managing Pests in Citrus" and "Rootstocks for Citrus in California" (in preparation) for specific recommendations. Consult your local University of California Cooperative Extension farm advisor or nurserymen for information that helps you select the appropriate cultivars for your orchard conditions.

Cultural Practices. Provide trees proper cultural care, including appropriate fertilization, pruning, and especially good irrigation and soil water management. As discussed below under "Root Rots," proper cultural practices can greatly reduce the occurrence and severity of many citrus diseases. Avoid planting too deeply. Carefully design, install, and operate the irrigation system to prevent asphyxiation of roots while providing sufficient water using methods that avoid direct wetting of trunks and waterlogging of the soil. Take steps to prevent injury to fruit, limbs, roots, and trunks from equipment, inappropriate chemical applications, poor

Sprinkler wetting of this citrus trunk has favored the development of extensive decay. Cutting off lower bark to expose tissue underneath has revealed healthy, yellowish wood (top center) and brown, rotted tissue (that nearest the soil). The gumming (also pictured on the first page of this chapter), location of the disease, and absence of staining in the wood indicate the cause is Phytophthora gummosis.

cultural practices, and adverse weather such as frost. Wounds directly damage citrus and increase its susceptibility to fruit decay, gummosis trunk diseases, and root rots.

For detailed cultural care recommendations, consult "Citrus Frost Protection" (in preparation); "Establishing the Citrus Grove" (in preparation); "Fertigation" (in preparation); "Irrigation" (in preparation); "Nutrient Deficiency and Correction" (in preparation); "Pruning" (in preparation); and "Soil/Water Analysis and Amendment Strategies for Citrus" (in preparation).

Sanitation. Use good sanitation to prevent the introduction and spread of soil- and waterborne pathogens, as described below under "Root Rots." During pruning and topworking (grafting or budding on new stock), scrub tools clean and chemically disinfect them before cutting a new plant to avoid spreading mechanically transmissible pathogens such as those causing stubborn and tristeza diseases. Topworking is not advisable if plants are already infected because many pathogens can move systemically between the scion and rootstock. Remove from the orchard any prunings of woody tissue and cuttings from hedging or skirting, or destroy them by burning or chipping, to prevent pathogens from continuing to reproduce and increase inoculum in the orchard.

ROOT ROTS

Phytophthora root rot and Armillaria root rot, as well as less-common diseases such as dry root rot and Rosellinia root rot, are caused by microorganisms that infect citrus roots or roots and the crown. Abiotic diseases, such as root asphyxiation (insufficient soil oxygen), mineral disorders, and phytotoxicity (all discussed under "Leaf and Twig Disorders") can also directly affect the health of roots. Because they damage the plant's nutrient- and water-absorbing system, the first visible symptom of root diseases is usually a yellowing of leaves and a slow decline of the scion.

Management

To help prevent root rots, avoid planting in a site likely to be infested with soilborne pathogens, including species of *Armillaria*, *Fusarium*, *Phytophthora*, and *Rosellinia*. Prepare and plant the site properly and choose stock that is free of, and resistant to, disease. Provide adequate soil drainage and good soil aeration to reduce the likelihood and severity of root disease. For example, before planting at sites where soil drains poorly or hardpans obstruct water flow, rip or backhoe through impervious soil layers, install a subsurface drainage system, or plant on mounds. If planting on mounds, avoid tipping trunks by properly preparing subsurface layers so roots become anchored. When establishing furrows, provide berms along the trees so that the crowns remain dry and out of the water. Note that in some areas where Armillaria root rot is established on native plants or in old citrus orchards where extensive root systems occur, deep plowing or ripping can increase the spread of infected roots throughout the orchard. In these areas site-specific methods that avoid spreading pathogens may be needed to break through impervious soil layers.

Use stringent sanitation and prevent soil or water movement from infected areas. Avoid working in groves when soil is wet because this is when soilborne pathogens are most readily spread and trees are more susceptible to infection. Begin harvest and any other orchard activities in areas where healthy trees occur and finish work in diseased areas to minimize pathogen movement. Bring only clean bins and equipment into groves.

During cultural operations, avoid injury to the underground portions of the crown, especially during the cool and wet season when fresh wounds are more susceptible to infec-

Yellowing of leaves and decline of the scion are symptoms of root and trunk diseases. Symptoms often become obvious after warm weather causes stress, as with this orange tree infected by *Phytophthora*.

prolong the life of an *Armillaria-* or *Fusarium*-infected tree. However, if trees are infected with *Armillaria* or *Rosellinia* or are unproductive due to disease, remove them completely, including the roots and especially the stumps. Let the infected trees dry thoroughly on site before disposing of them. With *Armillaria* and *Rosellinia* species, also remove the nearby, apparently healthy trees; once symptoms appear on a tree, these pathogens have probably already spread to the roots of surrounding trees.

Check regularly for signs of root rots and their damage. Also look for and remedy problems that contribute to root rots, such as injury from vertebrates or inappropriate soil or water conditions. For more disease-specific recommendations, including fungicides, consult the online UC IPM *Pest Management Guidelines: Citrus*.

Phytophthora Root Rot

Phytophthora species

Phytophthora spp. are the most important soilborne pathogens infecting citrus. Phytophthora root rot is common in all growing areas, thriving in areas of excessive moisture and poor drainage. Trees of any size and age can be affected, but young trees are more severely affected because of their small or limited root systems when they are first planted. *Phytophthora* spp. may also infect the trunk (causing Phytophthora gummosis) and fruit (causing brown rot). *Phytophthora citrophthora* and *P. parasitica* are the most common of these pathogens in California citrus, but other species of *Phytophthora* can also be present.

Although they are not true fungi (technically, they are Oomycetes, or water molds, related to brown algae), *Phytophthora* species have many funguslike attributes.

Symptoms and Damage

Phytophthora root rot destroys the feeder roots of susceptible rootstocks and causes a slow decline of the tree. The leaves turn light green or yellow and can drop prematurely. Twigs die back, there is less than normal flush of new shoots and leaves, and fruit production declines. *Phytophthora* spp. infect the root cortex (the layer beneath the epidermis) of pencil-sized and smaller roots. The cortex softens and separates from the stele (the vascular tissue core of roots). If the destruction of feeder roots occurs faster than their regeneration, the uptake of water and nutrients is severely limited. The tree frequently fails to grow and develop normally (becomes "unthrifty"), its stored energy reserves become depleted, and the tree may die. The cause of disease is often difficult to determine because nematodes, salinity, and waterlogging of soil cause similar damage. Only a laboratory analysis can provide positive identification of Phytophthora root rot.

tion by many pathogens. To avoid causing phytotoxicity, follow label instructions for applying fertilizers, fungicides, herbicides, insecticides, and nematicides at rates no higher than recommended; burning of root tissue may occur when excessive amounts of these materials are used. Chemical injury to roots increases tree susceptibility to pathogen infection and disease. Before fertilizing young trees, wait at least 6 weeks after planting or until the trees show new growth; otherwise, phytotoxicity may occur.

Careful irrigation is essential for preventing root rots and minimizing their damage. Schedule irrigations according to the physical and climatic conditions of the site and trees' changing needs for water. Keep irrigation water away from trunks and prevent waterlogging around root crowns. The health of infected trees can sometimes be improved by modifying irrigation, such as increasing intervals, reducing run time, or switching to alternate middle row irrigation or a different irrigation system (such as minisprinklers).

Excavating the tree root crown and allowing it to dry may

Seasonal Development

The pathogen can be easily spread, and orchardists can move it from a few trees to the entire orchard. *Phytophthora* species spread through the movement of contaminated nursery stock, on equipment and shoes, and through any human or animal activity that moves moist soil or water (rainwater or irrigation) from one place to another. In addition to its various life stages that can be moved by activities and people, when conditions are wet, *Phytophthora* species produce motile zoospores that are specialized to spread ("swim") in water in the rhizosphere (soil near roots). The rhizosphere contains root-associated microorganisms and root secretions attractive to *Phytophthora* zoospores. Phytophthora spores encyst (attach themselves to) and infect roots within a few hours. Mass flow of water carries inoculum such as zoospores from tree to tree (Figure 17).

If a tree looks stressed, dig up some soil and check the feeder roots. Healthy roots are firm, whereas roots affected by root rot have soft outer tissue that easily slides off the central core when pinched. If roots appear diseased, collect small roots and the surrounding soil for laboratory analysis to obtain positive identification. Sample for the two predominant species, but other species of *Phytophthora* may be involved. *Phytophthora parasitica* primarily occurs from July through September, whereas *P. citrophthora* predominately occurs from January through March.

Dry Root Rot
Fusarium solani

Dry root rot is a disease complex that often takes years before severe symptoms develop. *Fusarium solani*, a common soilborne fungus, is often isolated from the diseased tissue. Exactly how dry root rot develops is unknown. Usually, it affects only a few scattered trees in a grove. However, it can develop into an epidemic in some orchards. Overall, dry root rot symptoms in the canopy are similar to those caused by *Phytophthora* species and other agents that damage the roots or girdle the trunk.

Symptoms and Damage

Symptoms of dry root rot include reduced tree vigor, dull green leaf color, poor new growth, and twig dieback. If extensive root damage occurs, leaves suddenly wilt and dry on the tree.

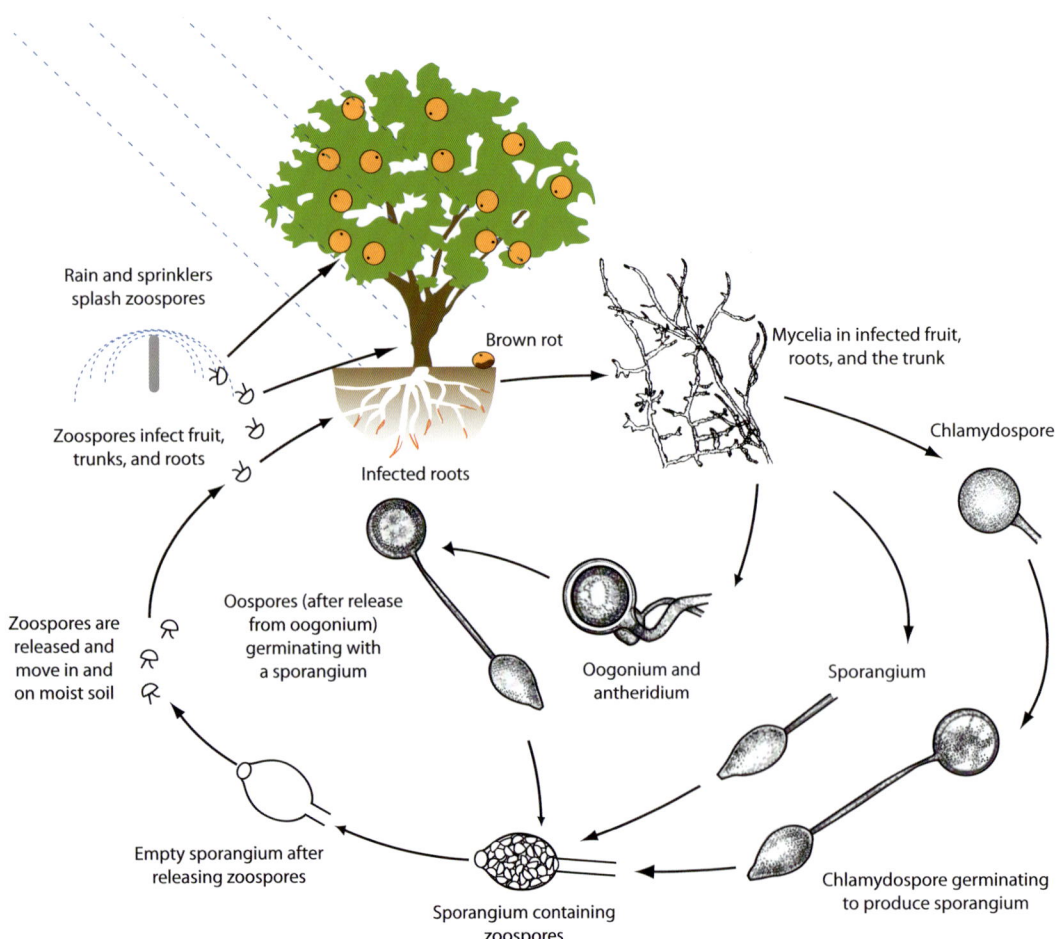

Figure 17. The disease cycle of *Phytophthora* species infecting citrus. *Phytophthora citrophthora* and *P. parasitica* persist primarily as mycelia in infected plant tissue and as chlamydospores and oospores. Mycelia produce chlamydospores and other spore-forming structures (sporangia, oogonia, and antheridia). Sexual reproduction occurs when antheridia (male parts) develop near the base ("neck") of an oogonium (female reproductive organ). Oogonia produce oospores and release them into the environment. When conditions are suitable, oospores form sporangia. Sporangia produce and release motile zoospores. Spores are splashed or move actively (e.g., zoospores "swim") in water. Spores and mycelia are also carried along with the movement of infected plants or contaminated soil or water. Upon contact (usually during wet conditions or through wounds), pathogen mycelia and spores infect citrus and cause brown rot of fruit, Phytophthora gummosis of trunks, and Phytophthora root rot. Adapted from Timmer and Duncan 1999, Wilcox 1992.

Removing lower bark from this citrus trunk reveals dark unhealthy wood next to healthy greenish white wood. The dry discoloration extends deeply into wood and also occurs in large roots (not shown). The cause is dry root rot from infection by *Fusarium solani*.

Wood discolored by dry root rot can be brown, grayish, or purple. Unlike cankers due to Dothiorella or Phytophthora gummosis, there is no gumming when dry root rot is the cause.

The disease usually starts in larger roots and spreads into the crown. Areas of bark on the underground portion of the crown show a moist, dark decay, which later dries and adheres to the wood. In some cases, dry bark may also be visible aboveground. The wood below the dead bark is hard, dry, and stained brown, grayish, or purple. Unlike Phytophthora gummosis, dry root rot does not produce gumming, and lesions extend deep into the wood. The initial infection can occur at planting or at any time during the life of the tree, but aboveground symptoms due to girdling of the crown may not appear until several years after the initial infection. Once the crown is girdled, foliage and the entire tree dies and the tree may fall over. The apparently sudden tree death often occurs after hot weather.

Seasonal Development

The causal agent of dry root rot, *Fusarium solani*, infects a tree through the crown or larger roots that have been injured by *Phytophthora* spp., water-saturated soil, mechanical damage, gophers, or root burn by an overdose of fertilizers or pesticides. All common rootstocks are susceptible to dry root rot.

This lemon trunk failed (tipped) due to dry root rot. Before its death, the tree produced an unusually abundant fruit load, then foliage turned yellow throughout the canopy.

Fusarium solani typically is a saprophytic fungus that develops in dead and dying wood. The fungus is common on citrus bark, leaves, and roots, usually without causing any damage. The development of dry root rot disease is not well understood, but stress and other injuries are believed to predispose the tree, allowing infection by *Fusarium solani* to develop into a disease that eventually kills trees.

Check regularly for dry root rot and problems such as Phytophthora root rot or vertebrate damage that provide entry sites for *Fusarium solani*. If you suspect dry root rot, dig all the way around the tree to look for disease because the decay may be underneath the crown roots or on one or more of the main lateral roots.

Armillaria Root Rot

Armillaria mellea

Armillaria root rot is a fungal disease of many crops and native and ornamental woody species, including avocado, citrus, and oak trees. The fungus persists in infested roots and wood in soil, infecting new plantings and spreading underground to infect nearby hosts. Once infection reaches the crown, it is very difficult to save a tree.

Symptoms and Damage

Symptoms caused by *Armillaria mellea*, also known as oak root fungus, may not be evident until after the disease is well established. The first symptoms of Armillaria root rot are poor growth or dieback of shoots and leaves that are small, yellow, and dropping prematurely. Symptoms often appear first on one side of the tree, then throughout the entire canopy. If infected trees are not removed, symptoms develop over time in nearby trees in a pattern of ever-widening circles as the pathogen spreads underground through the orchard. In winter, the *Armillaria* fungus often forms clusters of mushrooms at the base of infected trees a few days after rain. The short-lived mushrooms always occur in groups. The viable mushroom caps vary in color from off-white to honey yellow. Each cap is about 1 to 10 inches (2.5–25 cm) in diameter.

Mushrooms of *A. mellea* have a ring on the stalk just under the cap and shed numerous microscopic white spores. These basidiospores are generally not considered to be important in disseminating the pathogen. When the mushrooms die or are no longer viable, they turn black and shrivel.

The most reliable sign of Armillaria root rot is a white growth of fungal mycelia in the cambial tissue. If trees exhibit aboveground symptoms of infection, cut under the bark of the root crown and major roots to check for mycelial fans, which are areas of fungal extension that have a strong mushroom odor. In early stages of colonization, *Armillaria mellea* mycelia typically grow in patches in the cambium between the outer bark and wood and later colonize the wood, causing a typical white rot wood decay. This distinguishes *Armillaria* from many other wood-rotting fungi, which have mycelia that are limited to the secondary xylem or wood and never invade the cambium.

Seasonal Development

The fungus spreads among trees by root contact or through rhizomorphs (black strings of fungal mycelia), which can grow short distances through the soil and contact and penetrate healthy citrus roots (Figure 18). From the initial site of root infection, the fungus invades lateral roots and the crown region. Within trees, *Armillaria mellea* spreads as white mycelial plaques in the cambium region and eventually girdles the crown region and kills the root system.

Armillaria root rot caused this pale and wilted foliage, sparse canopy, and limb dieback. Unless this tree is removed, including its major roots and stump, the pathogen is likely to spread underground and kill nearby trees.

During the fall and winter, a few days after rain, short-lived mushrooms often grow around the base of *Armillaria*-infected trees. They vary greatly in color and size, but *Armillaria* mushrooms always occur in groups, never singly.

Armillaria root rot is native to California's woody plants. It affects many tree crops planted on hillsides, in former riverbeds, floodplains, and in other areas subject to water flow. The fungus can survive for many years in dead or living roots of fruit and nut trees and on ornamental and native tree species, often along former streambeds and near creeks. The fungus requires cool, moist soil conditions for development and spread; it is therefore rarely a problem in desert areas. *Armillaria* must build up a significant amount of inoculum (infective parts) on roots or buried wood before it can successfully infect healthy trees. Therefore, disease is

The white fungal mycelia of *Armillaria mellea* often occur in small, scattered patches in cambium, as exposed here by cutting under bark around this root crown.

Armillaria mycelia sometimes form extensive fan-shaped plaques (flat sheets) between the outer bark and wood. When first exposed by peeling back bark, the fungus emits a strong mushroom odor.

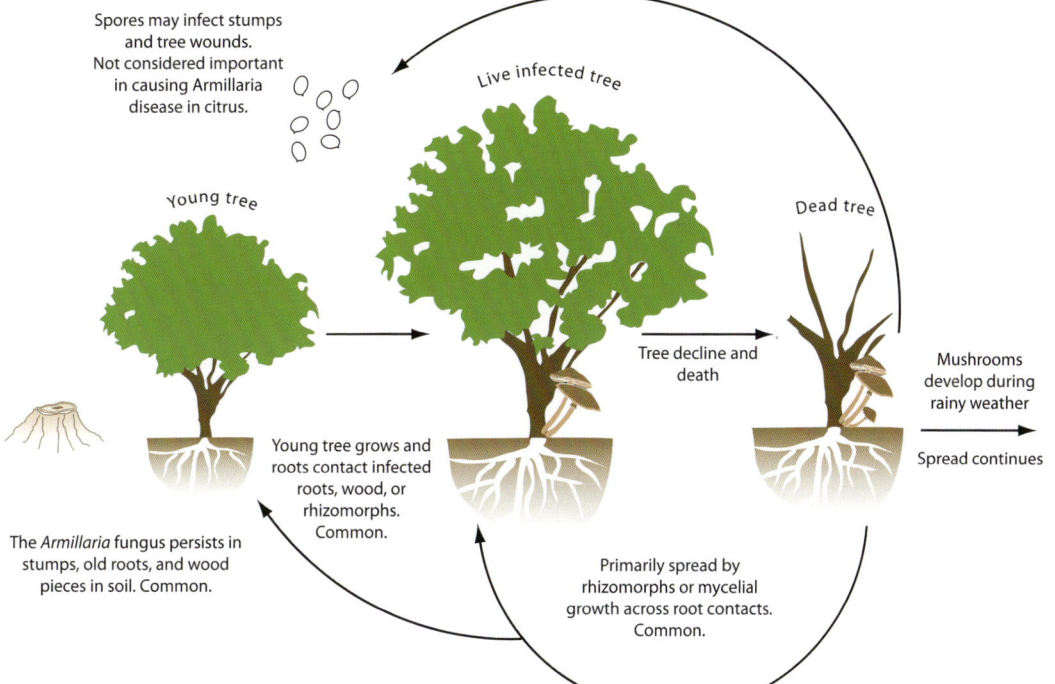

Figure 18. Armillaria root rot development cycle and spread. *Armillaria mellea* persists for years in infected roots and wood in soil. New infections occur when roots grow and contact infected roots or wood and when fungal rhizomorphs growing short distances in soil contact nearby plants. During fall and winter when conditions are wet, *Armillaria* mushrooms grow from infected trees. These mushrooms produce *Armillaria* spores, but spores are not considered to be an important source of Armillaria root rot in California citrus. Eliminating infected trees and removing old stumps, large roots, and wood pieces from soil can break the disease cycle.

most likely to develop in trees weakened from other causes and in trees nearest infected trees and stumps.

The pathogen *Armillaria mellea sensu lato* (in the broad or previous sense, as discussed above) is actually a poorly understood complex of several *Armillaria* spp. Disease varies in severity depending on the species of host plant and *Armillaria* involved. *Armillaria mellea sensu stricto* (in the narrow or strict sense or as recently defined) is the predominate *Armillaria* species occurring in California.

Rosellinia Root Rot

Rosellinia (=Dematophora) necatrix

Rosellinia root rot is a minor citrus disease in California. Symptoms and development resemble those of Armillaria root rot, including yellow foliage, little or no new leaf and shoot growth, premature leaf drop, and eventual tree death. Symptoms occur because of extensive root death or girdling of the basal trunk.

Rosellinia necatrix produces white mycelial growth underneath the bark of infected roots. Under the bark, the whitish mycelial patches can resemble those of *Armillaria mellea*. However, *R. necatrix* mycelia lack the characteristic mushroomlike odor, and they are often limited to smaller patches as compared with *Armillaria mellea*. Later, mycelium on infected roots becomes dark brown to black and produces dark, round sexual fruiting bodies (perithecia), each about 1/12 to 1/25 inch (1–2 mm) in diameter.

Rosellinia necatrix attacks a wide range of woody hosts. For example, grape is a host, so citrus planted in former vineyards is at risk for Rosellinia root rot. Its biology and management are much the same as for Armillaria root rot. Like *Armillaria*, *Rosellinia* persists in larger wood pieces and stumps in soil, from which it spreads to adjacent living trees.

TRUNK DISEASES

Common trunk disease symptoms include wood decay and bark with shelling and scaling, discolored lesions, and gumming. Serious damage to the trunk disrupts transport of water and nutrients and results in overall decline of the tree. Phytophthora gummosis is the most common citrus trunk disease. Other pathogenic trunk diseases include Dothiorella gummosis, exocortis, Hendersonula tree and branch wilt, Hyphoderma gummosis, and psorosis. Growths on bark caused by vein enation (also called woody gall) are discussed under "Leaf and Twig Diseases." Trunk diseases due to non-

Armillaria spreads through root contact or by rhizomorphs. Rhizomorphs, shown here growing on a large root, are dark strings of fungal mycelia. They resemble small roots and can grow short distances through the soil. In advanced stages, rhizomorphs can grow in decayed wood (white rot).

The *Rosellinia* fungus produces white mycelial growth visible around this excavated root crown. Unlike with Armillaria root rot, the pale mycelia of *Rosellinia* can grow on top of bark and soil, as well as under the bark of infected trunks and roots.

As *Rosellinia* develops, fungal mycelia on bark and wood may turn dark brown to black and produce these small, round sexual fruiting bodies (perithecia).

Gnarled bark on trunks (called woody gall) is caused by a pathogen spread by aphids and during grafting. The pathogen also causes bumps on leaves (vein enation), as pictured in the "Leaf and Twig Diseases" section. Except in old plantings, this disease is rare, as it is eliminated during propagation with certified disease-tested budwood.

infectious causes and those with major symptoms throughout the tree are discussed below in "Trunk Disorders" and "Diseases Affecting Growth Habit and Yield."

Management

Review "Prevention and Management" in the introduction to this chapter, as those methods apply to many diseases of trunks. Use good sanitation as described earlier for "Root Rots" to reduce the spread of pathogens such as *Phytophthora* spp.

Avoid planting an orchard with trees that are already unhealthy or likely to become infected by carefully checking the lower trunk and rootstock of new trees and planting only those free of disease symptoms and injury. Prevent the introduction of the pathogens that cause exocortis and psorosis by planting only certified disease-free stock, such as that from the Citrus Clonal Protection Program. Planting new trees correctly can reduce the development of trunk disease. Planting trees on a berm or high enough so that the first lateral roots are just covered with soil, without allowing bud unions to get buried or soil to build up around the crown, lowers the likelihood of *Phytophthora* spp. infection and gummosis disease.

Correct any soil problems, such as poor drainage and insufficient aeration, and provide trees with appropriate irrigation and other cultural care. As discussed for "Root Rots," these measures reduce crop stress that increases trees' susceptibility to *Phytophthora* spp. and the pathogens that cause Dothiorella gummosis and Hendersonula (*Nattrassia mangiferae*) tree and branch wilt.

In established orchards, protect trees from injuries that allow pathogens to enter and infect the tree. Control bark-chewing pests, such as ants and vertebrates, and avoid wounding bark when operating equipment near trees. The exception is pruning off dead and dying branches, which helps to stop the spread of Dothiorella gummosis and Hyphoderma gummosis. When pruning, avoid the mechanical spread of pathogens. After working on each tree, thoroughly scrub and clean propagation and pruning tools such as clippers, knives, and saws and chemically disinfect them.

Monitoring orchards is an important part of management because diseases such as Phytophthora gummosis can be controlled only by prompt action at early stages of disease development. Inspect trees several times per year for trunk damage, disease symptoms, and disease-promoting cultural practices and environmental conditions. Promptly correct any problems as discussed above.

Tree removal may be the only recourse with some diseases, such as exocortis and psorosis. However, cutting the tree down may not prevent infection of other trees when there is root contact. In this case, remove the entire tree, including the stump and major roots. For more disease-specific recommendations, consult the online *UC IPM Pest Management Guidelines: Citrus*.

Phytophthora Gummosis

Phytophthora citrophthora and *P. parasitica*

Trunk disease from infection by *Phytophthora* species is called Phytophthora gummosis. All scion cultivars are susceptible under environmental conditions that promote disease. If the rootstock is susceptible to *Phytophthora* infection, the disease can spread into the crown and woody roots (sometimes called foot rot). These same pathogens also cause Phytophthora root rot and brown rot of fruit.

Symptoms and Damage

A tree with Phytophthora gummosis initially has firm bark, which later dries and eventually cracks and sloughs off. Lesions enlarge around the circumference of the trunk, slowly girdling the tree. Decline can occur rapidly within a year, especially under conditions favorable for disease development, or can occur over several years. Sap oozes from small cracks in the infected bark, giving the tree a bleeding appearance.

Gumming due to *Phytophthora* spp. may be obvious only seasonally. Gum can be washed off during heavy rain, and other diseases sometimes cause gumming. Thus, gumming alone is not definitive for diagnosis. Late stages of Phytophthora gummosis are distinct, but early symptoms are often difficult to recognize. Yet early detection and prompt management actions are essential for saving a tree.

Secondary pathogens often infect through lesions created by *Phytophthora* spp. These secondary infections kill and discolor the wood, in contrast to infections caused only by *Phytophthora* spp. that do not discolor wood.

Seasonal Development

Phytophthora species are present in almost all citrus orchards and are readily spread when infested soil or water are moved, as discussed above under "Phytophthora Root Rot." Under moist conditions, the pathogens produce large numbers of motile zoospores, which are carried or disseminated in water to trunks (Figure 17). Phytophthora gummosis develops

Gum oozing from small cracks in trunk bark is often the first external symptom of Phytophthora gummosis. However, gumming alone is not definitive for diagnosis: oozing can be seasonal, gum washes off during heavy rain or sprinkler wetting of trunks, and several pathogens cause gumming.

Cracked, dry bark on the basal trunk of this navel orange tree is due to Phytophthora gummosis caused by *Phytophthora citrophthora*.

Removing cracked bark exposed cankered wood on this lower trunk. Unlike with dry root rot, this discoloration by Phytophthora gummosis is relatively shallow and does not extend deeply into wood.

rapidly under moist, cool conditions. Hot summer weather slows disease spread and helps drying and healing of the lesions.

Inspect your orchard several times per year for disease symptoms. Look for signs of gumming on the lower trunk and crown and for conditions that promote disease. Check the trunk and base of tree for ant feeding, which can produce gumming that resembles that of Phytophthora gummosis. Regularly inspect trees and growing areas to ensure that the sanitation and cultural practices described earlier for *Phytophthora* management in the "Root Rots" section are being followed because many of these methods also help prevent gummosis.

Dothiorella Gummosis (Gummosis and Canker, Dothiorella Blight)

Botryosphaeria ribis

Dothiorella gummosis is usually a minor problem in citrus. A wound or injury is generally required before disease can develop from infection by the fungus *Botryosphaeria ribis*. This weak pathogen has an asexual (anamorph) stage named *Neofusicoccum ribis* and was formerly referred to as *Dothiorella gregaria*, hence the common name Dothiorella gummosis.

Dothiorella gummosis results in distinct sunken cankers beneath dead outer bark on limited portions of branches or the trunk. Cankered bark may have a grayish cast or a surface color that is lighter than normal. Dead bark remains tightly attached and may exude gum. Underneath the bark surface, the cambial layer and underlying wood may be discolored brown to yellowish. The canker can spread up and down the cambium and wood in shallow, discolored grooves. Leaves and twigs on scattered branches or the entire tree may decline and die, usually with fruit and leaves remaining attached. This leaf and shoot dieback disease is called Dothiorella blight.

Dothiorella gummosis can cause rapid decline and death of a tree. Disease damage may not be obvious until the onset of warm weather causes trees to become water stressed.

This lemon tree has dropped most of its leaves because the limbs and trunk have developed cankers (not shown) due to Dothiorella gummosis.

This bark is discolored due to Dothiorella gummosis. Cankered cambial tissue and wood underneath have long, shallow grooves. These symptoms developed during spring on this mandarin after limbs were injured earlier by freezing winter weather.

Dothiorella gummosis caused this yellowish and brown discoloring of wood revealed by cutting off bark. Unlike damage from Phytophthora gummosis, with Dothiorella gummosis bark stays tightly attached to wood and usually does not develop large cracks.

Limb dieback and viscous gumming from bark caused by Hendersonula tree and branch wilt on a navel orange in the San Joaquin Valley.

Young trees are especially likely to die if the affected tissue is not removed. With Dothiorella gummosis, often it is not immediately obvious that bark and wood are cankered and dead. To help diagnose Dothiorella gummosis, carefully inspect trunks and limbs for grayish, lighter-colored, or sunken infected bark and oozing dark liquid, especially on young trees and trees with scattered dieback.

Hendersonula Tree and Branch Wilt (Sooty Canker)

Nattrassia mangiferae (=*Hendersonula toruloidea*)

Hendersonula tree and branch wilt causes bark cracking and peeling, or dead bark can remain tightly attached to dead limbs. Cracked bark can bleed profusely and contain viscous or dried gum. Removing affected bark often reveals black, sooty growth underneath, so this disease is sometimes called sooty canker. This dark discoloration occurs under the cambium layer (inner bark) but extends only shallowly into the wood. On limbs infected by *Nattrassia mangiferae*, leaves suddenly wither, turn brown, and dry up. Dead leaves typically remain attached to the twigs.

The fungus causing this disease is a relatively weak pathogen of citrus. It usually infects the tree through wounds caused by freezing, fertilizer burn, sunburn, or mechanical injury, such as from cultivation or pruning. Severe damage is more likely on trees lacking good cultural care and a proper growing environment.

These limbs were killed by Hendersonula tree and branch wilt. Because this disease can kill quickly, leaves suddenly wither and dry up but usually remain attached to the dead twigs.

When Hendersonula tree and branch wilt is the cause of dieback, peeling back bark on dead limbs often reveals dark, sooty growth that extends shallowly into wood.

Hyphoderma Gummosis

Hyphodontia sambuci

Hyphoderma gummosis was first reported in California citrus in 2001 and has been found only on lemon. The disease occurs in the San Joaquin Valley and coastal growing areas. The wood decay fungus *Hyphodontia sambuci* (formerly *Hyphoderma sambuci* and hence the disease name) causes branch and trunk cankers as the pathogen invades the cambium. This results in profuse gumming around infected areas. Extensive scaffold branch and trunk colonization causes wilting and dieback of the tree canopy that ultimately results in tree death. The pathogen, like other wood decay fungi, cannot infect its host through intact bark, but once inside it infects the cambium like *Armillaria* spp. To initiate infections, it requires injuries such as wounds from pruning or freeze damage that expose wood. Airborne basidiospores colonize exposed woody tissues during moist conditions. An extensive crust of pink to white develops into the fruiting bodies of *Hyphodontia sambuci*, but this growth can be easily overlooked during early stages of development. Fruiting bodies develop around infection sites after wet weather.

Wood Decays

Chlorotic, undersized, sparse leaves and branch dieback are common symptoms of wood decay fungi infecting roots, the basal trunk (root crown), or limbs. These fungi include *Armillaria mellea* and *Hyphodontia sambuci* (discussed above) as well as *Ganoderma*, *Oxyporus*, *Phellinus*, and *Schizophyllum* spp. These fungi are called white rots because they often cause decayed wood to become soft and white or yellow. In this type of decay, all the major components (cellulose, hemicellulose, and lignin) of the wood are degraded. Brown rots, such as those caused by *Antrodia sinuosa* and *Coniophora* spp., primarily decay cellulose and hemicellulose. They leave behind the brownish wood lignin, which is usually dry and crumbly.

Wood decay fungi produce fruiting bodies on the bark, tree crown, or stumps; these grow from the colonized underlying wood. Fruiting bodies on the soil surface grow through the soil from the wood of roots near the trunks. Fruiting bodies may be obvious toadstool- or umbrella-shaped mushrooms like those of *Armillaria mellea* or large and shelflike brackets or conks such as those of *Ganoderma* and *Phellinus* species. *Antrodia*, *Oxyporus*, and *Schizophyllum* spp. produce bracket-shaped, seashell-shaped, or thin flat fruiting bodies. Some decay fungi, such as *Hyphodontia* spp., form relatively

This limb wilt due to Hyphoderma gummosis has been reported only on lemon. Infection always begins where bark has been injured, such as by the large pruning cuts at the base of these dying limbs.

These fungal fruiting bodies (spore-producing structures) of *Ganoderma* are perennial; they remain present continuously for several years. Where fruiting bodies are attached to limbs or trunks, the wood underneath is decaying.

inconspicuous crusts on infected bark. Fruiting bodies produce numerous tiny spores (basidiospores) that are spread by the wind.

Many species of decay fungi initiate infections when their spores contact injured tissue on living trees, such as wounds that expose wood from pruning, vertebrate chewing, or infection sites of *Phytophthora* or other pathogens. Decay fungi can colonize stumps and infect through root grafts to adjacent trees. Spores landing on wood-exposing injuries initiate infections that colonize wood of adjacent parts of the tree. Fruiting bodies develop on infected tissues usually after extensive colonization has occurred. In a strict sense, most wood decay fungi are saprophytes that grow on the nonliving woody tissues of the tree. Still, the structurally weakened wood predisposes the infected tree to cultural (e.g., crop load) and environmental factors (e.g., wind storms) that can lead to extensive tree damage and death.

Psorosis

Citrus psorosis virus

Psorosis is a graft-transmissible viral disease that is most often found in old citrus plantings, especially sweet orange. Infected trees usually are not killed but slowly decline and eventually become unproductive.

The most distinguishing field symptom is scaling and flaking of the bark on the trunk and limbs of grapefruit, mandarin, and sweet orange. Patches of bark on the trunk or scaffold branches develop small pustules, which later enlarge and break up into loose scales. Large strips of bark can slough off from trunks and limbs. Xylem or wood underneath becomes darkly discolored with or without gumming. In advanced stages, most of the xylem becomes plugged with gum that blocks water and nutrient transport and causes severe decline of the tree. Stained wood is easily observed by cutting cross sections through infected limbs and trunks.

Young leaves may develop chlorotic flecking or clearing (a lack of green color) on portions or over the entire leaf. Distinct clearing can occur on or adjacent to leaf veins and may coalesce into large patches. Symptoms are most visible on leaves nearing full expansion in spring and summer. Leaf symptoms disappear as leaves mature.

The causal agent of the psorosis disease is the *Citrus psorosis virus* (CPsV). CPsV is classified in the genus *Ophiovirus* (*ophio* is the Greek word for "snake"), indicating the elongated twisted and coiled appearance of the virus particles (visible only when highly magnified, as with an electron microscope). The severity of psorosis symptoms observed in the field is related to different pathogenic strains of the virus.

CPsV is transmitted in infected budwood and with contaminated grafting and pruning tools. Seed of some citrange

This bark scaling and flaking is characteristic of infection by *Citrus psorosis virus*. As disease progresses, large sections of bark may slough off, and the tree will produce few fruit.

cultivars are known carriers of the virus. Positive identification is made by grafting budwood from suspected hosts onto indicator seedlings (commonly sweet orange or Dweet tangor) then observing new leaf growth for symptoms. Specialized molecular (targeting the virus genetic material) and serological (targeting the coat protein of the virus particle) laboratory tests are used by researchers but may not be commercially available.

Where an old tree shows characteristic psorosis symptoms, scraping away the infected bark area to stimulate the formation of wound callus can result in temporary recovery. Generally, a psorosis-infected tree will be less productive than healthy trees, and replacement is the best option.

Exocortis viroid stunts tree growth on sensitive rootstocks and causes a modest reduction of fruit yield. The viroid spreads to other trees on pruning clippers and saws unless tools are scrubbed clean and chemically disinfected after working on infected trees.

Exocortis commonly occurs in older groves established with budwood not tested for citrus viroids. Dry, cracked bark that peels off the root crown (bark shelling) is the characteristic symptom of exocortis.

Exocortis

Citrus exocortis viroid

Exocortis is of minor importance in most groves in California because strict regulations on budwood sources have kept new plantings largely free of this and other viroid diseases. Exocortis viroid is easily disseminated on infected budwood and contaminated propagation and pruning tools.

Exocortis disease develops only in certain susceptible rootstocks, most commonly trifoliate orange and its hybrids. Rangpur lime and some citron and lemon rootstocks are also affected. Characteristic symptoms of exocortis disease include bark shelling (peeling) of susceptible rootstocks and potentially stunting of the entire tree. The disease causes the bark to crack longitudinally, and bark may lift in thin strips. Droplets of gum often appear under the loose bark. Shelling sometimes develops slowly and is restricted to a small area for several years; at other times, the disease develops rapidly over the entire rootstock. Infected trees rarely die, but growth is slowed and productivity gradually declines.

Before working on a new plant, disinfect propagation and pruning tools with a chemical solution; heat or open flame does not kill viroids. Use certified viroid-free plants, such as those from the Citrus Clonal Protection Program.

TRUNK DISORDERS

Noninfectious (abiotic) trunk diseases include bud union disorders, genetic disorders (shell bark and dry bark), as well as environmental injury from sunburn and frost. In addition to directly damaging citrus trees, these disorders increase tree susceptibility to pathogens discussed above in "Trunk Diseases." To prevent these problems or reduce their damage, choose appropriate planting stock, provide good cultural care, and use methods discussed in "Prevention and Management."

Shell Bark and Dry Bark

Shell bark is an inherited disorder that occurs on lemons. Nucellar seedlings are less severely affected than old lines of lemons. Symptoms begin to appear after about 10 years on Eureka lemon and even later on certain Lisbon lines. Small areas of the outer bark crack and loosen in long strips. This bark injury, called shelling, usually develops from the bud union upward but sometimes starts in the crotch of scaffold branches. The bark lesions enlarge very slowly. On Lisbon, the disease usually does not cause noticeable decline. On Eureka, however, shelling can be so extensive that the required regeneration of bark tissue weakens the tree.

If disease extends into the inner bark and cambium, it is called dry bark; the affected areas dry out and crack slightly but do not shell because no new tissue is generated underneath. Secondary decay organisms can invade the lesions and aggravate the decline. If dry bark extends around most of the tree's circumference, nutrient transport is severely limited and the tree will die.

As a preventive measure, use only budwood from tolerant parent trees. In old orchards, severe pruning of diseased trees helps stimulate new growth and regeneration of affected bark tissue, allowing at least temporary recovery.

Two small squares cut away in bark reveal a dark line where the joining of an incompatible rootstock and scion has resulted in uneven growth. This delayed bud union disorder is sometimes called crease.

Intense sunlight produces heat that causes cracks and cankers in bark and wood. When sunburn is the cause of damage, the injury occurs to unshaded tissue on the south and west side of trees exposed to afternoon light.

Whitewash trunks whenever previously shaded wood is exposed to direct sunlight, as in this topworked citrus orchard.

Bud Union Disorders

Certain scion-rootstock combinations show an incompatibility reaction that can appear shortly after grafting or can take 10 to 20 years to develop. An example of delayed incompatibility is bud union crease. A crease or fold forms at the bud union due to uneven growth. With increasing overgrowth, the food-conducting tissue is compressed, resulting in a slow girdling of the tree.

When planting or replanting, avoid susceptible (incompatible) combinations. For example, bud union overgrowths develop in Frost nucellar navel on Trifoliate, lemon on Cleopatra mandarin, Eureka lemon or Satsuma mandarin on Troyer and Trifoliate, and certain scion lines on sour orange.

First decide what scion cultivar will be used. Next identify the rootstocks that are compatible, well adapted to local conditions, and resistant to problems such as disease. Because rootstock influences scion quality, choose compatible combinations that provide the desired fruit quality, as summarized in Table 5 in the chapter "Managing Pests in Citrus." For more information, contact your University of California Cooperative Extension farm advisor or consult sources such as "Rootstocks for Citrus in California" (in preparation).

Sunburn

Bark can be injured by heat produced during direct exposure to intense sunlight. The bark of young trees, and of trees of any age that are severely pruned, is easily damaged. Sunburned bark appears dried and often cracks and sheds or separates from wood. Intensive sunlight and prolonged heat can also blemish fruit and leaves. The side of fruit continuously exposed to the sun develops yellow or brownish leathery areas, and fruit become slightly lopsided. Fruit albedo or pulp may dry. Sunburned leaves develop yellow or brown blotches, and leaves may be smaller than normal or misshapen. Sunburned tissue becomes susceptible to infection and disease from fruit, leaf, and trunk pathogens.

Drought stress increases tree susceptibility to sunburn. Provide appropriate irrigation and good cultural care to keep roots healthy. Whitewash exposed bark.

Bark is peeling from dead limbs on this citrus tree injured by freezing weather several years earlier.

Small, lopsided fruit with dark seeds and a rind that does not color properly are symptoms of Huanglongbing (citrus greening). Bud mite, chimera, and stubborn and tristeza diseases (all pictured elsewhere) can also cause misshapen fruit. Because many pests cause similar damage, seek expert help if you are uncertain of the cause.

Frost and Freeze

Cold temperatures can injure the limbs and trunk. Damage to major woody parts usually occurs only on young trees that often subsequently die. Limb and trunk damage to older citrus trees may not be apparent until long after cold weather, when cankers develop in bark and wood or bark peels off. Except when trees are young, cold weather usually damages only exposed, immature shoots and current-season fruit, as discussed later in "Fruit Disorders" and "Leaf and Twig Disorders."

FRUIT DISEASES

Many pathogens and environmental conditions damage fruit. The most important fruit disease in the field is brown rot caused by *Phytophthora* species. Alternaria rot, anthracnose tearstain, Septoria spot, bacterial blast, blue and green mold, and Botrytis rot (in approximately this order of importance) also damage citrus fruit in California orchards. After harvest, fruit decays caused by *Penicillium* spp. may cause extensive crop losses. Damage due to adverse growing conditions and inadequate cultural care are discussed below under "Fruit Disorders." Many of these disorders cause injuries that increase fruit susceptibility to infection and disease development from pathogens.

Also be alert for symptoms of exotic diseases that do not occur in California and promptly report any suspected new problems to agricultural officials. Bacterial canker (citrus canker), Huanglongbing (citrus greening), citrus variegated chlorosis, and citrus leprosis are among the exotic diseases that cause fruit to distort, develop lesions or poor flavor, or fail to achieve mature color and size. See their discussion in the "Exotic Diseases" section.

Management

Cultural practices and environmental modifications that improve tree health and reduce fruit injury and wetting help prevent fruit diseases. For example, protect fruit and trees from freeze damage and minimize chemical or mechanical injuries that become decayed. Manage irrigation and soil problems appropriately, as discussed in the "Root Rots" section, to reduce tree stress. Less tree stress greatly reduces the incidence of split fruit, which commonly become infected by decay pathogens. Prune out dead limbs to remove tissue where decay pathogens produce spores. Pruning canopies also improves air movement, thereby reducing fruit rots favored by high humidity. Prune tree skirts 24 inches (60 cm) or more above the ground to reduce brown rot caused by *Phytophthora* spores splashing from the soil surface.

Harvest fruit at optimal maturity, but do not pick fruit when it is wet; fruit picked at the wrong time or when wet is more likely to become diseased during storage. Handle fruit carefully during harvest to reduce injuries that enable pathogens to enter. Promptly transport fruit to the packinghouse where it will be processed to preserve fruit quality. Preharvest fungicides are available to reduce brown rot and certain other fruit diseases. Consult the online UC IPM *Pest Management Guidelines: Citrus* for disease-specific recommendations.

Brown Rot

Phytophthora spp.

Damage from brown rot can occur each winter in most citrus-growing regions of California. Trees can show symptoms of brown rot alone or in combination with Phytophthora gummosis or Phytophthora root rot.

Brown rot develops primarily on mature or nearly mature fruit. Lesions are brown or tan colored and firm and leathery unless colonized by secondary decay organisms. Infected fruit usually drop and have a pungent odor. At high humidity, infected fruit become covered with white mycelia. Fruit are readily colonized by secondary pathogens, such as blue and green molds (*Penicillium* spp.), which cause the rind to soften. Occasionally, blossoms, leaves, and twigs are infected and turn brown and die.

Brown rot is caused by several *Phytophthora* species and mainly damages fruit growing near or touching the ground. The disease develops when conditions are cool and wet and after *Phytophthora* spores from the soil are splashed onto the tree skirts during rainstorms or irrigation (Figure 17). At temperatures of 57° to 73°F (14° to 23°C), spore-contaminated fruit can become infected after a continuous wet period of 3 hours. Fruit in the early stage of the disease may go unnoticed at harvest and infect other fruit during shipping and storage.

Alternaria Rot

Alternaria citri

Alternaria rot affects mainly lemons and navel oranges. Fruit infected by *Alternaria citri* frequently turn a mature color prematurely and develop dark decay that is generally softer on lemons than on oranges. Infections typically occur in the grove, but disease often does not develop until after harvest; thus, most Alternaria rot damage occurs during storage. On navel oranges, the disease is also called black rot because it causes dark brown to black firm blotches at the stylar (bottom) end or in the navel. If you cut infected fruit in half, you can see the rot extending into the core.

There are several strains of *Alternaria citri*. The strain that causes Alternaria fruit rot does not produce a toxin. Toxin-producing strains that produce other diseases of citrus have not been reported in California. Strains on mandarin causing brown spots have been referred to as *A. alternata* pv. *citri*.

Under cool, wet conditions, brown rot often develops on fruit. The disease starts as firm, water-soaked spots, usually on the bottom of fruit or on sides where fruit touch. At the early stage of disease development, on the lower left of this orange, brown rot is easily overlooked.

Advanced brown rot disease causes large brown to tan blotches on fruit. This decayed tissue emits a pungent odor. Although disease begins in the grove and is sometimes visible on the tree, advanced or obvious decay often develops only after harvest.

Alternaria infects mainly oranges and lemons, especially navels that have split from stress. This disease also causes uneven and premature fruit color change.

Anthracnose

Colletotrichum gloeosporioides

Anthracnose affects mainly mature grapefruit, as well as navel and Valencia oranges and occasionally lemons. The disease typically occurs on stressed trees with old, dead wood, where the fungus produces abundant spores. Anthracnose symptoms include twig dieback, premature leaf drop, staining on fruit, and postharvest fruit decay. Infected leaves initially develop yellow tips that turn brown and die. One phase of the disease is called tearstain or tearstreak because streaks of discoloration form on the rind. Infections typically occur in the grove, but fruit rot mainly develops on harvested fruit.

Fruit disease often first appears as tiny red to brown or black speckling on the rind. Speckles typically develop where water has dripped onto fruit and spores collect and germinate. Germinating spores produce dark, microscopic hyphal structures (appressoria) that appear as dull green to reddish streaks on immature fruit and as brown to black streaks on mature fruit rinds. The streaks cannot be washed off. When the appressoria germinate, usually after harvest, they penetrate the rind and cause dark, sunken decay lesions.

Disease is most common during warm springs with prolonged wet periods and when significant rains occur later in the season than normal. Dying leaves and twigs become covered with dark fungal spores that spread the pathogen. During wet or foggy weather, anthracnose spores drip or splash onto fruit, germinate, and begin the fruit disease.

Anthracnose often occurs together with Septoria spot. The *Septoria* fungus and possibly certain environmental conditions can also cause tearstaining discoloration.

Septoria Spot

Septoria citri

The *Septoria* fungus causes spotting of Valencia oranges, late-season navel oranges (navels harvested in mid- to late spring, e.g., fruit treated with growth regulators), and occasionally of grapefruit and lemons. The disease occurs in the San Joaquin Valley and in coastal and southern interior districts of California during cool, moist weather.

Infection occurs primarily on cold-injured fruit and after high rainfall. The early symptoms of Septoria spot on fruit are small, light tan to reddish brown pits, $\frac{1}{25}$ to $\frac{1}{12}$ inch (1–2 mm) in diameter. Lesions usually do not extend deeper than the oil-bearing tissue (flavedo) but can become larger

These brown to reddish green discolored streaks on the rind were caused by anthracnose. Septoria spot and occasionally other fungal diseases cause similar discoloring (tearstaining).

Anthracnose caused this fruit decay. Brown rot and Septoria spot are among the other causes of this type of damage.

and deeper pits, ⅙ to ¼ inch (4–6 mm) in diameter, that extend into the albedo (white portion of the peel). Dark brown to black fruiting bodies (pycnidia) often develop in the small depressions. Septoria spot can be confused with copper injury, except that the pits caused by copper do not have fungal fruiting bodies. Septoria spot lesions generally develop after the fruit change color coinciding with winter rains. As disease progresses, such as on fruit in storage, lesions can develop into a tearstaining pattern or merge into larger, dark, sunken blotches.

Septoria citri also causes spotting or lesions on leaves or twigs that are injured or weakened, such as by frost or pests. On leaves, lesions ¹⁄₂₅ to ⅙ inch (1–4 mm) in diameter develop as raised blisterlike black spots surrounded by a yellow halo. The halo turns pale brown in older infections.

The pathogen is naturally present on leaves and twigs in many orchards. Spores are spread throughout the tree and onto fruit by splashing irrigation water or rain. The disease develops slowly over several weeks, and reports indicate that the fungus may initially cause infections without visible symptoms (i.e., latent or quiescent infections). Infected fruit develop disease after adverse conditions, such as after cold or rapidly fluctuating temperatures, higher than average rainfall, or under a prolonged cool and moist environment (either in the grove or in storage).

Rind-damaged fruit are downgraded or culled in the packhouse. Fruit with Septoria spot may be quarantined and rejected by certain foreign markets.

Bacterial Blast (Citrus Blast, or Black Pit)

Pseudomonas syringae

Bacterial blast, also called citrus blast, on oranges and grapefruit usually kills only leaves and twigs. On lemons, fruit are the most susceptible to infection, and the disease is called black pit.

Bacterial blast occurs mainly in the Sacramento Valley, where wet, cool, windy conditions often prevail during winter and spring, favoring the development and spread of this bacterial disease. See "Leaf and Twig Diseases" for more information.

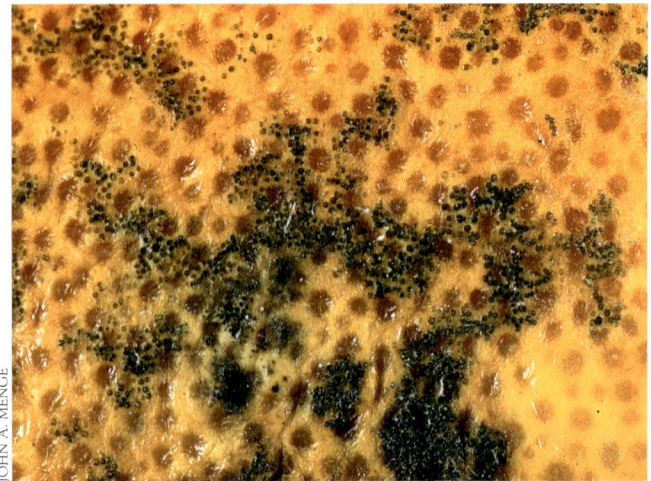

Dark brown to black fruiting bodies grow in the pits of rinds infected by *Septoria citri*. This fungal growth, shown here under magnification as when viewed through a hand lens, distinguishes Septoria spot from similar rind injury caused by copper sprays.

Septoria infection usually occurs in the field during wet weather, but disease often does not become apparent until fruit are in storage. As disease progresses, the initially small, shallow infection pits can merge into these large, dark, sunken blotches.

Septoria spot causes light tan to reddish brown shallow pits in infected fruit rinds.

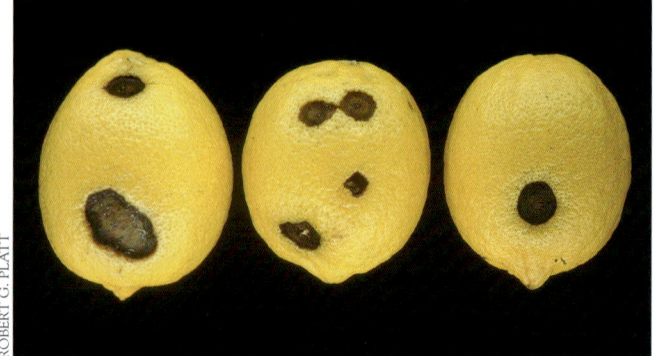

Bacterial blast on lemons is called black pit because it causes dark, sunken lesions. On grapefruit and oranges, the disease usually kills only leaves and twigs in a scattered ("blast") pattern throughout the windward side of the canopy.

Blue and Green Mold (Clear Rot)

Penicillium spp.

Blue and green mold is mainly a problem during storage but can also occur in the orchard on injured fruit. Disease initially causes a water-soaked, soft area on fruit called clear rot. This early symptom is easily overlooked. Disease becomes obvious once whitish mycelia and the blue or green spores develop on the fruit surface, usually after harvest.

Blue mold *(Penicillium italicum)* and green mold *(P. digitatum)* usually occur together. These fungi are common in soil and on decaying plant materials. Blue and green molds cannot infect intact fruit rinds.

Botrytis Rot (Gray Mold)

Botrytis cinerea

Botrytis rot is a problem on coastal lemon and occasionally on Valencia orange. "Gray mold" refers to the fuzzy fungal growth that occurs on infected parts and is the name for the postharvest disease when *Botrytis* decays fruit during storage.

During prolonged wet periods, *Botrytis cinerea* infects tissue through injuries or delicate tissue like senescing flower petals. Infected twigs can die back several inches. Infected blossoms often result in fruit drop or rind injury. Petals infected with the fungus spread the disease when they contact small fruit. At the point of contact, the fungus stimulates abnormal cell growth, so that ridges or raised scars develop on fruit. Mature Valencias develop a soft decay at infection sites. Larger branches can develop ooze resembling that of Phytophthora gummosis.

Sooty Molds

Sooty molds are black, somewhat feltlike fungal growths caused by *Capnodium* spp. and other fungi. They can be rubbed off the surface of contaminated fruit, leaves, or stems. Sooty molds generally grow on honeydew excreted by juice-sucking insects, including aphids, mealybugs, soft scales, and whiteflies. Commonly caused in citrus by *Capnodium citri*, sooty molds do not infect trees or fruit but grow only on the surface. Sooty molds usually cause no damage to the crop unless the plant becomes so heavily covered

Blue and green mold damage starts as a small, watery lesion called clear rot (left fruit). As disease progresses, the *Penicillium* spp. fungi that cause this decay produce white mycelia and blue and green spores.

Botrytis rot is also called gray mold because under humid conditions it develops gray, fuzzy growth. This fungal growth consists of numerous tiny-stalked spore-forming structures, which can develop on infected young fruit (shown here) and on dead blossoms and leaves.

These brown, raised, elongate scars on an immature lemon were caused by Botrytis rot. When very young fruit are infected and conditions remain wet, fruit may turn entirely brown and die.

Botrytis infection occurred when this lemon was young. The rind developed raised bumps and remained green at maturity.

This dark, felty sooty mold growth can be scraped off of plant surfaces, unlike fruit rots that extend into the rind and flesh. Where sooty mold occurs, look for aphids, citricola scale, cottony cushion scale, and other phloem-sucking insects that excrete honeydew on which these dark fungi grow.

that photosynthesis is reduced and the shading of surfaces causes foliar chlorosis or delayed fruit ripening.

Postharvest fruit washing can remove sooty mold, but contaminated fruit may be downgraded at the packinghouse. Manage sooty mold by controlling the insects that produce honeydew, as discussed in the chapter "Insects, Mites, and Snails."

FRUIT DISORDERS

Weather-related causes of fruit damage include frost, rind disorders, sunburn, and wind. Also discussed here are split fruit related to cultural practices and water deficit, pesticide spray injury (phytotoxicity), rind injury (oleocellosis or oil spotting), genetic mutations (chimeras), and poorly understood maladies (peteca of lemons and puff and crease of oranges and mandarins). Some of these maladies can be confused with other causes of rind damage, as summarized in Table 3 in the chapter "Managing Pests in Citrus."

Protecting trees from injury and providing proper cultural care as discussed in "Prevention and Management" at the beginning of this chapter help prevent many fruit disorders and the associated rots discussed above in "Fruit Diseases." See the "Leaf and Twig Disorders" section later for more photographs and discussion of many of these maladies.

Chimeras

If genetic mutation occurs in a branch or twig and that tissue survives, it can produce new shoots (called a chimera or sport) with characteristics different from the rest of the tree. Mutations can affect the color of the rind or pulp or the shape of the fruit. Leaves on these twigs can have a different shape or size or variegated color. Mutation can cause the development of multiple buds, creating bunchy growth or "witches' broom." A chimera can produce an improved crop: some of today's cultivars were propagated from chimeras. Usually sports are of inferior quality and should be avoided as propagation material. Prune sports that obstruct normal growth or interfere with harvest.

Wind Injury

In areas with persistent winds, fruit on the outside canopy can be scarred where twigs or thorns rub against the rind. Fruit can also drop from the tree. Strong, persistent wind causes water stress by dehydrating the leaves, causing leaf necrosis or twig dieback or stunting of the growth of young trees. Drying winds such as the Santa Anas cause bronzing, pitting, and curling of leaves, mostly on the outside of the canopy exposed to the wind. In certain coastal districts, chilling winds blowing from the ocean inhibit the normal growth of citrus.

A raised section in fruit, typically in a wedge-shape, is usually from chimera. This genetic mutation is of minor importance.

Pale to brown scabby scars develop on the rind where fruit rubs against twigs or thorns.

Appropriate irrigation and windbreaks can help reduce wind damage. A natural windbreak can be a row of fast-growing tall trees. Individual shelters of burlap or wood frames may be suitable for young trees.

Sunburn

Heat produced during direct exposure to intense sunlight can blemish fruit, cause chlorotic or necrotic leaf blotches, and produce bark cankers. On sunburned fruit, the side exposed to the sun develops brown, leathery areas. Some fruit become slightly lopsided. See the "Trunk Disorders" section for more discussion of sunburn and its prevention.

Hail

Impact from hail tatters leaves and causes sunken scars on the upper and outer exposed surfaces of fruit and twigs. If hail impacts young fruit, scars can become large and distinctive as the rind enlarges.

Frost and Freeze

Because it is a subtropical plant, citrus is particularly susceptible to cold damage (frost and freeze). Susceptibility to cold varies with the age and physiological state of the tree and with the cultivars of the rootstock and scion. For example, Eureka lemon and grapefruit are among the most cold-sensitive cultivars, whereas mandarin, Meyer lemon, and sweet orange are more cold hardy.

Immature or young shoots and fruit and young trees are the most susceptible to cold damage. Young trees can be killed. When trees are older, economic damage usually occurs only when fruit are injured. Cold ruptures oil cells in the fruit rind, resulting in oil leakage and damage to the rind surface, visible as watery, brownish specks or pits called ice marks. Fruit pulp underneath the ice marks ultimately dries, and the ice marks provide entry sites for decay organisms. Cold-damaged fruit may drop suddenly from the tree, or severe fruit damage can occur without fruit drop or obvious rind markings. This fruit is unmarketable, because the flesh is dehydrated. The damage may not be apparent until fruit is cut in cross section or processed at the packhouse.

Cold also can cause bark cankers and shoot dieback. Damaged leaves and twigs become water soaked, withered, and dark brown to black. Frost damage mainly appears on outer and upper shoot terminals and on the outside of fruit exposed to radiation frost.

Frost and freeze occur under different conditions but produce the same damage. See "Citrus Frost Protection" (in preparation) for more information.

Hail impact causes sunken scars on the exposed upper or outer surfaces of fruit and twigs.

When yellow to brown leathery areas occur on the unshaded surface of fruit in the south and west sides of the tree, the cause is probably sunburn. Nearby leaves may also have chlorotic or necrotic sunburned blotches.

This brown rind scarring sometimes develops on oranges after cold weather. Other times after cold damage no obvious external symptoms are visible on fruit even though flesh inside has been injured and the fruit is unmarketable.

Cutting open these navel oranges revealed dehydrated, shrunken flesh in the left fruit damaged by cold temperatures.

This rind damage, called rind stipple or concentric rind stipple, sometimes develops on grapefruit after cold, wet weather. The brown discoloration occurs in numerous small pits.

Rind Stipple of Grapefruit

A unique pattern of rind discoloration develops on grapefruit during prolonged periods of cold, wet weather. The rind develops small, brownish pits (ice marks). The pits are often surrounded by concentric rings that are formed as water droplets dry. The pits may coalesce into larger, irregular lesions. If green fruit is injured, a green halo appears around pits and persists until the fruit turns yellow. This distinctive rind damage pattern has been reported only on grapefruit. The symptoms occur mostly on exposed fruit on the north side of the tree. Stippled fruit is downgraded at the packinghouse. No treatment is available.

Rind Disorder (Mandarin Rind Disorder)

A rind disorder that develops on mandarins sometimes results in severe crop loss in the Sacramento and San Joaquin valleys. Brown, water-soaked blotches or dark, sunken areas develop on exposed surfaces of fruit in the outer canopy. Fungi develop in the discolored areas, and the flesh underneath softens and rots. Damage is most severe on Satsuma mandarins but occasionally occurs on other mandarin varieties such as Clementine.

Mandarin rind disorder occurs during high rainfall in the fall. Once fruit begin to develop mature color, rind injury becomes apparent within days after rain. The disorder is a physiological problem that develops during adverse environmental conditions when the rind becomes water soaked and the oil glands rupture. The injured rind is then colonized by various secondary decay fungi. Treatments include one or two foliar applications of oil or antitranspirants that repel water, applied before forecasted rain or around fruit color break.

Peteca of Lemon

Coastal lemons sometimes develop round depressions in the rind that become discolored or brownish. Called peteca, this necrosis and collapse of albedo cells becomes obvious after harvest. Less-obvious damage is sometimes visible before harvest as slightly brown, sunken spots on the skin. The cause is unknown.

Once fruit develop mature color, dark, sunken blotches can develop on mandarins within a few days after rainy weather. This rind disorder most often occurs on Satsuma mandarins.

Lemons growing near the coast sometimes develop depressions in the rind that become discolored or brownish. This disorder is called peteca.

Puff and Crease

The innermost spongy white layer of the rind (albedo) sometimes separates from the outer surface of the fruit segments. This internal damage is visible externally as an uneven appearance of the rind surface. Some portions of the rind surface appear inflated (puffy) and other areas are indented (creased), hence the name "puff and crease." Navel and Valencia oranges and Satsuma mandarins are susceptible to this tissue separation.

Symptoms become increasingly prevalent as fruit mature. There may be physiological differences between damage development on oranges versus mandarins. Puff is sometimes considered a separate malady from crease, but their overall damage appearance is similar. The cause of rind separation from flesh and how to prevent it are not well understood. The application of plant growth regulators and various cultural practices and environmental conditions (irrigation, nutritional status, temperature, and fruit load) may influence the extent to which puff and crease occurs. Consult "Citrus Physiology and Phenology" (in preparation) for more discussion.

Oleocellosis (Oil Spotting)

A phytotoxic reaction occurs from peel oil when glandular cells in the rind are injured. Oleocellosis (also called autophytoxicity, autotoxicity, or oil spotting) occurs on rinds after abrasion, rough handling, or thorn punctures. The rind of oranges and lemons is especially susceptible to this disorder. Wet conditions from high rainfall or excessive irrigation often result in a more turgid rind that can worsen the condition if harvest immediately follows these conditions. Occasionally, when turgid fruit are exposed suddenly to low temperatures, peel oil may be released, causing oleocellosis.

Split Fruit

After tree stress, rinds often split at the bottom of fruit. Causes of tree stress include unseasonably high temperatures or other extreme weather, inadequate irrigation, and potassium deficiency. Decay fungi usually colonize wounded fruit, resulting in secondary diseases such as Alternaria rot and blue and green mold. Even when only a few fruit are affected on the tree, pathogen spores developing in split fruit can disperse and heavily contaminate healthy fruit surfaces. After harvest, this pathogen contamination can cause postharvest fruit rot.

This uneven appearance and cracking on the outer surface of Valencia orange rinds is called crease. The top of the right fruit also has a raised or inflated ridge. Internally, the rind has separated from flesh. Other names for this puzzling malady include crease, crease and split, puff, or puff and crease.

Rind split, usually at the bottom of fruit, can occur after tree stress. Extreme weather and poor irrigation are common causes of split fruit.

Spray Injury (Phytotoxicity)

Pesticide sprays, such as fungicides, growth regulators, herbicides, and oils, can injure citrus trees, causing phytotoxicity. Most agricultural chemicals can cause damage if used incorrectly or applied during adverse environmental conditions. See "Leaf and Twig Disorders" for more discussion and photographs of spray injury.

These shrunken, dark blotches on lemon rinds are spray injury due to manganese and zinc nutrient application.

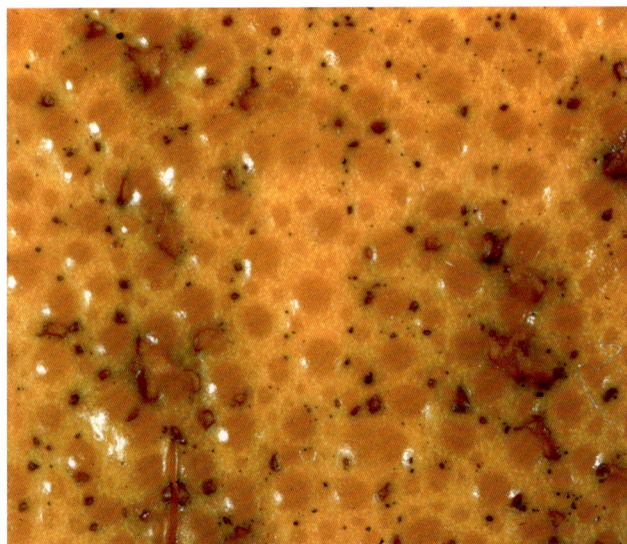

Dark rind pitting can occur after a copper spray. This phytotoxicity resembles Septoria spot, except there is also dark fungal spore growth in discolored tissue when *Septoria* (pictured earlier) is the cause.

LEAF AND TWIG DISEASES

Several pathogens and abiotic disorders cause leaves and twigs to wilt, die back, distort, or become discolored, often without affecting other parts of the tree. Most leaf and twig diseases do not cause serious long-term losses in citrus. However, pathogens and other pests affecting roots, trunks, and overall tree growth often cause symptoms on leaves and twigs as well. Be sure to diagnose the cause of leaf and twig symptoms correctly by inspecting the whole tree and comparing your findings with the appropriate sections of this book. For example, anthracnose and Botrytis rot are pictured below but are discussed mainly in the "Fruit Diseases" section. Other causes of damage are presented under "Leaf and Twig Disorders," "Exotic Diseases," and "Diseases Affecting Growth Habit and Yield."

To minimize leaf and twig diseases, provide good cultural care to reduce tree stress. Prune out dead and diseased limbs and twigs in spring after the rainy period to reduce pathogen spore production and disease. Proper fertilization timing can reduce disease, so schedule fertilization and pruning during spring or early summer and avoid these activities later in the season. This timing minimizes excessive new growth during fall (this stage is particularly susceptible to bacterial infection and freeze damage). Cold-damaged tissue then becomes susceptible to anthracnose and other diseases. Windbreaks can reduce the incidence of wind-associated bacterial blast. Use bushy cultivars with relatively few thorns to help avoid injuries that become sites of pathogen infection. For more recommendations including fungicides, consult the online *UC IPM Pest Management Guidelines: Citrus*.

Bacterial Blast (Citrus Blast, or Black Pit)

Pseudomonas syringae

Bacterial blast, also called citrus blast, kills leaves and twigs. When fruit are damaged, the disease is called black pit. Bacterial blast occurs mainly in the Sacramento Valley, where wet, cool, windy conditions during winter and spring favor development and spread of the blast bacterium. Leaves and twigs of oranges and grapefruit and the fruit of lemons are the most susceptible to infection.

Infections caused by *Pseudomonas syringae* usually start as small water-soaked or black lesions in the leaf petiole and progress into the leaf axil. Once the petiole is girdled, leaves wither, curl, and eventually drop. Entire twigs may die back. Diseased areas on twigs are reddish brown and can become scabby or callused. Dieback scattered throughout the canopy creates the "blast" appearance for which the disease is named. Bacterial blast damage may be confused with symptoms from Dothiorella blight (discussed under "Dothiorella Gummosis"), frost, or wind.

The bacterium infects through small injuries, including those from thorn punctures, wind abrasions, and insect feeding. Bacterial blast damage is most severe on the south side of the tree, which is exposed to the prevailing winter wind-driven rains.

Botrytis Rot

Botrytis cinerea

Botrytis infects and kills twigs and small branches, primarily on coastal lemon trees. Botrytis rot damage may be confused with that from anthracnose, citrus blast, frost, water deficit or stress, and wind. A distinctive difference is that under cool, wet, or humid conditions, *Botrytis*-infected tissue becomes covered with gray fuzzy growth, called gray mold. It also damages fruit; see "Fruit Diseases" for more information.

Bacterial blast (citrus blast) starts as dark lesions in the leaf petioles and progresses into leaf axils. This disease occurs during winter and spring after wet, cool, windy conditions.

Bacterial blast causes leaves to curl, dry, and drop prematurely. Disease affects shoots scattered throughout the canopy side that is exposed to wind-driven rains.

Botrytis rot can kill shoots and fruit if they are infected when they are young. Botrytis rot is most common on coastal lemons, but this dieback occurred on navel orange in San Bernardino County.

Anthracnose killed these Valencia leaves. This necrosis might be confused with that caused by bacterial blast, Botrytis rot, or injury from water stress or wind. The black, sooty fungal spores visible on damaged leaves help to identify anthracnose as the cause.

The small bumps on veins (vein enation) on the upper surface of leaves and pits in the lower surface of the hand-held leaf were caused by an aphid- and graft-transmissible pathogen, most likely a virus. The pathogen also causes distorted bark, as pictured earlier under "Trunk Diseases."

Anthracnose

Colletotrichum gloeosporioides

Anthracnose causes leaf tips to become yellow, then turn brown and die. Dead leaf and twig tissue becomes covered with dark mycelium and orangish fungal spores, which provide inoculum that spreads to cause additional infections. Anthracnose is a problem primarily when it causes dark blemishes on fruit rinds. The disease mainly affects stressed trees with dead wood, where the fungus produces abundant spores. Damage is most common during springs with prolonged wet and warm periods and when significant rains occur later in the season than normal. See the "Fruit Diseases" section for more information.

Vein Enation (Woody Gall)

An aphid- and graft-transmitted pathogen causes gnarled bark (woody gall) and small bumps on citrus leaves (vein enation). The pathogen is most likely a virus, but the cause and development of vein enation disease is poorly understood.

The vein enation pathogen causes swellings on the veins on the lower leaf surface of cultivars such as Mexican lime, rough lemon, and sour orange. Bark outgrowths often develop around thorns or wounds on small-fruited acid limes (e.g., Mexican lime) and on the sensitive rootstocks rough and volcamer lemon.

The pathogen causing vein enation appears to interact with other graft-transmissible agents. For example, when the vein enation pathogen occurs together with the yellow vein virus or citrus viroids, the symptoms of each of these diseases are more severe. Conversely, when the tristeza virus and vein enation occur in the same plant, the symptoms of vein enation are suppressed.

Vein enation and woody gall are not commercially important except when severe galling occurs on young trees budded on sensitive rootstocks. The disease is rare in California because citrus is often propagated on nonsusceptible rootstocks, strict regulations on budwood sources have kept new plantings free of this disease, and secondary spread by the aphid vectors has been minimal.

These trees are dying from root asphyxiation (aeration deficit). Flooding caused their root zone to be excessively wet for an extended period and deposited several inches of soil beneath canopies, depriving roots of sufficient oxygen to survive.

LEAF AND TWIG DISORDERS

Mineral deficiencies and toxicities commonly cause a variety of leaf and twig symptoms. Aeration deficit, chimera, fire, frost, spray injury, and wind also damage leaves and twigs, as discussed below and (for some of these) also above in "Fruit Disorders."

In addition to their direct damage, many of these disorders increase tree susceptibility to pathogens, as discussed above in "Leaf and Twig Diseases." Provide trees with proper cultural care as discussed in "Prevention and Management" at the beginning of this chapter to help avoid or minimize damage from many of these disorders.

Aeration Deficit or Root Asphyxiation

Too little oxygen in the soil (aeration deficit) causes smothering (asphyxiation) of roots, a serious problem that can kill trees. Symptoms of aeration deficit usually are evident throughout the entire tree, including leaf yellowing and wilting, premature leaf drop, and stunted growth. If aeration deficit is extreme or prolonged, the leaves will suddenly collapse and turn brown. Fruit wither or soften and then usually drop. Branches typically die back from the tips, but only a few branches or one limb may die.

Asphyxiation occurs mainly in soils that are compacted, fine textured (clay soils), shallow with an impervious subsoil (hardpan), or otherwise have poor drainage (slow water infiltration). When problem soils are overirrigated or wet from prolonged rainy periods, water displaces the gas in pore spaces between the soil particles. Flooding from water runoff or where the water table is high also causes waterlogged soil. Saturated soil impedes oxygen diffusion (movement through the soil) and leaves few water-free pore spaces for oxygen to occupy. Insufficient aeration, excess soil moisture, and diseases such as Phytophthora root rot often act in combination to damage or kill trees.

Diagnose aeration deficit by assessing whether site conditions and cultural practices are conducive to root asphyxiation. Look for groups of trees that exhibit similar symptoms, since damage often appears in clusters of trees that share an area of problem soil. Examine small roots to determine whether they appear to be healthy (abundant, firm, and pale inside) or unhealthy (sparse and dark). Dig 1 to 2 inches (2.5–5 cm) below the surface and check the soil for the disagreeable odor of sulfides (rotten eggs), which indicates anaerobic (oxygen-deprived) conditions. Send samples of the soil to a laboratory to determine its composition (soil aeration decreases as the percentage of clay increases). Check for the presence of compact or cemented hardpan in the root zone, which may impede water penetration.

Improve soil conditions and cultural practices as described in the "Root Rots" management section. For example, keep furrows or tree basins open during rainy weather to facilitate drainage. Where an impervious soil layer is the cause of poor aeration, consider methods such as deeply ripping the soil alongside each tree (during cool times of the year to minimize tree stress) and installing subsurface drains.

Water Deficit or Stress

Water stress occurs when a tree's moisture demands exceed its supply. Afternoon wilting of shoot tips and young leaves is an immediate symptom. After severe or prolonged moisture deficient, older leaves curl, wilt, and drop. Fruit quality may decline.

The afternoon wilting of young leaves is an immediate symptom of extreme water stress. If the cause is unhealthy roots or adverse soil conditions, wilting can occur even if soil is saturated with water.

DISEASES

Frost damage, leaf mesophyll collapse, twig dieback, and wind injury (all discussed below in this section) and split fruit (discussed above under "Fruit Disorders") are caused or aggravated by water stress. Abnormally low soil moisture (insufficient irrigation) is an obvious cause of water stress. Extreme heat, sudden changes in temperature, high wind, and low humidity contribute to moisture stress. Phytophthora root rot and anything that causes adverse soil conditions or unhealthy roots can cause water stress even when readily available soil moisture is present.

To minimize water stress, provide trees with appropriate cultural care, especially proper irrigation. Create a good growing environment through appropriate soil conditions and windbreaks. Avoid cultivation near trunks, which can injure roots. Avoid chemical applications to foliage during hot or windy conditions, as this can interfere with the physiological functions of the tree, thereby increasing water deficit damage.

Twig Dieback

Periodically, twig dieback is a problem on citrus trees in most growing districts. Causes include a poor root system, a mild scion-rootstock incompatibility, and moisture- and weather-related stress.

Sudden twig dieback sometimes occurs in the San Joaquin Valley in the spring. Terminal leaves turn brown, die, and remain attached to twigs or drop from the tree. Roots are mostly inactive at this time because the soil is cool. However, increasing daytime temperatures increase the tree's water loss and the canopy's water demand, resulting in water deficit or stress even when there is ample soil moisture. Gum may form in the conducting tissue of twigs, further inhibiting water transport. As a result, leaves and twigs may die immediately or become more likely to die some weeks later, often during another hot period. On navel orange trees, this type of dieback often occurs after fruit have been harvested.

Twig dieback in the San Joaquin Valley can also occur in late summer. The brown, dead terminal leaves usually remain attached to the tree. This dieback typically occurs after hot temperatures and may be related to moisture stress or poor root health.

Frost and Freeze

Frost- or freeze-damaged leaves and twigs become water soaked, wither, and turn brown to black. Cold weather can kill young or regrafted trees but usually not mature trees. Damage varies according to the age and physiological stage of the tree, cultivar of scion and rootstock, location, and extent of cold conditions. Fruit loss is more common than mature tree death. Certain cultural practices, such as avoiding pruning or fertilizing during late summer and providing frost protection during critical injury times, can reduce the impact of cold temperatures on branches and leaves. See "Fruit

Frost and freeze cause upper and outer leaves and twigs to wither and turn dark brown to black. Unless they are young, citrus trees usually survive cold temperatures.

This twig dieback occurred during spring in the San Joaquin Valley. Cool soil prevented inactive roots from meeting the tree's sudden increase in water demand during hot weather. This damage was unavoidable, but be sure to diagnose the cause. Improper irrigation, unhealthy roots, and various pathogens cause similar damage.

Cold can cause tree canopies to appear as if they have been scorched by fire. Citrus growing on flat land and young trees are more likely to be damaged during radiation frost or freeze (during clear nights when plants loose heat to the sky). During an advective freeze (when a cold air mass moves in), trees on higher ground are damaged as well.

Soft tissue between leaf veins can collapse and become translucent or pale. This injury, called mesophyll collapse, occurs when trees are unable to provide leaves with enough water.

Disorders" earlier in this chapter and "Citrus Frost Protection" (in preparation) for more information.

Wind Injury

Strong, persistent wind or hot or cold winds of shorter duration can dehydrate leaves, stunt the growth of young trees, and scar fruit. Drying winds can bronze, pit, and curl leaves. Clusters of leaves on shoot tips turn brown and die, mostly on the outside of the canopy exposed to the wind. To reduce wind injury, irrigate appropriately and plant a windbreak row of fast-growing, tall trees. Individual shelters of burlap or wood frames may protect young trees.

Hail

Impact from hail causes sunken scars on the upper or outer exposed surfaces of twigs and fruit. Leaves are torn, shredded, and may drop prematurely after hail impact.

Mesophyll Collapse

The soft interior leaf tissue (mesophyll) between the veins sometimes collapses and becomes translucent. Affected leaves may dry completely and turn brown. Mesophyll collapses when trees are unable to supply enough water to the leaves, often because of low soil moisture or hot, dry winds. A poor root system, saline soil, and heavy mite feeding also stress trees and increase the occurrence of mesophyll collapse.

Fire

Orchards planted near wildlands are more likely to be exposed to wildfires. Fire kills leaves and fruit, causing them to blacken or brown. If tree canopies are singed without injury to major limbs or trunks, prune away the dead branches after foliage regrows. Provide optimal cultural care to encourage tree recovery.

For orchards near wildlands, it can be difficult to keep trees from burning during a wildfire. Advance preparation is the best approach. Consider creating firebreaks (areas of little or no vegetation) to help keep fire from moving into orchards. A well-managed border of low-growing vegetation reduces the risk of orchard invasion by wildfire and also can reduce off site movement of agricultural chemicals and soil. If fire threatens, irrigate to increase orchard wetness and reduce the potential of burn damage.

The small scars on twigs of this citrus occurred on the exposed upper and outer sides of branches. The cause was impact from hail.

Hail shredded and tore these citrus leaves and caused discolored scars on the stems.

Fire killed these fruit and leaves, causing them to turn black or brown. Especially where citrus grows near wildlands, manage border vegetation to minimize the likelihood that fire will burn into orchards.

Pale specks or blotches develop on the upper side of citrus leaves after exposure to high levels of ozone air pollution. Ozone pollution is usually too low to cause this obvious leaf injury. But tree growth and fruit yield are reduced by the air pollution levels common in California.

Chlorotic or necrotic blotches on leaves can be caused by sunburn. If damage is not too severe and roots are healthy and adequately irrigated, leaves yellowed by sunburn often recover their normal color once conditions become less stressful.

Sunburn

Heat produced during direct exposure to intense sunlight can cause chlorotic or necrotic blotches on leaves, blemishes on fruit rinds, and cankers on bark. To help prevent sunburn, keep roots healthy, irrigate appropriately, and whitewash exposed bark. See the "Trunk Disorders" section for more sunburn photographs and prevention tips.

Air Pollution

Air pollution, called smog when air is visibly dirty, can stunt tree growth and reduce fruit yield. High concentrations of ozone cause pale specks or blotches on the upper side of citrus leaves. Ozone forms when volatile organic compounds (VOCs) and nitrogen oxides combine in the presence of sunlight. Tree growth and fruit yield are reduced by ozone pollution levels common in California even when there are no obvious leaf damage symptoms. Consult "Ozone Air Pollution" (in preparation) for more discussion.

Spray Injury (Phytotoxicity)

Chemical injury to plants (phytotoxicity) can occur if pesticides, foliar nutrient sprays, and other agricultural chemicals are used incorrectly. Improper or careless mixing or application, and drift or other movement from nearby crops, are usually the cause. Certain growth stages are more susceptible, and some stressful environmental conditions increase the risk of phytotoxicity. For example, phytotoxicity from spray oil is most likely to occur during certain stages of fruit development, such as green lemon. Almost any material can injure trees if applied during hot and dry conditions and in orchards with a soil water deficit.

Oils can cause phytotoxicity because a persistent film of oil interferes with the physiological functions of the tree. Oil spray can discolor fruit, cause chlorotic or necrotic blotches on leaves, and in severe cases result in fruit and leaf drop and dieback in the tree canopy. In comparison with older oils, the narrow-range oils mostly used today pose a low risk of phytotoxicity when properly applied.

Growth regulators such as 2,4-D* cause leaf curling and chlorosis if applied at the wrong concentration or the wrong time (e.g., when immature growth is present). Trunk contact with the higher amine concentrations of 2,4-D* used as an herbicide in other crops can girdle citrus trunks.

Fungicides containing copper may pit leaves and fruit. On leaves, discolored pitting begins on the lower surface and may extend through to the upper surface. Certain areas in southern California and Kern County are more likely to suf-

2,4-D applied for growth regulation can cause citrus leaves on immature shoots to cup or curl. In some cases, plant growth overall can be undesirably stunted. Drift from 2,4-D use as an herbicide on other crops can seriously damage citrus.

*Restricted-use material. Permit required for purchase or use.

fer copper injury than other citrus-growing regions, which may be due to the air pollution in those areas.

Postemergence and preemergence herbicides can injure citrus by direct contact with foliage or fruit or by uptake through roots, depending on the material and specifics of its application. Glyphosate applied to nearby weeds commonly injures citrus. This systemic herbicide is absorbed through foliage and other green tissue but also sometimes through tender bark. Glyphosate can deform new growth, darken foliage, and abort flowers. Shoots may develop narrow, distorted leaves, multiple buds, and stunted, bunchy tissue (rosetting).

Leaves or fruit may discolor when a contact herbicide is sprayed onto the tree skirts. If paraquat is accidentally sprayed on or drifts onto citrus trees, leaves may drop. When roots take up too much of a preemergent herbicide, symptoms appear above ground mainly on the leaves. Damage varies depending on the herbicide and amount absorbed, such as fading of leaf color or causing entire leaves to turn brown and die.

Fumigants kill beneficial mycorrhizal (root-associated) fungi, sometimes resulting in stunting and poor growth of citrus seedlings. Other materials can cause outbreaks of mites, disrupt biological control of pests, or pollute air or water. See the chapters "Weeds" and "Insects, Mites, and Snails" for more discussion.

See the "Fruit Disorders" section for more photographs of spray injury. For more detailed discussion of phytotoxicity, consult "Environmental, Physiological, and Cultural Injuries and Genetic Disorders" (in preparation) and the UC Davis Weed Research and Information Center Web site (http://wric.ucdavis.edu). For recommended pesticides and growth regulators, including material-specific tips on how to avoid phytotoxicity, consult the online *UC IPM Pest Management Guidelines: Citrus*.

Glyphosate herbicide causes small, puckered, needlelike leaves, resembling zinc deficiency. Glyphosate-injured shoots may develop fewer leaves than normal, and buds may only partially open.

Glyphosate spray curled and yellowed these citrus leaves. Some small fruit and blossoms were killed. Except when trees are young, citrus usually recover well from glyphosate injury when spray contaminates a limited portion of the canopy.

Leaf veins turn yellow or white (vein clearing) after excess citrus root uptake of the preemergent herbicides diuron (shown here) or bromacil.

Simazine caused this chlorosis between leaf veins, symptoms that resemble certain nutritional deficiencies. Damage increased in severity in proportion to the amount of herbicide absorbed by the tree's roots.

This color variegation and misshapen, undersized leaves are due to genetic mutation. Bunchy shoot growth (witches' broom) and misshapen or discolored fruit may also develop.

Chimeras

A change in a plant's genetic material (mutation) sometimes causes abnormal leaf shape, size, or color (variegation) on one or several twigs. Shoots may develop bunchy growth called witches' broom. Fruit may be misshapen or discolored.

Chimeras (also called sports) are usually of inferior quality and should be avoided for propagation. Prune sports that obstruct normal growth or interfere with harvest.

Mineral Deficiencies and Toxicities

Mineral deficiencies occur because of low amounts in the soil or adverse conditions that reduce the plant's ability to absorb nutrients. Excess soil moisture, low temperatures, low or high soil pH, and root diseases or injuries inhibit plants' nutrient absorption. Excess of some minerals in the soil or irrigation water can inhibit absorption of other minerals or may be toxic to the plant. The lack or excess of certain minerals usually appears first as leaf symptoms but eventually affects fruit size, quality, or yield.

In California citrus, nitrogen and zinc are most commonly deficient. Magnesium, manganese, potassium, and iron can also be lacking. Knowing your orchard soil conditions and water quality helps alert you to what mineral disorders are likely to occur. Iron, manganese, and zinc deficiencies are common on alkaline soils. Zinc deficiency is most severe on sandy and sandy loam soils. Iron deficiency often results where soils drain poorly or have a high lime content. Toxicities from minerals are less common than are deficiencies. Boron and sodium cause toxicity where soils or irrigation water contain high levels of these minerals.

Deficiency symptoms are most apparent during the winter and early spring. An exception is magnesium deficiency, which usually becomes apparent in late summer or fall. When a deficiency is mild, the symptoms often disappear later in the season.

The location of symptoms on the tree and distinct patterns of leaf discoloration and abnormal growth often help diagnose the cause of nutrient disorders. For example, zinc and manganese deficiencies are most pronounced on young shoots, with the effects of zinc deficiency most noticeable on the south side and those of manganese deficiency on the north side of trees. Magnesium deficiency is most prevalent on mature leaves in late summer or fall. Iron deficiency manifests typically on the lower half or skirt of the tree, affecting leaves on the outside as well as the inside of the canopy.

Although many common deficiencies or toxicities produce distinct symptoms, the cause of nutrient disorder symptoms is often difficult to diagnose. Symptoms of certain disorders resemble each other. Nutrient disorder symptoms are often secondary, developing as a result of other problems occurring at the same time, so symptoms of one disorder are masked by those from other disorders or biotic diseases.

Correctly sample leaves at the proper time of year to help diagnose the cause of symptoms. Use leaf analysis and other knowledge of the local situation to verify your field identification and to provide information on the general nutritional status of the tree. Consult "Fertigation" (in preparation) and "Nutrient Deficiency and Correction" (in preparation) for details on the diagnosis and management of nutrient disorders affecting citrus.

A white crust on this soil surface indicates root exposure to high salt concentrations. Poor water quality and inadequate irrigation management can expose citrus to excess salinity, causing yellow leaves, reducing tree growth and yield, and increasing tree and fruit susceptibility to various diseases.

Foliage that is pale overall indicates a nitrogen deficiency. Even when trees are healthy and receiving proper cultural care, nitrogen deficiency symptoms are common during early spring.

Nitrogen (N) Deficiency. Foliage is pale overall when nitrogen is deficient. Nitrogen deficiency is frequently observed in early spring, when soils are too cold or wet to allow roots to absorb adequate amounts of nitrogen for new growth. However, pale foliage does not mean that a true deficiency of nitrogen is the cause. If deficiency symptoms develop even though trees have been fertilized as recommended, investigate other causes, including overly wet or cool soil and whether roots are healthy. Nitrogen is usually provided as nitrate or ammonium for absorption by roots or as urea, which is readily absorbed by leaves.

Zinc (Zn) Deficiency. Leaves deficient in zinc remain small ("little leaf") and occur on shoots with shortened internodes, resembling glyphosate herbicide damage. Zinc-deficient leaves also have irregular pale to white, chlorotic blotches between the veins. Dark green, round spots often occur scattered throughout the leaf.

Zinc deficiency occurs where soil is sandy, or topsoil has been eroded away and pH is low, such as in granitic soils. Zinc also becomes unavailable where soil pH is high, such as when irrigation water is alkaline or soil is calcareous. Quickly remedy zinc deficiency by foliar application of zinc sulfate or chelate. Soil application of zinc chelate compounds or modifying the pH of irrigation water or soil can slowly remedy zinc deficiency.

Magnesium (Mg) Deficiency. Leaves deficient in magnesium develop chlorotic blotches between veins. The blotches enlarge progressively along the leaf tip and edges, leaving only a center V-shaped, dark green area along the basal portion of the main vein. Leaves may hang downward or droop. In severe cases, leaves turn entirely yellow and drop prematurely. Magnesium deficiency appears primarily in older foliage.

Magnesium salts can be applied to either soil or leaves to remedy a deficiency. However, first consider the status of other nutrients. Increasing magnesium can decrease absorption of calcium or potassium. Conversely, excess calcium or potassium decreases magnesium availability.

This midvein chlorosis is caused by a mild zinc deficiency. Inadequate zinc uptake is usually due to adverse soil or water pH, other stressful soil conditions, or unhealthy roots.

Advanced zinc deficiency causes extensive chlorosis between veins, resembling phytotoxicity from certain preemergent herbicides. Zinc-deficient leaves may remain smaller than normal. Shoots may be stunted.

Magnesium deficiency causes foliar yellowing that begins at the leaf tip and margins and moves inward. A somewhat V-shaped dark green patch of tissue often remains along the basal portion of the main vein.

Manganese (Mn) Deficiency. Manganese deficiency initially causes pale to yellowish blotches between green veins. Leaves may become chlorotic overall, with the veins retaining some pale green color. Symptoms usually occur on young leaves and often disappear as leaves mature. Manganese and zinc deficiencies sometimes occur together, and their symptoms may not be different enough to distinguish which is the cause of chlorosis.

Manganese deficiency occurs most often where soil pH is 7 or higher (alkaline). Applying manganese salts to leaves or soil can remedy the deficiency. However, manganese is toxic at all but low concentrations, and toxic levels can become available to roots at low soil pH. Excess manganese also induces zinc deficiency.

Potassium (K) Deficiency. Potassium deficiency is most obvious in older leaves, where leaf edges often curl downward. Deficient leaves may be pale overall or have yellow

Interveinal chlorosis from iron deficiency appears as yellowing between the small, darker green veins. This net-vein pattern occurs primarily in young leaves of the fall growth flush.

Leaves turn yellowish overall but larger veins remain slightly green where manganese is deficient. Iron or zinc deficiency can causes similar symptoms. Symptoms of several deficiencies can be common on young leaves of the fall growth flush as soils cool and root activity diminishes.

mottling and chlorotic or necrotic edges. Where potassium is deficient, cold hardiness, disease resistance, and fruit flavor and yield may be reduced. Water-stress-related maladies may become more common. Deficiency is remedied by applying potassium to soil or leaves. Avoid excess potassium concentrations, which reduce magnesium availability.

Iron (Fe) Deficiency. When iron is deficient, young leaves slowly lose their green color, eventually becoming yellowish to cream colored. Leaf veins remain green except in severe cases. Fruit may turn yellow.

Iron deficiency is common in alkaline and calcareous soils where the high pH inhibits iron absorption by roots. Iron deficiency can result from anything that causes adverse soil conditions or makes roots unhealthy. Iron deficiency is difficult to correct. Where waterlogging is the main cause, change irrigation practices to alleviate excess soil moisture. Iron chelate application, especially to soil, may help alleviate chlorosis. Acidifying soil, such as by switching nitrogen

Yellowish leaves with edges bent downward, especially around the leaf tip, are symptomatic of insufficient potassium.

fertilization from nitrate to ammonium compounds, and practices that improve root health may slowly remedy iron deficiency.

Boron (B) Deficiency or Toxicity. Boron deficiency symptoms include corking (swelling) of the veins on the underside of the leaf and pinholes developing in leaves and shoot tips. Twigs may die back. In fruit, pockets of gum can develop in juice vesicles close to the core.

Excess boron causes leaf tips and margins to turn yellow. Brown, dried gum spots can appear on the underside of leaves. Eventually the leaf tips and margins turn brown, wither, and die. Symptoms are most apparent in older leaves in late summer and fall. Excess sodium or overall high salinity also cause chlorosis and necrosis resembling boron toxicity. Lemons and grapefruits are more susceptible to boron toxicity than are oranges.

Tiny amounts of boron are essential for plant growth, but there is a narrow "acceptable" range between sufficient and excessive concentrations. Boron can be deficient in decomposed granitic soils. Too much boron is a problem in some southern California and San Joaquin Valley waters and soils with naturally high boron concentrations and where citrus is irrigated with low-quality groundwater or reclaimed wastewater.

Test irrigation water quality if you suspect boron toxicity. If the boron level is high (more than 0.5 ppm), the best remedy is to switch to another source of water or blend water sources. Where this is not possible and the injury is severe, a more tolerant crop should be grown. Where injury is moderate, frequent, heavy irrigation can wash the excess boron below the root zone. Additional nitrogen fertilizer, especially with calcium nitrate, and some liming of acid soils may also alleviate boron toxicity but may have a negative impact on fruit quality.

EXOTIC DISEASES

Citrus bacterial canker, Huanglongbing (citrus greening), citrus variegated chlorosis, and citrus leprosis are serious citrus diseases elsewhere in the world. Look out for and promptly report to agricultural officials these or other pests you suspect to be new pests. Collaborate with officials attempting to detect and eradicate exotic pests. Obey quarantine regulations to help keep exotic pests out of California. Do not bring into California from other states or countries citrus or other plants belonging to the Rutaceae family. Propagate and plant only certified, pathogen-tested stock obtained from the Citrus Clonal Protection Program or a reputable, local supplier. For more information, consult the exotic and invasive pest resources listed at the back of this book under "World Wide Web Sites."

Citrus Bacterial Canker (Citrus Canker)

Xanthomonas axonopodis

Citrus bacterial canker is caused by a bacterium originating from southeast Asia. Bacterial canker now occurs in Florida and elsewhere in the world but not in California as of 2010.

Leaf spotting and rind scarring are the most common damage symptoms. Premature leaf drop, twig dieback, and fruit drop also occur when conditions favor disease development. On leaves, lesions usually appear first on the underside as slightly raised blisterlike spots $1/12$ to $2/5$ inch (2–10 mm) in diameter. As disease progresses, lesions enlarge and damage may be visible on the upper side of leaves and on fruit and twigs. Older lesions are circular to irregularly shaped and brown or tan with a paler yellow margin or chlorotic halo. Older lesions become corky or spongy and often sunken or craterlike on leaves but more raised and blisterlike on twigs and fruit.

The bacteria readily spread from infected trees to nearby

Excess boron in soil or irrigation water caused this yellow mottling and spotting and tip dieback on leaves. Severe boron toxicity causes premature leaf drop and dieback of twigs. Virtually identical symptoms are caused by excess sodium or overall high salinity.

Citrus bacterial canker causes circular scabby lesions on leaves, which may drop prematurely. Help keep citrus canker and other exotic diseases out of California by purchasing plants only from reputable, local sources and by obeying quarantines.

Citrus bacterial canker lesions usually appear first on the underside of leaves. Lesions are typically slightly raised or sunken blisterlike spots. Damaged spots are circular to irregularly shaped and brown or tan with a paler yellow margin or chlorotic halo.

Fruit infected with bacterial canker develop rind lesions and scars and often prematurely drop from the tree.

citrus in wind or rain, entering the citrus tissue through natural openings or wounds. The citrus leafminer has been shown to exacerbate citrus canker epidemics in Florida because the mines that it produces create additional entry points for the bacteria. All varieties of citrus are susceptible, although mandarins show some resistance. Laboratory tests such as ELISA and PCR can be used to detect *X. axonopodis* strains. See *Citrus Bacterial Canker & Huanglongbing (Citrus Greening)* (Polek, Vidalakis, and Godfrey 2007) and the website of the California Department of Food and Agriculture (www.cdfa.ca.gov) for more information.

Huanglongbing (HLB), or Citrus Greening

Candidatus Liberibacter spp.

Huanglongbing (HLB) occurs in Asia, India, Saudi Arabia, South America, the Caribbean, Belize, Mexico, and the southeastern United States, but it has not been found in California as of 2010. Huanglongbing is thought to be caused by species of the phloem-limited bacteria *Candidatus* Liberibacter (proposed genus). The HLB pathogens reside in the phloem vascular system, limiting nutrient flow and causing damage to fruit, leaves, roots, and stems.

The characteristic early symptoms of the disease are pale shoots in individual limbs or sectors of the canopy (hence the Chinese name *huanglongbing*, which means "yellow shoot disease") and asymmetrical blotchy yellow mottling of leaves. Fruit are affected later in the disease cycle. Fruit become lopsided and fail to turn a mature color (hence the South African name for the disease, "citrus greening"). Internal segmentation is asymmetrical, the seeds are aborted, and the juice may develop a highly acidic or bitter flavor that makes the fruit unmarketable for either fresh produce or juice. Leaves and fruit drop off the tree easily, and the tree shows twig and limb dieback. Diseased trees can die within 3 to 5 years.

HLB infects all commercial citrus varieties, as well as some closely related plants in the Rutaceae family such as *Murraya paniculata* (orange jasmine) and *Clausena* spp. (such as wampee, or wampi). Symptoms may take more than a year to manifest themselves after a tree has become infected. Symptoms alone are not diagnostic, and laboratory testing is required for identification. Damage from HLB can resemble

The leaf yellowing in about one-quarter of this orange canopy is due to Huanglongbing, also called citrus greening. The cause is phloem-inhabiting bacteria spread by the aphidlike Asian citrus psyllid.

Mottling and yellowing of foliage that crosses leaf veins is a symptom of Huanglongbing. Note (where marked by arrows) that mottling is asymmetrical: there is a yellow area on one side of the midvein and a dark green area opposite. If this exotic tree-killing disease or another new pest is suspected to be present, promptly report it to agricultural officials.

that caused by stubborn disease, tristeza, and certain nutrient problems such as zinc deficiency.

The *Candidatus* Liberibacter asiaticus and *Ca*. L. americanus species are spread by the Asian citrus psyllid (*Diaphorina citri*) pictured in the chapter "Insects, Mites, and Snails." The Asian citrus psyllid also damages plants directly by injecting a toxin during feeding. New shoot growth that is heavily infested by psyllids does not expand or develop normally and may easily break off. The Asian citrus psyllid occurs in all citrus-growing states in the United States and Mexico. The *Ca*. Liberibacter africanus strain is transmitted by the African citrus psyllid (*Trioza erytreae*), and its distribution is limited to Africa.

In addition to the psyllid vector, the pathogen can be spread by grafting and possibly by citrus seed. For more information, see *Citrus Bacterial Canker & Huanglongbing (Citrus Greening)* (Polek, Vidalakis, and Godfrey 2007), *Asian Citrus Psyllid* (Grafton-Cardwell et al. 2006), and the USDA-ARS Web site *Identification, Characterization, and Detection of Foreign and Newly Emerging Domestic Bacteria* (Sechler et al. 2009, online at http://www.ars.usda.gov/research/publications/Publications.htm?seq_no_115=230192).

Citrus Variegated Chlorosis

Xylella fastidiosa

Citrus variegated chlorosis (CVC) is caused by the bacterium *Xylella fastidiosa*, which inhabits and clogs tree xylem (the water-conducting tissues). A number of genetically separate strains of *X. fastidiosa* cause disease symptoms in other crops, including separate strains that cause almond leaf scorch, oleander leaf scorch, and Pierce's disease of grape. All of these *Xylella* strains and diseases are found in California but not the CVC strain of *X. fastidiosa* (as of 2010).

Citrus variegated chlorosis occurs in South America and has extensively damaged Brazilian citrus. Trees that are infected when they are young develop severe yellowing between veins, resembling zinc, boron, and potassium deficiencies. Affected leaves have chlorotic spots, which have brown gummy lesions on the lower side. Affected fruits are often small and hard with high acid content; they are not acceptable for either fresh market or juice. Disease symptoms are often most pronounced on older tissues. Typically, damage first appears on only one branch and then spreads throughout the tree. All commercial citrus cultivars are believed to be susceptible to CVC, but damage is especially severe on sweet orange. Field diagnosis of this disease is difficult. Symptoms vary and resemble damage caused by nutritional deficiencies and certain other pathogens.

CVC can be transmitted by grafting and by a number of different species of sharpshooters (a subgroup of leafhoppers). Sharpshooters ingest the bacterium when they feed on xylem fluids from infected plants and then spread the pathogen when they move and feed on other plants. The *X. fastidiosa* strains in California are most efficiently vectored by the glassy-winged sharpshooter (*Homalodisca vitripennis*).

Citrus variegated chlorosis occurs in South America and is caused by a leafhopper-vectored bacterium that clogs xylem tissue. Infected trees develop severe yellowing between veins, which typically appears first on one branch and then spreads throughout the tree.

Citrus variegated chlorosis symptoms are often most pronounced on older leaves.

Pale blotches between veins and brown gummy lesions visible on the lower side of leaves are characteristic symptoms of citrus variegated chlorosis.

Citrus Leprosis

Citrus leprosis virus

Citrus leprosis virus (CiLV, two types, cytoplasmic and nuclear) is prevalent in Central and South America but has not been found in California as of 2010. Citrus leprosis causes lesions on fruit, leaves, and twigs that are initially yellow then become brown and the tissue dies in the center. On fruit and leaves, lesions may have concentric patterns that are often roundish. Older lesions may contain or exude brownish gum. Stem lesions may coalesce, followed by flaking or peeling of bark. There may be fruit drop, premature leaf drop, and twig dieback. Citrus trees often die within a few years after becoming infected. Leprosis primarily affects oranges, but also mandarins.

Citrus leprosis virus is transmitted by flat mites, also called false spider mites (*Brevipalpus* spp.). In the absence of any virus, feeding by these mites causes scabby discoloring on fruit or leaves, injury that can resemble leprosis damage. Effectively managing mite populations and good cultural care of trees greatly reduce disease severity; however, there is no cure for leprosis-infected trees.

Citrus leprosis virus causes lesions on leaves, fruit, and twigs. Lesions are initially yellow, may have concentric patterns, and are often roundish. Report suspected findings of this disease to agricultural officials. In Central and South America, citrus trees often die within a few years after becoming infected by this virus.

Pale, roundish lesions on fruit are early symptoms of citrus leprosis. Leprosis primarily affects oranges, but also mandarins.

These brown lesions with dry, dead tissue in their centers are advanced symptoms of citrus leprosis. Infected fruit may also prematurely drop.

DISEASES AFFECTING GROWTH HABIT AND YIELD

Three serious diseases affect overall growth habit and yield in citrus, often without producing distinctive symptoms on the trunk or roots. Stubborn disease is caused by a spiroplasma. Tristeza is caused by a virus. Lemon sieve tube necrosis is an inherited disorder affecting lemon trees. Other causes of tree dwarfing and reduced yield include cachexia (also discussed below) and exocortis and certain exotic diseases like Huanglongbing (discussed above).

Management

To prevent the introduction and spread of these pathogens, obey quarantines, use good sanitation, and remove infected trees. Plant, replant, and topwork using certified disease-tested stock. See the Citrus Clonal Protection Program Web site (http://ccpp.ucr.edu), the "Prevention and Management" section, or contact your county agricultural commissioner or University of California farm advisor for sources of disease-tested and true-to-type budwood.

Do not bring into California from other states or countries citrus or other plants belonging to the Rutaceae family. You may inadvertently introduce diseases to your orchard. Conversely, prevent disease present in California from spreading to other localities. To avoid spreading tristeza, do not ship plants or plant parts from infected southern California districts to areas where tristeza is not present or is localized, such as the San Joaquin and Coachella valleys.

Monitor established orchards to detect disease and make treatment decisions. Observe trees carefully for any characteristic signs of these diseases. Consider submitting appropriate plant samples to a diagnostic laboratory if the suspect pathogen can be verified by commercial testing. Replace diseased and unproductive trees. In southern California, where tristeza is widespread, you may want to remove infected trees only when they become unproductive.

Learn how to identify the potential invertebrate vectors

in your orchard, especially how to distinguish the black citrus aphid (already established in California) from the brown citrus aphid (a quarantined pest and more damaging pathogen vector). For photographs and an illustrated key to aphid species in citrus, see the "Aphids" section in the chapter "Insects, Mites, and Snails." Promptly report to agricultural officials any suspected brown citrus aphids and diseases or other pests suspected of being new to California.

Certain crops that are alternate hosts of citrus-feeding aphids or leafhoppers may not be compatible for growing near citrus. For example, cotton, or melon, aphid (the vector of tristeza) often develops high populations on pomegranate and many vegetable crops. High cotton aphid populations migrating from crops nearby may facilitate spread of tristeza in citrus.

Stubborn Disease

Spiroplasma citri

Stubborn disease occurs throughout California's warm inland growing areas. It affects primarily grapefruit, mandarin, sweet orange, and tangelo trees, and the disease is more severe if trees are infected when young. The pathogen is a spiroplasma spread by leafhoppers (primarily beet leafhopper) and by grafting.

Symptoms and Damage

Stubborn disease is often difficult to diagnose in the field, especially when other disorders are present or in the early disease stages when symptoms are subtle. Stubborn disease does not kill trees, but it stunts growth and reduces fruit quality and yield, sometimes severely.

If young trees are infected, the entire tree may remain small and unproductive. If trees become infected when they are mature, only a single branch may show symptoms, and symptoms may or may not spread slowly throughout the canopy. Even though leaf and shoot growth may be only mildly affected on mature trees, the size and number of their fruit and fruit quality can still be significantly reduced.

The most obvious symptoms of stubborn disease are undersized and few fruit or the absence of fruit, and stunted or feathery growth of the canopy due to shortened internodes on shoots. The leaves are small and grow abnormally upright and close to the stems. Leaves may also be cupped, thicker than normal, and have variable chlorotic patterns resembling nutrient deficiencies. The trees usually develop out-of-season growth flushes and blossoms on portions of the canopy. The few fruit produced remain small and are lopsided. The best way to see the off-center core and uneven sides of a fruit is to cut it in half, from the stylar (bottom, or blossom) end through the stem.

Some possible symptoms depend on the cultivar and maturity of the fruit. You may see stylar end greening: the blossom end of the fruit remains green while the stem end becomes colored. Fruit of seedy cultivars have dark-colored, small seeds aborted early in their development. Fruit may have an insipid or bitter flavor. In some cultivars, fruit become acorn shaped.

This unseasonable growth flush and bloom is symptomatic of stubborn disease. On affected shoots, leaves may be small and grow abnormally upright and close to the stems.

Stubborn disease often causes fruit to become lopsided and remain small. Bud mite, chimera, tristeza, and the exotic Huanglongbing disease also cause misshapen fruit. Look throughout the tree for several symptoms to help you diagnose the cause.

Young trees affected by stubborn disease are stunted and produce undersized and few fruit, or no fruit at all.

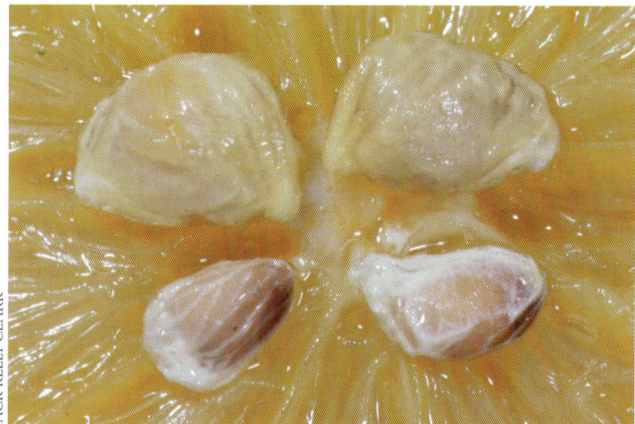
Seeds of citrus fruit affected by the stubborn disease (bottom) are smaller and darker or more pink than normal seeds.

Seasonal Development

Spiroplasmas are single-celled microorganisms that belong to a group of soft-skinned bacteria that lack an outer cell wall. They are named for their helix (spiral) shape formed during certain stages of their development. In addition to citrus, *Spiroplasma citri* occurs in weeds such as London rocket (*Sisymbrium irio*), many mustards (*Brassica* spp.), and wild radish (*Raphanus raphanistrum*). Leafhoppers become infective after feeding on these spiroplasma-infected weeds, then spread the pathogen when they move and feed on citrus, primarily during summer after weeds dry. Treating leafhoppers in the field does not prevent the spread of this spiroplasma.

Positive identification of stubborn disease can be time consuming, and such laboratory tests may not be commercially available. In commercial orchards, deciding whether a tree is infected and should be removed is usually based on the presence of multiple characteristic symptoms in the field. Observe the trees carefully for any signs of stubborn disease in late fall or early winter. For example, a sparse crop is a useful diagnostic symptom and becomes apparent during the time fruit develop their mature color. Map or flag trees suspected of being infected and recheck the orchard several times during the year to confirm your diagnosis.

Tristeza

Citrus tristeza virus

Citrus tristeza virus (CTV) is one of the most important pathogens affecting citrus in all production areas around the world. *Tristeza* means "sadness" in Spanish and Portuguese, and the name describes the impact it has had on many citrus industries. In California, CTV is widespread throughout southern California, but it is more limited in distribution in the San Joaquin Valley and desert valleys because of environmental conditions and localized efforts to remove infected trees. The occurrence of many different strains of CTV have made studying its biology and management difficult. Strains have a wide range of virulence, from mild symptoms that cause no visible damage to severe symptoms such as stem pitting and loss of productivity, or (if it is grafted on sour orange) they may cause rapid death of the tree (quick decline). Thus, the symptoms produced by the virus depend on the scion and rootstock cultivar combination and the virulence of the virus strain. Furthermore, trees may be infected with more than one strain, with one strain masking the expression of symptoms of another strain, and many tristeza symptoms resemble damage caused by other diseases and disorders.

Two orange trees with quick decline due to *Citrus tristeza virus*. The tree on the left is in an advanced stage of decline, and the tree on the right is in the early stages of decline. Tristeza occurs primarily in old trees where sweet orange scion is grafted to a sour orange rootstock.

Symptoms and Damage

Symptoms include quick decline, stem pitting, and yellowing. Quick decline can occur when sweet orange varieties grown on sour orange rootstock become infected with certain strains of CTV. Quick decline occurs because the phloem sieve tubes just below the bud union die, preventing movement of starch from the scion downward to the root system. During periods of heat stress or drought stress, the tree rapidly dies, often within a few weeks. In the past, growers commonly used sour orange rootstock because it provides resistance to fungi and is productive in a wide range of soil conditions. Sour orange rootstock is no longer widely used because of the potential for quick decline caused by CTV.

Prior to tree death, quick decline is sometimes detectable by peeling bark off the bud union and looking for thickening in bark below the graft and by inspecting wood around the union for honeycombing or a band of discoloration. When citrus trees are weak and stressed, the saprophytic fungus *Fusarium solani* (discussed earlier) can become pathogenic, causing dry root rot and tree death that looks similar to CTV quick decline. Thus, laboratory testing for CTV prior to tree death is important for confirmation of CTV as the causal agent.

Specific strains of CTV can cause stem pitting, in which the vascular tissue in trunks and branches does not produce normal cells, leaving elongated sunken areas or pits. Stem pitting is considered a severe symptom in citrus because it affects the health of tree branches and limbs. Most citrus types are susceptible to stem pitting disease, even if they are grafted onto resistant rootstocks such as trifoliate and its hybrids. Stem pitting can be recognized by peeling bark off branches or the trunk and looking for pits in wood and protrusions or ridges on the underside of bark. Especially sensitive varieties are grapefruit, tangelo, and tangor, while many mandarin varieties are resistant. Laboratory testing can be used to confirm that a stem pitting isolate of CTV is involved.

The underside of a section of citrus bark (photo left) removed to inspect for tristeza damage. Prior to tree death, quick decline is sometimes detectable by peeling bark off the bud union and looking for thickening in bark below the graft and by inspecting wood around the union (photo right) for honeycombing or a band of discoloration.

Visual diagnosis of tristeza is difficult and usually inconclusive. Historically, detection of CTV was accomplished by graft inoculating Mexican lime seedlings and looking for symptoms including vein corking, vein clearing, stunting, stem pitting, and leaf cupping over a period of 6 to 9 months. Laboratory tests targeting the coat protein of the virus using ELISA or tissue blotting and molecular methods using PCR (targeting CTV-specific genes) are now used for a rapid diagnosis of CTV and may indicate the severity of the strain(s) involved.

For tristeza monitoring and management information, consult the online UC IPM *Pest Management Guidelines: Citrus*.

Cachexia (Xyloporosis)

Cachexia is caused by citrus variants of the *Hop stunt viroid*, Citrus viroid IIb & IIc (CVd-IIb & -IIc). This viroid disease is widespread in older citrus orchards. Cachexia is absent from most young groves in California because strict regulations on budwood sources have kept new plantings largely free of this and other viroid diseases. In addition to their potential to be introduced in new stock, viroids can also be transmitted mechanically on contaminated pruning and grafting tools.

Cachexia is symptomless in many citrus cultivars. In some mandarins and mandarin hybrids (tangelo and tangor), the viroids cause distinct (sometimes severe) gumming, discoloration, and stem pitting. In Rangpur lime, rough lemon, and sweet limes, the viroids cause fine wood pitting without discoloration or gumming (symptoms called xyloporosis).

Severely affected trees are stunted with chlorotic foliage. Trees may decline and infrequently die. Peeling back bark may reveal rounded pits in wood and protrusions or ridges on the underside of bark that fit into the pits in wood. The

Susceptible rootstock scion combinations infected with CTV may show light green foliage, poor growth flushes, and some leaf drop. Infected young trees may bloom early and begin producing fruit 1 to 2 years before uninfected trees. Small fruit may also be produced due to poor root growth. Girdling at the graft union reduces starch transport to roots, and feeder roots decline from the periphery inward. Yellowing of leaves and shoot symptoms occur in the summer when water needs cannot be met by a declining root system. Yellowing symptoms caused by CTV are difficult to diagnose and may resemble symptoms caused by other diseases.

Seasonal Development

Citrus tristeza virus is spread through budding and grafting or by aphids feeding on infected citrus and passing the virus on when they fly to and feed on other trees. Aphids that vector CTV include the cotton, or melon, aphid (*Aphis gossypii*), spirea aphid (*A. spiraecola*), and the black citrus aphid (*Toxoptera aurantii*). The cotton aphid is the primary vector for the CTV strains found in California. The rate of spread of the disease depends on the aphid species, the particular virus isolate, and the receptivity of the tree. Peak periods of transmission occur in spring and fall when the trees are flushing, virus titer in infected trees is high, and the aphids are flying.

Elsewhere in the world, brown citrus aphid (*Toxoptera citricida*) is a more efficient vector of CTV than the cotton aphid, and tristeza is a more damaging disease where this aphid is found. Brown citrus aphid is currently found in Florida, Mexico, Hawaii, and elsewhere in the world, but not in California. CTV could become a more serious disease in California if new aphid species or virus strains are introduced.

Cachexia causes gumming and brown discoloration of inner bark or phloem of citrus trunks. Peeling back bark may also reveal rounded pits in wood and protrusions or ridges on the underside of bark that fit into the pits in wood (not shown).

These Eureka lemon trees near the coast have chlorotic canopies and prematurely dropping leaves. The cause is lemon sieve tube necrosis, an inherited disorder that results in a cyclic decline, then temporary recovery, of coastal lemons, followed by premature tree death.

inner bark or phloem can have gumming and brown discoloration. When infection is suspected in symptomless trees, grafting budwood onto sensitive hosts (indicator plants) and laboratory molecular tests can be used to confirm cachexia.

Lemon Sieve Tube Necrosis

Lemon sieve tube necrosis is an inherited disorder of lemon trees. Eureka budlines and Frost Lisbon lemons are affected. The disease is a problem in coastal growing areas. In the San Joaquin and Coachella valleys, lemon sieve tube necrosis does not result in noticeable decline.

Trees with lemon sieve tube necrosis go through a cyclic decline. About 4 or 5 years after planting, the older food-conducting sieve tubes near the bud union die. Several years later, younger sieve tubes also die, severely restricting food transport to the roots. As a result, many feeder roots die, fruit ripen prematurely, shoots grow poorly, and some leaves turn yellow and drop.

The sieve tube dieback stimulates new cambium and phloem production, and the tree recovers temporarily. Once the new sieve tubes also become necrotic, the decline process starts again and the above symptoms reappear. Only a microscopic analysis can reveal the collapsed sieve tubes.

Only certain budlines are affected by this inherited disorder. Eureka lemons that have this disorder could die in 8 to 15 years. Before planting this variety, obtain the most recent cultivar recommendations from your University of California farm advisor.

Insects, Mites, and Snails

A large number of invertebrates occur in citrus, but only a few are likely to cause damage in any given orchard. Some species are beneficial parasites (parasitoids) and predators. Others have no effect on crop production. Learn to identify the major invertebrate pests for the various growing regions of citrus, their natural enemies, and the damage that they cause. Understanding the biology of invertebrates and the principles of monitoring and management will help you minimize pest damage and maximize the effectiveness of biological pest control.

The species of pests that are important vary among locations according to climate and geography (Figure 11). For example, citrus thrips is a key pest in the San Joaquin Valley, desert valleys, and hot areas of interior southern California, but it is only an occasional pest in coastal California, where greenhouse thrips is the more common pest thrips.

Like citrus pests, the importance of natural enemies that control pests also differs by region. For instance, parasites provide substantial to complete biological control of California red scale and citricola scale in southern interior and coastal California. In the San Joaquin Valley, many of these same parasite species are present, but they are less effective because extremes of summer heat and winter cold synchronize scale stages, limiting the periods that the scale stages susceptible to attack are available to parasites.

Some pests have an effective complement of natural enemies in all regions. Brown soft scale, cottony cushion scale, purple scale, and several species each of mealybugs and whiteflies are under excellent biological control if undisturbed by adverse weather, ants, excessive dust, or insecticides.

Pest and natural enemy importance is also affected by factors such as the management practices and variety of citrus in the orchard. For example, orchards that are heavily sprayed with broad-spectrum insecticides tend to have citrus red mite and caterpillar problems because of the loss of natural enemies. Another example is that bud mite is a pest only on coastal lemons and (less frequently) on lemons in the southern interior, so it is not of concern in other regions of the state.

Exotic pest introductions cause major problems for citrus IPM programs because they arrive without their native complement of natural enemies. Pests such as Asian citrus psyllid and glassy-winged sharpshooter that are vectors of plant pathogens require eradication programs that use broad-spectrum pesticides and which can eliminate the natural enemies of other pests. New pests are likely to be introduced in the future, disrupting established IPM programs and contributing to the ongoing changes in pest management practices.

For more discussion of how pests and natural enemies differ by growing region, consult Figure 11, Table 7, and the introductions to the individual pest sections later in this chapter, and especially "Citrus IPM" (in preparation). For species-specific monitoring and management methods, including an up-to-date list of recommended pesticides, consult the UC IPM Pest Management Guidelines: Citrus and Citrus Year-Round IPM Program (both online at www.ipm.ucdavis.edu).

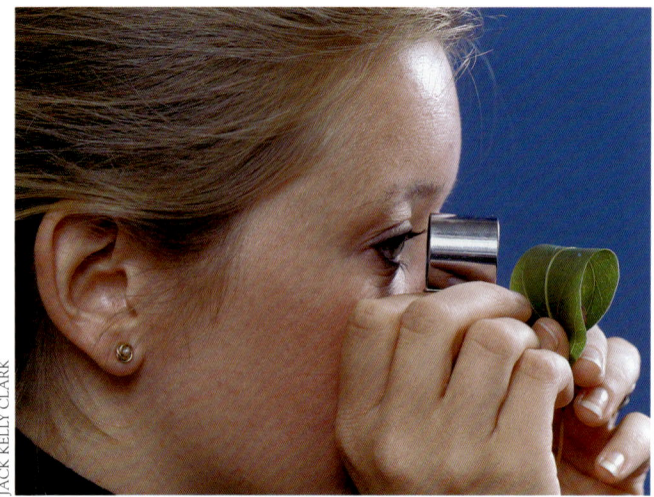

Inspect foliage with a hand lens when the presence of mites or tiny insects is suspected. Hold the lens close to your eye and move the leaf being viewed until it is in focus.

Invertebrate Damage

Invertebrate pests damage citrus during feeding by chewing or sucking on plant parts. Invertebrate feeding can result in damage or stress in the tree or reduce fruit quality or yield. Fruit rinds can be affected by many different invertebrates, certain pathogens, and disorders such as mechanical injury or wind scarring, as summarized later in Table 3 in the chapter "Managing Pests in Citrus." For example, phloem-feeding Homoptera such as soft scales secrete honeydew, which promotes the growth of a mold that produces a blackish sooty appearance on fruit rinds. Superficial injury or fouling of the rind surface or the mere presence of invertebrates or their excrement on fruit is unacceptable to many consumers. This aesthetic or "cosmetic" injury significantly reduces the marketable value of fruit, causing substantial economic loss to growers even when the injury does not affect internal fruit quality. Some high-value foreign markets, such as Japan, accept only unblemished citrus, and many countries prohibit (quarantine) the import of invertebrate-contaminated produce.

Identifying Insects and Mites

Use the photographs and descriptions in this chapter to help you identify common insect and mite pests, their damage, and natural enemies in citrus orchards. Illustrated keys are provided to help you distinguish among similar-looking species of ants, aphids, armored scales, caterpillars, mealybugs, thrips, weevils, and whiteflies.

Correct identification is vital. Unless you correctly diagnose the cause of damage you cannot know which management actions will be effective. Even closely related pest species may require different control measures. For example, fire ants prefer solid protein baits while the Argentine ant and other important ant species are attracted to sweet liquid baits. Also recognize whether the species you encounter are known to be established in California so you can promptly report to agricultural officials any suspected new pests.

Monitoring Insects and Mites

A good monitoring program for insects and mites is essential for implementing integrated pest management. California red scale, citricola scale, citrus red mite, citrus thrips, and caterpillars (orangeworms) are among the invertebrates with special monitoring methods, as summarized in those pest sections in this chapter.

Monitoring for certain pests requires materials such as an insect sweep net or special traps. Often a hand lens and sometimes a binocular dissecting microscope are necessary to identify a pest or natural enemy correctly. As you monitor for specific pests, be on the lookout for other pests. Spread your monitoring efforts over the entire orchard.

Monitoring for some pest problems can be conducted at the same time as other routine orchard practices. For example, inspections to ensure that trunk barriers are effectively excluding ants and snails from trees and that tree skirts are not touching the ground or weeds can be conducted concurrently by the person who regularly inspects the irrigation system for leaks and other problems.

You must monitor for individual species only at certain times of the year, as summarized in Figure 11 in the chapter "Managing Pests in Citrus." A well-illustrated guide on what to monitor during each crop stage, current monitoring method recommendations, and sample recordkeeping forms are provided in the UC IPM Pest Management Guidelines: Citrus and Citrus Year-Round IPM Program. For additional help and new developments, consult your University of California Cooperative Extension farm advisor or PCA.

In addition to monitoring the pest and (where appropriate) the level of predator or parasite activity, note weather conditions and keep records of previous cultural and pest management actions. Use your monitoring results, together with information such as citrus variety, orchard location, and pesticide spray history, to determine the need for management actions. These data are essential for making sound management decisions, and they provide valuable historical data for future seasons.

Monitoring Methods

Monitoring methods for invertebrates vary according to the pest species, life stage, and the purpose for monitoring. Sometimes multiple methods are combined to complement each other: for example, California red scale is monitored using both pheromone trapping and visual checking for scale-infested fruit. See the individual pest sections for more explanation and photographs of the monitoring techniques summarized below.

Visual Search. You can assess a pest population by counting the number of individuals present on a predetermined number of fruit, leaves, or twigs. This method is used for estimating citricola scale nymphs and citrus red mites on leaves. The percentage of plant parts infested can also be used. For example, the citrus thrips treatment threshold is based on the percentage of new fruit that are infested. Timed counts (the number of pest individuals found during a predetermined time interval) are used for certain caterpillars and katydids.

Traps. Traps can detect an insect population, reveal its flight activity, and help you decide whether and when a pest population needs treatment. For example, the number of citrus red scale males captured during the fourth (September–October) flight helps you estimate the percentage of infested fruit.

Sticky traps are white, yellow, green, or blue cards coated with a clear sticky material. White is not especially attractive to most types of insects and is commonly combined with a synthetic lure as described below. Other colors attract different species of insects. Generally, green attracts bean thrips, blue attracts flower thrips, and yellow is attractive to a wide variety of flying pest insects including aphids, psyllids, sharpshooters, thrips, and whiteflies, as well as natural enemies such as parasitic wasps and vedalia beetle. The insect is attracted to the color and adheres to the card on contact. The cards can be changed weekly to monitor population densities and are invaluable in detecting the introduction and presence of exotic pests.

Pheromone traps generally use a sticky card combined with a synthetic female sex pheromone lure that attracts males, which then get stuck on the card. Pheromone-baited sticky traps are commercially available for the adult moths

Sticky traps are white, yellow, green, or blue cards coated with a clear sticky material. Different colors attract different species of insects. This yellow trap is attractive to a wide variety of flying pest insects including aphids, psyllids, sharpshooters, thrips, and whiteflies, as well as natural enemies such as parasitic wasps and vedalia beetle. This trap is being covered with clear plastic so its contents can be examined later.

of a number of pest caterpillars (citrus cutworm, citrus leafminer) and scale insect pests (California red scale). The date of the first pheromone trap catches are used as the biofix (starting point) for calculating degree-day units and predicting when life cycle events will happen.

Pheromone traps are also used as an indicator of pest densities. They come in many shapes and sizes. They may be small rectangular cards as for California red scale or triangular traps as for citrus leafminer. The citrus cutworm pheromone trap does not use a sticky card but consists of a bucket with a toxicant in it that kills the moth when it enters the bucket, as pictured earlier in the chapter "Managing Pests in Citrus."

Double-sided sticky tape can be wrapped around twigs to trap scale crawlers, such as California red scale or citricola scale, as they wander on bark in search of a feeding site. The tapes can be replaced weekly to determine when crawler emergence is occurring, as discussed later in the sidebar "Sticky Tape Traps for Timing Scale Sprays" (p. 106).

Liquid bait traps used for fruit flies contain a food odor to attract the fly to the trap. The insect drowns in the liquid.

The Tedders trap (named after its designer) is placed on the ground, where it takes advantage of weevils' tendency to climb objects they encounter. The trap's base guides weevils to climb through an increasingly narrow opening (an inverted funnel), so that weevils enter a collecting chamber where it is difficult for them to find the small hole to walk

Pheromone traps, such as this triangular trap for citrus leafminer, typically use a sticky card combined with a synthetic female sex pheromone lure that attracts males, which then get stuck on the card. Traps can detect an insect population, reveal its flight activity, and help you decide if and when a pest population needs treatment.

Readily dislodged insects such as Diaprepes root weevil, European earwig, Fuller rose beetle, and certain predators can be monitored by shaking citrus foliage (branch beating) over a collecting sheet.

out. Diaprepes root weevil adults emerging from the soil or migrating into the orchard will climb the trap as if it is a tree trunk and become trapped inside, as pictured later in the "Diaprepes Root Weevil" section.

Shake and Beat Sampling. Sweep net and beating-sheet sampling are useful for monitoring citrus cutworm larvae and katydids that are difficult to search for visually because their green coloring blends in with the leaves. They can be dislodged by shaking or beating branches over a tray or cloth or by inserting branches into a sweep net and vigorously shaking it. Earwigs and the adults of Diaprepes root weevil and Fuller rose beetle can also be easily dislodged by beating or shaking branches. The color of the beating cloth or tray should be the opposite color of the insect monitored. A dark cloth is useful for cutworms, but a light cloth would be better for earwigs.

GENERAL PREDATORS

General predators such as lacewings, minute pirate bugs, and certain lady beetles feed on a variety of insects and mites. Often, each individual predator consumes large numbers of prey. Because they prey on many different pest species, general predators are often present in and around orchards before harmful pest populations develop.

Many specialized predators and parasitic flies and wasps feed only on certain pest species; these are discussed later in the sections that describe their hosts. The importance of these specialized natural enemies varies by growing region as summarized in Table 7. See the chapter "Managing Pests in Citrus" for discussion of the types of biological control (augmentation, classical, and conservation) for using beneficial pathogens, parasites (parasitoids), and predators.

The more common general predators that help control insect and mite pests in citrus are described and illustrated below. For more information, consult resources such as the *Natural Enemies Gallery* (online at www.ipm.ucdavis.edu/PMG/NE/index.html) and *Natural Enemies Handbook* (Flint and Dreistadt 1998).

Green Lacewings

Chrysoperla and *Chrysopa* spp.

Green lacewing (family Chrysopidae) larvae feed on mites and small insects, including aphids, leafminer, peelminer, psyllids, scales, small caterpillars, and whiteflies. The adult stage of *Chrysoperla* consumes honeydew, nectar, and pollen. *Chrysopa* spp. adults are predaceous and additionally consume pollen and sugary liquids.

Lacewings develop through four life stages (Figure 19). Adult lacewings are bright green and about ⅔ to ¾ inch (15–20 mm) long. They are slender with four delicate wings that fold over the body like a tent when they are at rest. Their wings have a netlike or lacy pattern of many fine cells. Females lay eggs on leaves, attached by a long, threadlike stalk that helps protect eggs from predation. *Chrysoperla* deposits its eggs singly and *Chrysopa* deposits its eggs in groups of about 10 to 70.

Larvae are mottled, often gray or yellowish gray, with a somewhat flattened body that tapers at each end. They have long, curved mandibles (mouthparts) for grasping prey. The larvae molt through three instars until they reach a length of about ⅜ inch (10 mm), then pupate in a round silken cocoon.

Green lacewings of several species are available commercially for release, both as eggs and as larvae. Augmentative

Table 7. Biological Control of Citrus Pests in Major Growing Areas.

Pest		Biological control in growing areas				
Common name	Scientific name	Coastal	Southern interior	San Joaquin Valley	Desert	Specialized natural enemies
Mealybugs						
citrophilus mealybug	Pseudococcus calceolariae	S	C	NA	NA	Coccophagus gurneyi, Hungariella (=Tetracnemoides) pretiosa
citrus mealybug	Planococcus citri	C	C	C	NA	Leptomastix dactylopii, Leptomastidea abnormis, Cryptolaemus montrouzieri[1]
Comstock mealybug	Pseudococcus comstocki	S	C	C	NA	Allotropa convexifrons, Allotropa burrelli, Pseudaphycus malinus
longtailed mealybug	Pseudococcus longispinus	C	C	NA	NA	Acerophagus notativentris, Arhopoideus peregrinus, and Anarhopus sydneyensis
pink hibiscus mealybug	Maconellicoccus hirsutus	NA	NA	NA	C	Anagyrus kamali
Scales						
black scale	Saissetia oleae	C-S	S	NA	NA	many species, including Metaphycus bartletii, Metaphycus helvolus, Scutellista caerulea (=S. cyanea)
brown soft scale	Coccus hesperidum	C	C	NA	NA	Chilocorus spp.,[2] Metaphycus luteolus, Metaphycus spp., Microterys nietneri, Rhyzobius lophanthae[3]
California red scale	Aonidiella aurantii	C-S	C-S	S[4]	N	Chilocorus spp.,[2] Aphytis melinus,[4] Aphytis spp., Comperiella bifasciata, Rhyzobius lophanthae[3]
citricola scale	Coccus pseudomagnoliarum	C	C	N	N	Coccophagus lycimnia, Coccophagus scutellaris, Metaphycus flavus, Metaphycus luteolus
cottony cushion scale	Icerya purchasi	C	C	C	C	Rodolia cardinalis,[5] Cryptochaetum iceryae[6]
purple scale	Lepidosaphes beckii	C-S	NA	NA	NA	Aphytis lepidosaphes
yellow scale	Aonidiella citrina	C[7]	C[7]	C-S	NA	Aphytis lingnanensis, A. melinus, Comperiella bifasciata
Whiteflies						
ash whitefly	Siphoninus phillyreae	C	C	C	C	Encarsia inaron, Clitostethus arcuatus[3]
bayberry whitefly	Parabemisia myricae	S	S	NA	NA	Encarsia spp., Eretmocerus debachi, Eretmocerus furuhashii
citrus whitefly	Dialeurodes citri	S	C	NA	NA	Encarsia lahorensis, Encarsia protransvena, Encarsia strenua, Eretmocerus spp.
giant whitefly	Aleurodicus dugesii	C	C	NA	NA	Encarsiella noyesii, Entedononecremnus krauteri, Idioporus affinis
woolly whitefly	Aleurothrixus floccosus	C	C	C-S	C-S	Amitus spiniferus, Cales noacki, Eretmocerus sp.

Species under biological control can become pests if their natural enemies are disrupted, such as by dust, abundant ants, or insecticide application. Pests for which biological control is known to be important and their specialized natural enemies are listed. The natural enemies are parasitic wasps (parasitoids, commonly called parasites) except as noted.

KEY
C = complete biological control
S = substantial biological control, thresholds sometimes exceeded
N = generally not under biological control
NA = not applicable (as pest is absent or uncommon) or it is unknown whether pest absence is due to biological control or other causes
1. Mealybug destroyer lady beetle.
2. Lady beetles, including twicestabbed lady beetle (Chilocorus orbus =Chilocorus stigma).
3. Lady beetle.
4. Periodic Aphytis releases are required to provide biological control.
5. Vedalia, a lady beetle.
6. Parasitic fly.
7. Biological control of yellow scale is due to differential parasitism of yellow versus red scale by introduced wasps coupled with the competitive displacement of yellow scale by California red scale due to yellow scale's greater vulnerability to parasitism by Aphytis spp. wasps.

Sources: DeBach, Hendrickson, and Rose 1978; Ebeling 1959; Grafton-Cardwell et al. in preparation; Nechols et al. 1995; Rose and Zolnerowich 1997; Schauff, Evans, and Heraty 1996; Smith and Armitage 1931.

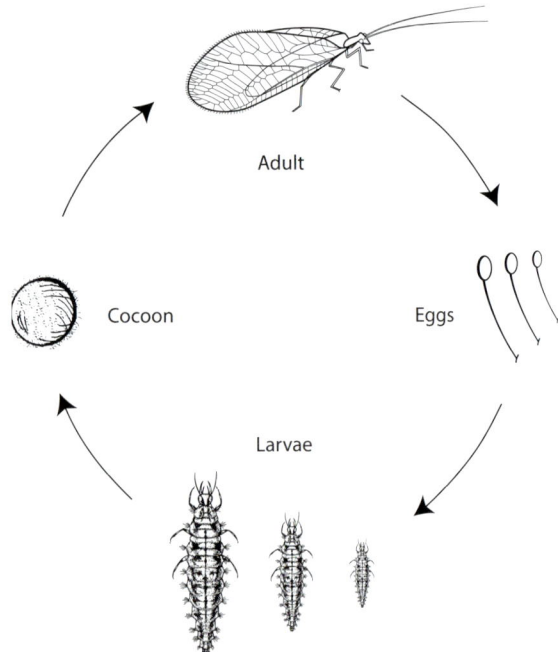

Figure 19. Green lacewing life cycle and stages. Females lay oblong eggs attached to plants by a silken stalk. Larvae develop through three instars, then pupate in loosely woven, spherical, silken cocoons attached to plants. The nocturnal adults emerge, mate, and seek prey, near which females lay their eggs.

Green lacewing adults, such as this *Chrysoperla carnea*, have golden eyes and slender, green bodies. Lacewings are named for their prominent wings, which have green, netlike or lacy veins. Adults are nocturnal but can be observed flying during the day if their resting place on a branch or foliage is disturbed.

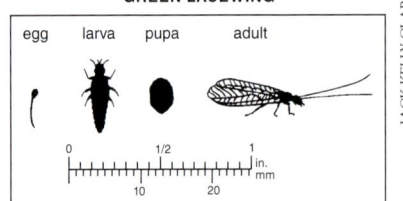

lacewing releases are not recommended in citrus, but natural populations can exert significant control of mites and small insects such as aphids.

Brown Lacewings

Hemerobius spp.

Brown lacewings (family Hemerobiidae) feed on mites and small insects such as aphids, leafminer, peelminer, psyllids, scales, small caterpillars, and whiteflies. Both adults and larvae are predaceous. Brown lacewings are similar in shape and wing pattern to green lacewings, but adults are brown and about one-half the size of green lacewings. Females lay eggs singly on the underside of leaves. Their oblong eggs resemble eggs of syrphid flies, but brown lacewing eggs have a tiny knob projecting at one end and have a relatively smooth surface. Syrphid eggs do not have a knob on top and appear to have fibers or lines crisscrossing their surface.

Larvae of brown and green lacewings are similar in appearance, but brown lacewing larvae are more slender and have a smaller head and jaws than green lacewing larvae. Unlike green lacewings, brown lacewing larvae often

Green lacewings lay oblong eggs, each with a silken stalk attached to the plant. Depending on the species, eggs are laid singly or in clusters, as with these eggs of *Chrysopa nigricornis*.

Lacewing larvae feed on mites and almost any small, soft-bodied insects. Larvae, like this third-instar *Chrysoperla carnea*, have long, curved mandibles for puncturing prey and sucking out their body fluids.

move their head rapidly from side to side when seeking prey. Mature brown lacewing larvae (second and third instars) lack the trumpet-shaped lobe (empodium) that green lacewing larvae have between the claws on the end of their legs. Brown lacewings are active at cooler temperatures and are more common at cooler growing locations than green lacewings.

Dustywings

Conwentzia and *Coniopteryx* spp.

Dustywings (family Coniopterygidae) feed as both larvae and adults on all stages of mites. When feeding on mites such as citrus red mite, each dustywing larva eats about 250 mites while developing through three larval stages. Larvae also feed on small insects such as aphids. Adults also consume honeydew and nectar.

Dustywings resemble brown and green lacewings in shape; all are in the order Neuroptera. Adult dustywings are sometimes confused with whiteflies because of their small size and the grayish white powder that covers their body and wings. Adults are about 1/8 inch (3 mm) long, about twice the length of most whitefly adults.

Female dustywings lay tiny, oval eggs, usually singly on the underside of leaves. Eggs are pinkish to yellow, about 1/50 inch (0.5 mm) long, and have a hexagonal (5-sided) pattern of narrow raised areas on the surface that causes

Brown lacewings, like this *Hemerobius pacificus* adult, are predaceous as both adults and larvae. Brown lacewing larvae resemble those of green lacewing, except their mature size is smaller.

Adult dustywings have appendages, bulging eyes, and a body shape that resembles that of small brown and green lacewing adults. Their coloration resembles whiteflies because dustywings are covered with pale powder.

Brown lacewing eggs, like this of a *Hemerobius* sp., are oblong and are laid singly on their side on plants without any attaching stalk. Brown lacewing eggs look similar to syrphid eggs, but brown lacewing have a tiny knob projecting at one end (photo right) and have a relatively smooth surface.

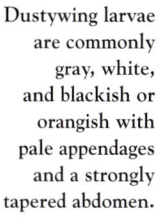

Dustywing larvae are commonly gray, white, and blackish or orangish with pale appendages and a strongly tapered abdomen.

Brown lacewing larvae, like this *Hemerobius pacificus*, are more slender and have a smaller head and jaws than green lacewings.

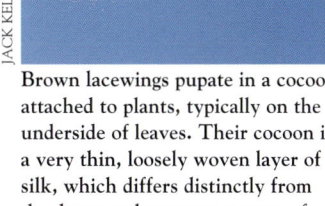

Brown lacewings pupate in a cocoon attached to plants, typically on the underside of leaves. Their cocoon is a very thin, loosely woven layer of silk, which differs distinctly from the dense and compact cocoon of green lacewings.

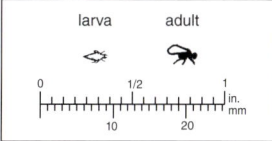

The dustywing cocoon is a flattened silk cover that is usually attached to the underside of leaves. The cocoons of brown lacewings and predaceous midges and the egg sacs of psocids (pictured with scale insects) and certain spiders look similar.

BROWN LACEWING

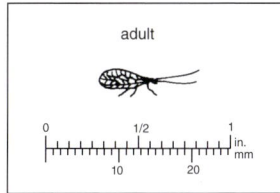
adult

DUSTYWING

larva   adult

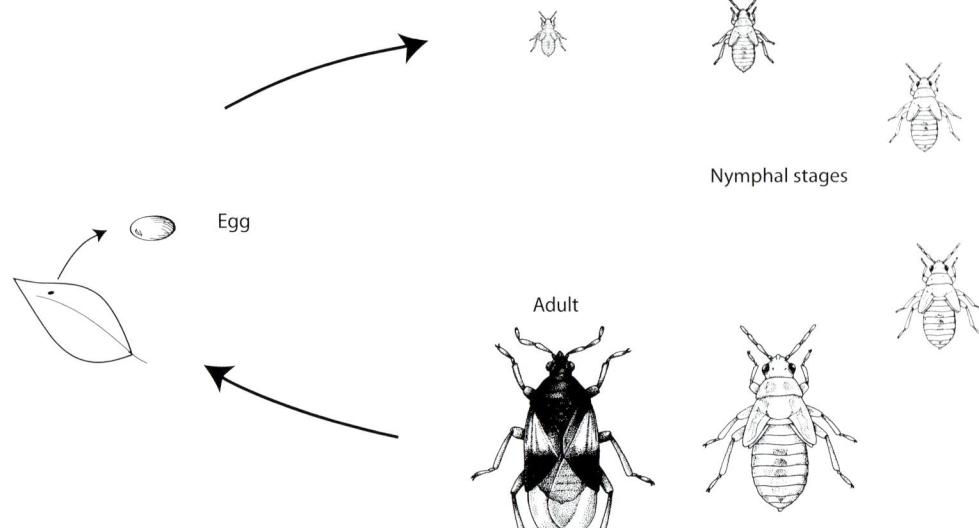

Figure 20. Minute pirate bug (*Orius* species) life cycle and stages. *Orius* hatches from an egg and develops through five nymphal instars before maturing into an adult. *Orius* and other true bugs, such as assassin bugs, have incomplete metamorphosis (gradual immature development with no pupal stage). Nymphs resemble adults, except for color, size, and the lack of wings. Adapted from drawings by Celeste Green in Smith and Hagen 1956.

eggs to appear to be covered with a fine mesh net. Larvae are similar in shape to lacewing larvae but are broader and smaller. Larval color varies, but usually it is a blotchy pattern of white and black or reddish orange.

Minute Pirate Bugs

Orius spp.

Minute pirate bugs (family Anthocoridae) feed on thrips, all stages of mites, insect eggs, aphids, mealybugs, psyllids, whiteflies, and small caterpillars. If feeding on mites, minute pirate bugs consume 100 or more in a lifetime. *Orius* spp. adults are about ⅛ inch (3 mm) long and black with black and pale wings. At rest, their overlapping wings form white or silvery triangular markings on their mostly black body. Eggs are often overlooked because females insert them into plant tissue, mostly along veins on the underside of leaves. *Orius* develop through five progressively larger nymphal instars (Figure 20), all of which are yellowish to orange with prominent reddish eyes.

Pirate bugs, such as this adult *Orius tristicolor*, have prominent, needlelike mouthparts with which they impale prey. At rest, the dissimilar parts of their folded wings overlap, forming white or silvery triangular markings on their mostly black body, or appear in an X-shaped pattern.

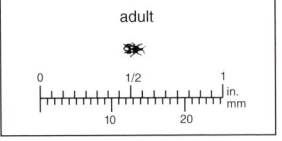

Orius nymphs are yellow to orangish pear-shaped predators. Older nymphs may be darker but also have distinctly red eyes.

Assassin Bugs

Assassin bugs (family Reduviidae) are predaceous as both adults and nymphs, mostly feeding on caterpillars, leafhoppers, and a variety of other small- to medium-sized mobile insects. Adults and nymphs stalk or lie in wait for prey, which they impale with their extended, 3-segmented, needlelike beak, and inject with venom.

Assassin bugs develop through the same life stages (egg, 5 nymphal instars, and adult) as illustrated for minute pirate bug (Figure 20). Adults are usually blackish, reddish, or brown, with a long narrow head and round beady eyes. Both adults and nymphs have an elongate body and longer legs than most other true bugs. Eggs are positioned on end and glued together in a group.

Lady Beetles

Adults and larvae of lady beetles (family Coccinellidae) are predators of mites and most soft-bodied, slow-moving or sessile (immobile) insects, including aphids, mealybugs, scales, and whiteflies. Lady beetles develop through four life stages (egg, larvae, pupa, and adult) as illustrated in Figure 21, below, and Figure 28 in the "Scales" section.

Over 500 different species occur in the United States. Selected species found in California citrus are listed in Table 8. The convergent lady beetle (*Hippodamia convergens*) and many other orange-colored lady beetle species, including the California lady beetle (*Coccinella californica*) and sevenspotted lady beetle (*Coccinella septempunctata*), feed primarily on whatever species of aphids are abundant. The vedalia beetle

Assassin bugs have elongate bodies and skinny legs that are longer than on most other types of true bugs. They impale prey with their needlelike mouthparts, as on the adult (left) and nymph of this assassin bug *(Zelus renardii)*.

Assassin bug eggs are laid on end, glued together in a cluster. These *Zelus* species eggs are barrel shaped and dark brown with a white cap and are often laid in the open on leaves.

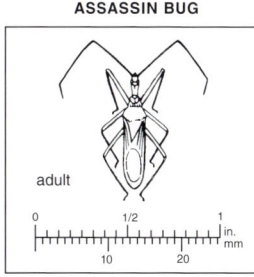

Figure 21. Convergent lady beetle life cycle and stages. Lady beetles develop from an egg through 4 larval instars, then into a pupal stage from which the adult emerges. Adapted from illustrations by F. H. Chittenden in Sanderson and Jackson 1912.

Table 8. Lady Beetles and Their Prey in Citrus.

Common name	Scientific name	Prey
ashy gray lady beetle	*Olla v-nigrum*	aphids primarily, also psyllids and mites
Australian lady beetle*	*Rhyzobius lophanthae*	scales
black lady beetle	*Rhyzobius forestieri*	scales primarily, also mealybugs
California lady beetle	*Coccinella californica*	aphids
convergent lady beetle*	*Hippodamia convergens*	aphids primarily
mealybug destroyer*	*Cryptolaemus montrouzieri*	mealybugs
multicolored Asian lady beetle	*Harmonia axyridis*	aphids, mites, psyllids, scales, and whiteflies
sevenspotted lady beetle	*Coccinella septempunctata*	aphids
spider mite destroyer*	*Stethorus picipes*	mites
steelblue lady beetle*	*Halmus chalybeus*	scales
twicestabbed lady beetle*	*Chilocorus orbus*	scales
vedalia beetle*	*Rodolia cardinalis*	cottony cushion scale
—*	*Delphastus* spp.	whiteflies

* See the discussion and photographs of these specialized lady beetles in the prey sections: "Aphids," "Mealybugs," "Mites," "Scales," and "Whiteflies."

(*Rodolia cardinalis*) feeds on only one species, the cottony cushion scale. The ashy gray lady beetle (*Olla v-nigrum*) and introduced multicolored Asian lady beetle (*Harmonia axyridis*) are examples of lady beetles that prey on a variety of pests, including aphids, mites, psyllids, and whiteflies. The specialized habits of various lady beetles and photographs of them are presented later in the sections "Aphids," "Mealybugs," "Mites," "Scales," and "Whiteflies."

Syrphid Flies

Syrphid flies (family Syrphidae) prey mostly on aphids during their larval stage. A single syrphid larva can consume hundreds of aphids during its development. Syrphid larvae also feed on other soft-bodied insects, such as mealybugs, psyllids, and whiteflies. Adult syrphids, also called flower flies or hover flies, are not predaceous. They eat pollen and nectar and often are observed visiting blossoms.

Syrphids are true flies (Diptera) and develop through four life stages (egg, larva, pupa, and adult). Adults of many species have black and yellow or whitish bands on top of their abdomen and resemble honey bees; however, adult syrphids cannot sting and have only one pair of wings. Females lay eggs singly near prey. Eggs are whitish to gray, oblong, and about 1/25 inch (1 mm) long. Eggs resemble those of brown lacewings pictured above, except that syrphid eggs do not have a knob on top and appear to have fibers or lines crisscrossing their surface. Syrphid larvae (pictured later in the "Aphids" section) are maggotlike, develop through 3 instars, and at maturity are up to 1/2 inch (13 mm) long. Larvae vary in color and commonly are brownish, green, yellowish, or a mottled mix of dark and light colors. Pupae are oblong, pear shaped, and often green to dark brown.

Predatory Mites

Although certain mites feed on plants and are pests, other species of mites are important predators of pests. In citrus, several *Euseius* species (family Phytoseiidae) are the most

The adult (right), larva (top), and pupa of the introduced multicolored Asian lady beetle (*Harmonia axyridis*) are shown here. The adult has more than 100 color forms, including orangish or red with no spots or up to 19 large or small spots. Both adults and larvae prey on a variety of pests. Lady beetle species that specialize more on certain prey are pictured later with those pest species.

Adult syrphids are commonly observed on blossoms. They have black and pale (yellow to whitish) alternating bands on top of their abdomen, resembling honey bees. A predatory syrphid larva is pictured later in "Aphids."

The adult ashy gray lady beetle has a mostly gray to pale yellowish abdomen and thorax and numerous black spots. A mostly black form with two large red spots also occurs and resembles the twicestabbed lady beetle pictured later in "Scales."

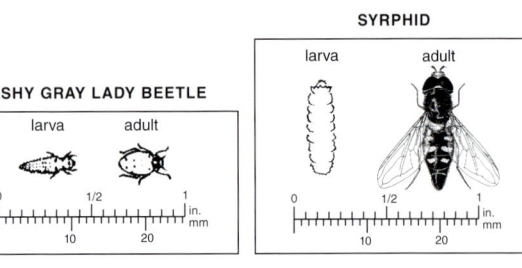

common predaceous mites, feeding on pest mites and insects such as scales, thrips, and whitefly nymphs.

After hatching from an oblong egg, predatory phytoseiid mites develop through one 6-legged larval stage and two 8-legged nymphal stages before becoming adults, as illustrated later for spider mites in Figure 42. *Euseius* move quickly when exposed to bright light, so examining the undersides of leaves in sunlight allows this shiny, fast-moving predator to be distinguished from the sluggish pest mites. See the "Spider Mites" and "Thrips" sections for more *Euseius* information and photographs.

Spiders

Spiders feed on a variety of insects, and their prey differs according to the type (family) of spider. Orb weavers (family Araneidae) and cobweb weavers (Theridiidae) eat small invertebrates that fly or walk into their sticky webs, such as moths and adults of aphids, leafhoppers, psyllids, and whiteflies. Jumping spiders (Salticidae) and wolf spiders (Lycosidae) mostly stalk or lie in wait and pounce on mobile prey such as immature thrips and small stages of caterpillars and katydids.

Spiders typically have 8 eyes that differ in size and arrangement according to taxonomic family. Orb weaver and cobweb weaver eyes are relatively close together, small, and all about the same size. Instead of relying on their vision, these spiders capture insects by sensing the vibrations of prey struggling in their webs. Jumping and wolf spiders have at least one pair of eyes that is noticeably larger than the rest, and their eyes are more widely spaced, increasing their field of vision. In comparison with web spinners, these visual hunters actively seek insects and capture their prey by grasping them with their legs and venomous mouthparts.

Spiders prey on many different insects. Whiteflies and thrips have been captured in this web of a cobweb weaver spider. The webbing is loose and apparently irregular, in contrast with the more uniform or regular webbing pattern of orb weavers.

An adult (left) and nymph of a predatory mite *(Euseius hibisci)*. *Euseius* species have a shiny, pear-shaped body that varies in color from reddish (when feeding on citrus red mite) to white or yellow (when feeding on citrus thrips, leaf sap, or pollen).

Jumping spiders are hairy, day-active predators that commonly have contrasting or bright-colored mouthparts (chelicerae) and adjacent sensory organs (palps) that make these front appendages appear especially prominent. Some species are iridescent or metallic colored.

Wolf spiders have long, hairy legs. Their body commonly has contrasting dark and light colors that make them difficult to see unless they are moving. Many wolf spiders resemble funnel weaver spiders (Agelenidae), but funnel weavers have four pairs of eyes that are about the same size. Wolf spiders have a middle pair of eyes that are distinctly larger than their other six eyes (not visible here).

Spiders typically lay eggs in a group within silk, as with this jumping spider egg mass. Spiders sometimes web leaves together or curl a leaf over their eggs. This predator's silk might be misidentified as caterpillar presence if you fail to closely examine the webbing, through which spiderlings or spider cast skins may be visible.

SCALE INSECTS

Scale insects suck plant fluids, reducing tree vitality. Armored scales (family Diaspididae), soft scales (Coccidae), and cottony cushion scale (Margarodidae) are pests in citrus.

Several characteristics distinguish the armored scales from the soft scales.

- The cover of soft scales is the actual body wall of the insect and cannot be removed. Armored scales secrete a flattened or slightly convex platelike cover that in some instars can be removed to reveal the actual scale body underneath.
- Soft scales and cottony cushion scale secrete sticky honeydew. Armored scales do not secrete honeydew.
- Except for mature females, soft scales and cottony cushion scale retain their antennae and legs throughout their life. If their feeding site becomes unsuitable, nymphs can walk slowly and relocate to feed on nearby plant parts. Armored scales are mobile for only a day or two after hatching. Once the mobile first instars (crawlers) settle to feed, armored scales lose their legs and antennae and remain at the same spot. An exception is the adult males of both types of scales, which have wings and are mobile.

California red scale and citricola scale are key pests in the San Joaquin Valley. Black scale is a common pest in southern California. In all growing areas, most other scale species are a problem only when their biological control is upset by ants, chemical treatments, excessive dust, or adverse environmental conditions. For more species-specific information, consult the online *UC IPM Pest Management Guidelines: Citrus*.

Immature males of most armored scales develop an elongate cover (top right). Underneath the cover, the male body gradually forms eye spots and appendages. At bottom left is a California red scale female with a parasite exit hole in its round cover.

California red scale feeding can discolor rinds and debilitate trees. California red scale is a key pest of citrus because it usually must be managed. How California red scale is controlled affects other pests and their natural enemies.

Scale Insect Management

Scales are managed by a combination of cultural, biological, and chemical control methods. Certain control methods are unique to the species of scale, such as pressure washing of fruit in the packinghouse to remove covers of California red scale.

Cultural control starts with maximizing tree health. Scale damage will be exacerbated if improper fertilization or poor soil and water management results in stressed trees. Proper pruning of the tree is very important for scale control. Pruning to open the tree canopy increases mortality of scales by lowering the humidity and increasing scale exposure to heat, especially in mandarins and grapefruit. In addition, parasites are more effective when pruning gives them better access to scales. Raising the skirts of trees and removing weeds that are tall enough to contact the citrus canopy reduce ant, Fuller rose beetle, and snail access to canopies. Apply a polybutene-based sticky material on top of a trunk wrap to provide an ant barrier. Minimize dust because excessive dust on leaves and fruit reduces the natural enemies' ability to locate pests.

Biological control for scales can be effective, but the effectiveness varies according to growing region and the species of scale. For instance, California red scale and citricola scale are under excellent biological control in southern California. In the San Joaquin Valley, however, natural enemies do not control citricola scale because of greater extremes of temperature, and effective biological control of California red scale requires more careful management, such as periodic introduction of *Aphytis* parasites. Augmentation (releasing purchased or field-collected natural enemies) is also sometimes used to control black scale, brown soft scale, and cottony cushion scale. Ant control is critical because ants severely disrupt natural enemies of scales.

Use pesticides only when needed. Follow a monitoring program to indicate whether and when a scale treatment is necessary, and, for some scale pests, treat only that portion of the orchard where populations exceed thresholds. Many chemical treatments for scales or other pests kill natural enemies. Conserve natural enemies by choosing chemicals least disruptive to biological control.

For more detailed and up-to-date management recommendations, including guidelines on what numbers of parasites or predators are considered sufficient and at what levels scales are a problem, consult the online UC IPM Pest Management Guidelines: Citrus.

Armored Scales

In California citrus, armored scale species include California red scale, yellow scale, and purple scale. Armored scales are categorized by a hard, platelike covering, and they do not excrete honeydew.

California Red Scale

Aonidiella aurantii

Yellow Scale

Aonidiella citrina

California red scale is usually the only armored scale observed in California citrus. It is one of the most important pests of citrus because of its widespread distribution and ability to cause serious crop damage. California red scale has many other occasional host plants, but it usually is a pest only on citrus.

California red scale males are attracted to a synthesized sex pheromone impregnated in the rubber cap attached above this 3-by-5 inch white sticky card. Trapping can identify when male flights occur, which orchard blocks have high scale populations, and where control is warranted.

The biology and management of California red scale and yellow scale are similar. An exception is that yellow scale is better controlled by parasites. It is also controlled by insecticides applied for California red scale, so yellow scale is rarely a problem.

Description and Biology

The development of California red scale and yellow scale is illustrated in Figure 22. In both species, each female produces a total of 100 to 150 crawlers (mobile first instars). About two or three crawlers a day emerge from under the female cover and settle in small depressions on fruit, leaves, or (with California red scale) bark. Before settling, crawlers can spread by wind or be moved on contaminated animals

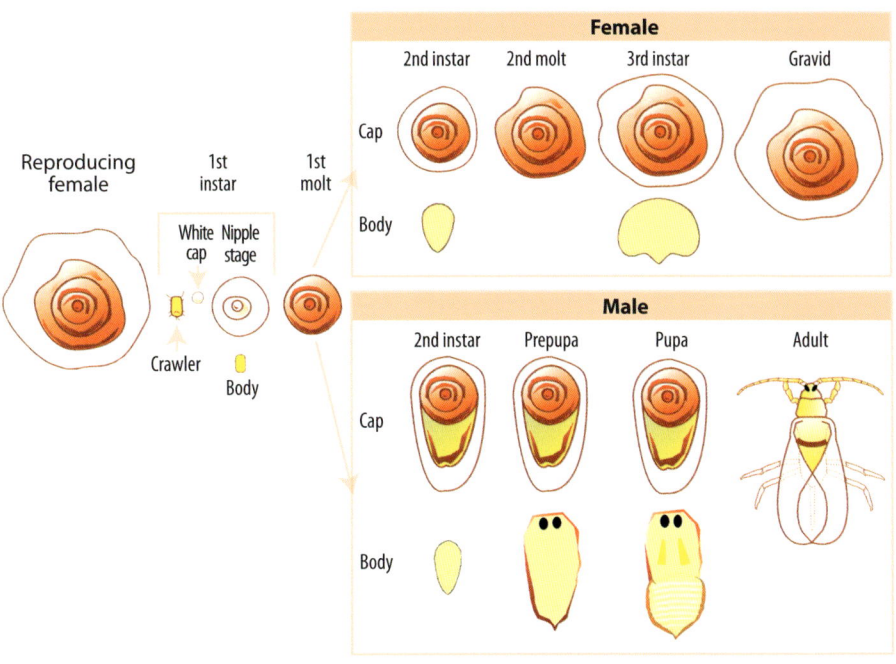

Figure 22. California red scale and yellow scale life stages and their development cycle. Female and male scales develop the same until after the first molt. During the middle of the second instar, the female extends her round cover (cap) as a gray margin, whereas the male develops an elongated cover. The females molt one more time, the males three more times. After the second molt, the female incorporates the cast skin into the gray margin of the cover from previous stage and begins to extend the margin further. The third-instar male begins to develop eye spots and appendages; and in the fourth instar, the male appendages are completely developed. After the fourth molt, the two-winged adult male emerges. By this time, the adult virgin female has extended her cover margin and is ready to be fertilized by the emerging male. The gravid (egg-carrying) female extends her margin further to reach the full size of the reproducing female.

(such as birds), equipment, or people. The first instars insert their sucking mouthparts into plant tissue and secrete a circular, white "cottony" cover (the white cap stage). The cover enlarges and develops a circular ridge.

As they mature, female covers remain round, and the scale body expands and fills the cover. During molts and after mating, the cover and female body become connected by a membrane on the underside that attaches the female securely to the plant, helping make the scales resistant to parasitism. Between molts (during the instar phase), covers of unmated females can be removed to reveal the body underneath; at this point, the scales are susceptible to attack by parasites. The female molts twice and adds a ring to the waxy cover each time she molts. When she attains the third instar, she is ready for mating.

Midway through the second instar, males begin to develop differently from females. Males form an elongated cover. The body develops eye spots, and during the prepupal and pupal stages males form the beginnings of legs and wings. After the fourth molt, the tiny, two-winged adult male emerges.

Third-instar female scales produce a sex pheromone that attracts males to walk or fly to them. Males are weak flyers and cannot fly in winds over 1 mile per hour. Males fly at a height of about 8 feet (2.5 m), mostly around dusk when temperatures exceed 60°F (15.5°C).

Adult male scales are minute, yellowish insects with one pair of wings. Unlike *Aphytis* parasites, this male red scale has long filamentous antennae and a dark reddish ring around the thorax near the base of the wings. The male red scale has a triangular-shaped head that is relatively narrow in front by the two red eyes.

Males of each generation emerge over several weeks in flights. The development of one generation requires about 1,100 degree-days (°F) above a threshold of 53°F (11.7°C). In the San Joaquin Valley, during most years there are four male flights (corresponding to four generations), one flight each peaking during May, June to July, August, and September to October. There are two or three generations per year in both the coastal and southern interior growing areas of California. The timing of the flights depends on average daily temperature. Using pheromone-baited sticky traps to monitor these scale flights helps to determine whether and when to take control actions.

California red scale females have circular, convex covers. The cover becomes orangish and develops concentric rings as the immobile females molt two times. A yellow crawler (mobile first instar) is visible on top of the lower female. Several recently settled first instars with a cottony cover (white cap) are also shown. Most of these scales are older, dark, first instars, called the nipple stage.

CALIFORNIA RED SCALE

Distinguishing California Red Scale from Yellow Scale. California red scale can be reliably distinguished from yellow scale only by properly slide-mounting adult females and examining them under a compound microscope. The two species differ in the number of sclerotized (hardened) areas on the underside of the terminal portion of their body (pygidium), as illustrated in Figures 23 and 24. However, in the field, appearance and distribution on the tree can indicate the species.

- Yellow scale is nearly always found only on leaves and fruit. California red scale feeds on twigs and mature wood as well as on fruit and leaves.
- On leaves, yellow scale feeding creates a more distinct yellow crescent of discolored leaf tissue around the scale body than does California red scale.
- The yellow scale cover is yellow-brown, and the California red scale cover is reddish brown. However, scale color is not a reliable method for distinguishing these species and may be affected by the background color of the host plant, parasitism, and maturity of the scale.

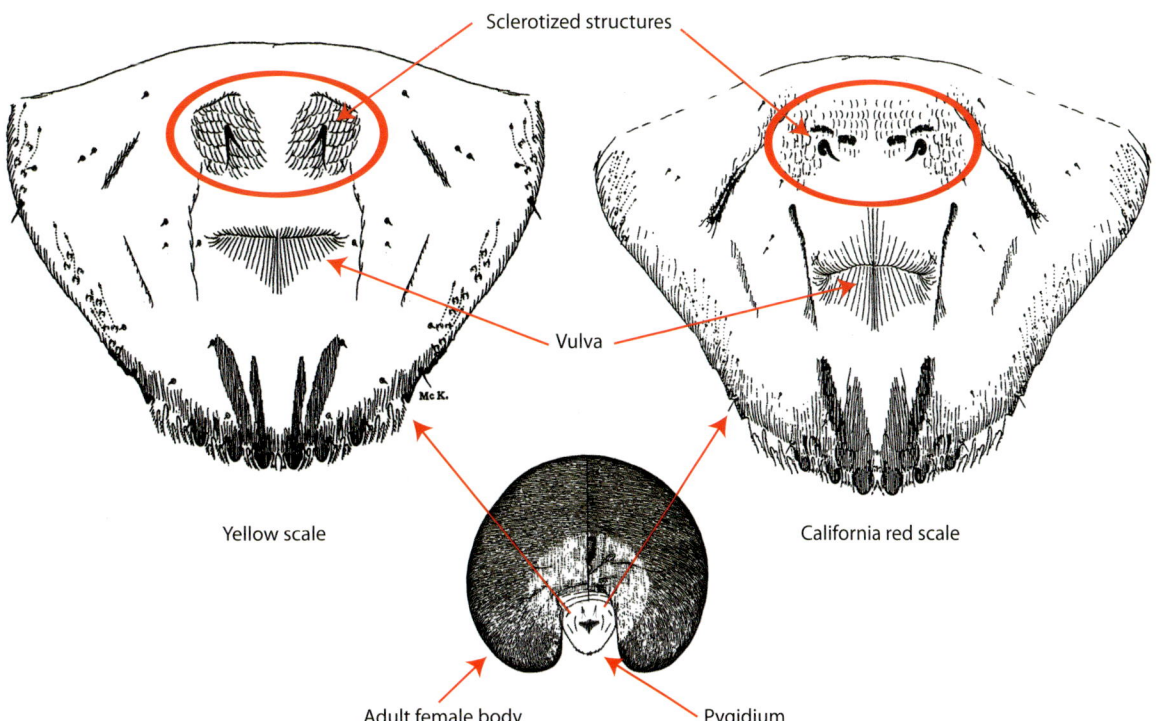

Figure 23. Features for distinguishing California red scale (*Aonidiella aurantii*) from yellow scale (*A. citrina*). In the field, appearance and distribution on the tree indicates the species, as described in the text. However, these species can be reliably distinguished only by properly slide-mounting adult females and examining the underside (ventrum) of the terminal portion of their body (pygidium) through a high-powered microscope. The most important character for differentiating *Aonidiella* species is the internal sclerotized (hardened) structures (apophyses) located in front of the egg-laying opening (vulva) of the pygidium, as illustrated above:
- Yellow scale (top left) has two sclerotized projections (apophyses), each resembling an inverted V shape.
- California red scale (right) has two groups of three sclerotized projections (apophyses).

Adapted from Gill 1997a, McKenzie 1956.

Yellow scale, *Aonidiella citrina*
One pair of sclerotized projections in front of vulva, each resembling an inverted V shape

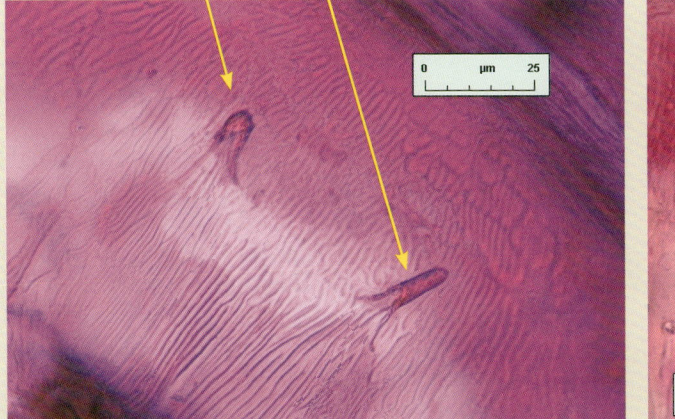

California red scale, *A. aurantii*
Two groups of three sclerotized areas in front of the vulva

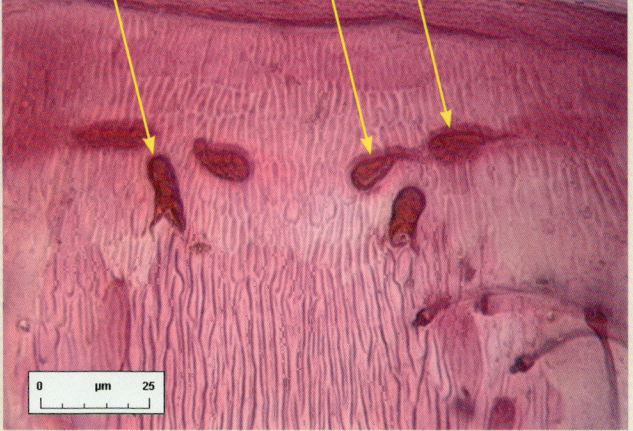

Figure 24. Photomicrographs for distinguishing California red scale from yellow scale. Sclerotized (hardened) areas on the underside of the terminal portion (pygidium) of the body of adult females are used to distinguish California red scale from yellow scale. See Figure 23 for explanation of these features and their location on the scale body. Photographs by GERMAIN J.-F., LNPV Montpellier, adapted from European Plant Protection Organization 2005.

Appearance and distribution on the tree help to distinguish the *Aonidiella* species infesting citrus. California red scale (on the left fruit) can infest branches, fruit, leaves, and twigs. Yellow scale (right) is more yellowish and nearly always found only on leaves and fruit. However, these species can be reliably distinguished only by microscopic examination.

Severe infestations of California red scale cause branch dieback and fruit and leaf drop. This damage is most likely to occur in late summer and early fall when scale populations are highest and moisture stress on the tree is greatest.

Damage

California red scale attacks all parts of the tree, including twigs, leaves, scaffolding branches, and fruit, sucking plant fluids with its long, filamentous mouthparts. More than 10 California red scale on a fruit rind can cause downgrading of fruit during packing. The most serious damage the scale causes is to tree health. Severe infestations reduce fruit yield and quality and cause premature fruit and leaf drop and tree decline and dieback. Lemon is the most susceptible host, followed by grapefruit, Valencia, navel, and mandarin orange. Tree damage is most likely to appear in late summer and early fall when scale populations are highest and moisture stress on the tree is greatest.

Natural Enemies

Resident natural enemies usually provide complete biological control of yellow scale and can be effective against the California red scale, depending on the climate of the growing region and pest management practices in the orchard. The most important natural enemies are several species of parasitic wasps. Predators, including the *Rhyzobius* lady beetle and lacewings, also exert significant control of California red scale.

The natural enemy species complement each other in providing biological control. For example, the small parasitic wasp *Encarsia pernisiosi* attacks and emerges from early instars, while *Aphytis melinus* and *Comperiella bifasciata* prefer later instars. *Encarsia* attacks primarily scales on twigs, while *Aphytis* and *Comperiella* prefer scales on fruit and leaves. *Comperiella* attacks a broader range of scale developmental stages and better tolerates summer heat than do *Aphytis* and *Encarsia*. All three species are common in southern California, while *Aphytis* and *Comperiella* provide most of the California red scale parasitism in the San Joaquin Valley.

Differences in biology and appearance help distinguish the species of parasite, as described below and illustrated in the photographs. See Figure 25 for these parasites' host stage preferences and *Life Stages of California Red Scale and Its Parasitoids* (Forster, Luck, and Grafton-Cardwell 1995) for more photographs and details on their biology.

Aphytis melinus. *Aphytis melinus* (family Aphelinidae) is the most important natural enemy of California red scale and yellow scale in California. However, in the San Joaquin Valley, *Aphytis* must be periodically released to ensure effective biological control. The *Aphytis melinus* adult has a yellow abdomen and thorax, three tiny red eye spots arranged in a triangle on top, and large eyes that bulge from the side of its broad head.

The *Aphytis* female inserts her ovipositor through the scale cover and injects the body with venom that prevents the scale from developing further. She then lays one to several eggs on the outside of the scale body, under the cover, and the scale is killed as the parasite larva feeds on it. The larva accumulates food in its gut until it molts into the prepupa, when it excretes the fecal pellets as meconium. The prepupa then molts into the pupa, which then molts into an adult.

Aphytis prefers to lay eggs on third-instar female scales because they are large and provide all of the food needed for its progeny to survive. But it also parasitizes male and late second-instar female scales, as illustrated in Figure 25, if the third-instar female scales are not available. *Aphytis* is an ectoparasite (external feeder) with larvae that feed attached to the outside of the scale body. At maturity, parasites leave behind a flat and dehydrated scale body and their cast skin and fecal pellets, which may be observed outside of the scale body by removing the scale cover. The adult parasite may emerge through a small, round exit hole in the scale cover or push out between the scale cover and plant substrate, so that parasitized scales often slough off.

The *Aphytis* female must also feed on the host to obtain protein to mature her eggs. During host feeding, the female punctures scale bodies with her ovipositor and feeds on the exuding fluid. *Aphytis* also mutilates scales by puncturing and killing them without using the host for feeding or egg

The *Aphytis melinus* adult has a yellow abdomen and thorax and relatively short antennae that widen toward their tip. The head has three tiny red eye spots arranged in a triangle on top. Large true eyes bulge from the side of the broad head.

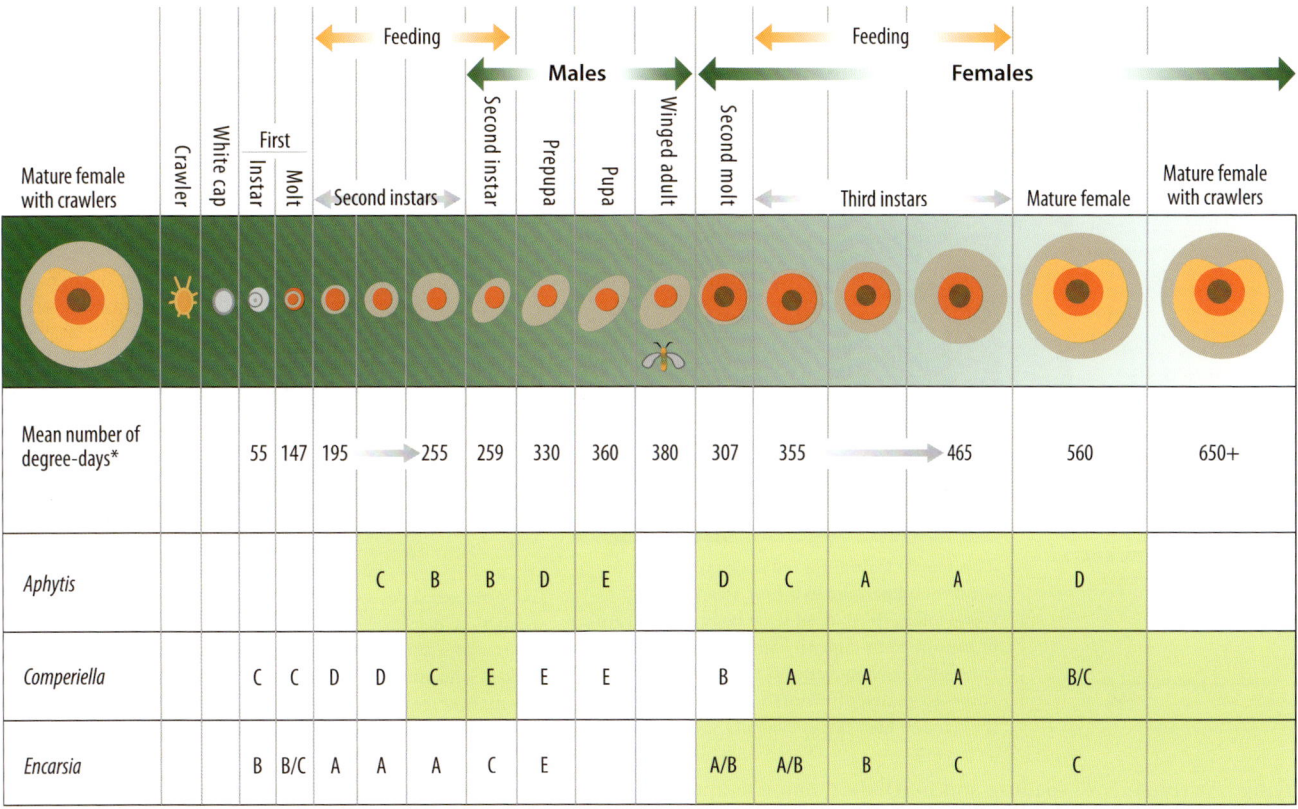

	Mature female with crawlers	Crawler	White cap	First Instar	First Molt	Second instars →	Second instar	Prepupa	Pupa	Winged adult	Second molt	Third instars →	Mature female	Mature female with crawlers
Mean number of degree-days*		55	147	195	→	255	259	330	360	380	307	355 → 465	560	650+
Aphytis						C	B	B	D	E	D	C	A	A	D
Comperiella		C	C	D	D	C	E	E	E		B	A	A	A	B/C
Encarsia			B	B/C	A	A	A	C	E		A/B	A/B	B	C	C

*Estimates: Varies across location and season

Stages in which parasitism is detectable with hand lens

A = most-preferred stage to attack → E = least-preferred stage to attack (rare)

Figure 25. California red scale can often be well controlled by natural enemies, primarily by *Aphytis melinus* and two other parasitic wasps (*Comperiella bifasciata* and *Encarsia perniciosi*). Each parasite (parasitoid) species prefers to lay eggs and develop as larvae on or in certain scale stages because the size and hardness of the scale body and the tightness of the body attachment to its cover change as the scale develops. When parasitism is most likely to be prevalent and observable during each scale generation, and the optimal times for augmentatively releasing *Aphytis melinus*, can be determined by monitoring degree-day accumulation as illustrated here. Illustration by G. Conville in Forster, Luck, and Grafton-Cardwell 1995.

Aphytis melinus inserts its ovipositor through the scale cover and lays eggs on the scale's body. Sometimes the parasite lays no eggs, but it still kills the scale by injecting it with a paralyzing venom or host feeding on scale body fluids.

Aphytis melinus lays these translucent, teardrop-shaped eggs on the armored scale body. When parasites compete for a relatively few desirable host stages, eggs may become damaged and flat. Unhealthy eggs often occur on the underside between the scale body and plant surface. Damaged eggs are difficult to observe, but they indicate that the scale will likely die even if the parasites fail to develop.

Maggotlike *Aphytis* larvae are feeding on top of this California red scale. As the larvae feed, golden yellow to brown material in their gut becomes visible through their pale, segmented body surface.

This mostly pale *Aphytis* prepupa is revealed by lifting the cover of its host. As *Aphytis* develops into this prepupal stage, it excretes its dark gut material. These brown, oblong fecal (meconial) pellets are visible to the top of this parasite.

Aphytis have no eye color soon after they pupate, as shown here. But during their week-long pupation period, eye color changes from initially clear to pink, then reddish brown and finally green. Oblong meconium (fecal pellets) are visible near the pupa's rear.

CALIFORNIA RED SCALE
adult female

This female California red scale cover with a small round parasite exit hole was flipped to reveal a flat, dried, dead scale body characteristic of parasitism by *Aphytis*. Some *Aphytis* adults emerge by pushing beneath the cover, causing covers to slough off plants so that parasitism by these individuals is not evident.

California red scales sometimes develop discolored brown mutilation wounds on their body caused by the ovipositor of an adult female *Aphytis*. Mutilated scale usually die even though they seldom contain healthy, immature parasites. Mutilation sometimes occurs during *Aphytis* host feeding, but host feeding is difficult to recognize in the field. Host-fed scales are usually just a collapsed body skin.

laying. At times, host feeding and mutilation cause as much mortality of scale populations as egg laying.

Aphytis has a short life cycle development time: 10 to 20 days from egg to adult, or two to three generations per scale generation. Some scale stages are unacceptable to *Aphytis*, including first instars and the molt stages (when the body and cover are tightly connected). Therefore, the parasite performs best against scale populations when all stages are present. Because of milder weather, multiple, parasite-susceptible host stages are more prevalent year round in growing areas in southern interior California and at the coast, increasing the naturally occurring biological control there.

Where resident natural enemies are insufficient, releasing purchased *Aphytis melinus* can be effective if ants are controlled and pesticides compatible with natural enemies are relied on in that orchard. How many and how often to release varies by location and grove history.

Comperiella bifasciata. *Comperiella bifasciata* is common in the Central Valley and southern interior districts and occurs at low densities in coastal growing areas. The adults are black with two white stripes on the head. Females have especially prominent head stripes and additionally have contrasting dark and pale areas on the wings. Males have clear wings.

Comperiella prefers to oviposit in third-instar and mated female scales, but it will oviposit in almost any scale stage except for female scales that are already producing crawlers (Figure 25). *Comperiella* is an endoparasite: the female lays eggs inside the scale body. Thus, *Comperiella* eggs cannot be observed. The parasite larva feeds inside the scale body. When the parasite larva molts into a prepupa, it excretes meconial pellets that become trapped inside the scale body, pushed to the sides like dark crescents. The prepupa develops into a black pupa and then into an adult. The emerging adult parasite chews a large, irregular exit hole in both the scale body and cover, leaving a parchmentlike, bloated scale skin with dark meconium inside. These dead scales (mummies) usually stay glued to the plant surface, though mummified scales may slough off as fruit grow.

Although *Comperiella* is a common parasite of California red scale, *Aphytis* is the commercially available species recommended for augmentative releases. *Comperiella* cannot compete against *Aphytis* because *Aphytis*, as the external parasite, consumes both the scale and the *Comperiella* larva. Unlike *Aphytis*-parasitized scales, which readily slough off of fruit, scales parasitized by *Comperiella* often remain attached to rinds, which is aesthetically objectionable.

This *Comperiella bifasciata* female parasitizes California red scale and yellow scale. Adults are mostly black with white head stripes. Unlike males, females also have contrasting dark and pale areas on the wings, giving them a fork-tailed appearance. *Encarsia perniciosi* is also dark. However, *Encarsia* does not have any white stripes and is smaller than *Comperiella*.

A *Comperiella bifasciata* adult male on the surface of an orange fruit infested with California red scales.

Encarsia pernisiosi. *Encarsia pernisiosi* is common in coastal growing areas. The adult is dark colored and smaller than *Comperiella*.

Encarsia pernisiosi is an internal parasite that develops similarly to *Comperiella*. Because it is smaller, *Encarsia* can oviposit and develop inside almost any scale stage, but it prefers second instars, either male or female scales (Figure 25). At maturity, the larva pupates inside the scale body. The adult chews a hole and emerges, usually from a scale stage that looks like a second molt or mummified third instar.

An older *Comperiella* larva and some of its greenish gray gut material are visible through this red scale's body surface after removing the scale cover. Because immature *Comperiella* are pale colored and inside their host, eggs and young larvae are not visible with a hand lens in the field.

During later stages of *Comperiella*'s development, as with this parasite prepupa, dark discoloration becomes visible through the scale cover. As the *Comperiella* larva matures into a pupa, it packs its meconium together, often forming dark parallel lines or a crescent shape.

A *Comperiella bifasciata* exit hole in the cover of a California red scale female. The dark discoloration that develops during later stages of parasite development remains visible through the scale cover after *Comperiella* emerges.

Through the host's body surface, a *Comperiella* prepupa and two groups of its dark meconium are visible inside this third-instar female California red scale.

A *Comperiella* pupa revealed by opening the body of its armored scale host. Pupae initially are opaque (pale yellowish or milky colored), then turn black.

Comperiella leaves a bloated (mummified) scale body with an exit hole. The emerging adult parasite also chews a large, irregular exit hole in the cover. *Comperiella* and *Encarsia* meconial pellets remain inside the empty scale body, in comparison with *Aphytis* meconia, which occur outside the host body that is usually flattened.

This mottled brown, gray, and white wasp (*Marietta* sp.) is a hyperparasite (secondary parasite). Its larvae feed inside of scales that are already parasitized by *Comperiella bifasciata*.

This lady beetle (*Rhyzobius lophanthae*) feeds on scales. A larger but otherwise similar-looking, blackish to dark reddish brown lady beetle (*Rhyzobius forestieri*) also feeds on all stages of scales and mealybugs.

Rhyzobius forestieri larvae are black and covered with spines. *Rhyzobius lophanthae* larvae look similar, except smaller. This *Rhyzobius* larva is eating eggs of a lecanium scale that does not occur on citrus but is about the same size and shape as a mature black scale.

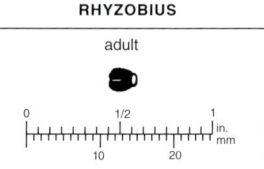

Marietta. A hyperparasite (secondary parasite) in the genus *Marietta* develops inside scales parasitized by *Comperiella bifasciata*. The *Marietta* adult is a mottled brown, gray, and white wasp. It lays its eggs inside the body of *Comperiella* and consumes the larva.

Rhyzobius lophanthae. The Australian lady beetle (*Rhyzobius =Lindorus lophanthae*) is the most important armored scale predator. The adult is 1/16 to 1/8 inch (1.5–3 mm) long and mostly dark brown or blackish with a lighter reddish brown pronotum. The larva's dark body is covered with numerous short, stout "spines."

Rhyzobius adults and larvae prey on all stages of armored scales where they are abundant in small patches (e.g., on interior wood), helping to prevent scale populations from increasing to the point where the scales become an economic pest. *Rhyzobius* can be numerous in some orchards with high scale infestations.

Monitoring California Red Scale

Multiple Monitoring Methods Exist. Use several methods in combination to decide whether biological or chemical controls are needed and when to use them. All growers should monitor California red scale using weekly pheromone trapping and inspections of fruit on the tree and in bins. Pheromone trap monitoring by flight (number of males per generation), sticky tape trapping for crawlers, and detailed evaluations of parasitism are additional methods that may be warranted. As discussed below, their use depends on the management practices in the orchard, such as the types of pesticides applied and whether biological control is relied on. For current monitoring recommendations and thresholds, such as how many fruit to examine and what levels of parasitism are considered sufficient, consult the online *UC IPM Pest Management Guidelines: Citrus*.

Pheromone Traps. Citrus growers use pheromone traps to monitor male scales in two ways: by changing cards weekly or by changing cards at the end of the first (May), second (June-July), and fourth (Sept.-Oct.) flights of male scales. The weekly card changes provide a biofix week of first scale activity from which degree-days (°D, or DD) can be calculated based on a lower developmental threshold of 53°F. Male flights occur approximately every 1,100°D, and crawler emergence occurs about 555°D from the peak of the male flights. Growers using a chemically based program can use degree-days to time sprays so that they contact the most susceptible stage of scale. Growers using a biologically based program can use degree-days to concentrate *Aphytis* releases just prior to flights when the virgin third-instar scales that the *Aphytis* prefers for oviposition predominate.

For growers on a chemically based management program, changing cards at the end of each flight gives the grower an indication of which orchards have heavy populations of scale. Generally, when an average of more than 1,000 scales are trapped during the fourth flight and fruit is infested with scale at harvest, treatment is planned for the next season. The goal is to maintain California red scale populations at levels that do not result in more than 10 scales per fruit at harvest. Pheromone cards do not provide reliable estimates of California red scale populations when insect growth regulators are used because the males are more sensitive to these insecticides than the females, and so the cards underestimate the scale population.

Growers on a biologically based program do not rely on pheromone card scale densities as an indication of scale populations because *Aphytis* prefers to parasitize female scales and the male scale numbers can be very high while the female population is low.

Weekly pheromone trap monitoring. Select 5 or 6 orchards that have a known population of California red scales to monitor every week so that you can determine when flights are occurring and time your sprays. Put out pheromone traps beginning in March before the first flight. Change the sticky cards weekly and the pheromone caps monthly through October. Use two to four pheromone traps per 10-acre (4.4-ha) block; add two traps for each additional 10 acres.

Weekly sticky tape traps for crawlers are recommended for use in combination with weekly pheromone traps for males. Although the California red scale crawler stage peaks at about 555°D after the peak in each male flight, the accuracy of predicting crawler peaks based on pheromone traps is affected by orchard management practices. Therefore, using crawler traps is recommended as described in the sidebar "Sticky Tape Traps for Timing Scale Sprays," (p. 106).

Pheromone trap monitoring by flight. In the remaining orchards (in addition to the four reference blocks), use pheromone traps to determine areas of heavy scale infestation.

- Hang the traps with a fresh lure just before the predicted first, second, and fourth flights: for the first flight this is March 1, for the second flight it is at 1,100°D after the biofix of the first male flight, and the fourth flight at 3,300°D from biofix. Monitoring the third flight is unreliable because it can appear to overlap with other generations.

- Use four pheromone traps per 10-acre block; add two traps for each additional 10 acres.
- Remove traps at the end of each flight and count scales (or estimate based on counting the scales inside the squares [20%] and multiplying by 5).
- Record results, such as by using the example recordkeeping form available online in the *UC IPM Pest Management Guidelines: Citrus*.

These traps will tell you which areas of the block have heavy infestations. If the fourth flight is heavy (more than 1,000 scales per card) and fruit is infested with scale at harvest, plan to treat during the next season.

Examining fruit. In all orchards, whether *Aphytis* wasps are released or not, conduct visual inspections of citrus fruit once a month during August, September, and October. Walk around 20 trees in each quadrant of the block and record the number of fruit examined along with the number of fruit with noticeable patches (10 or more) of scales. Calculate the percentage of fruit with more than 10 scales.

Bin counts. At harvest, look at the fruit on the surface of at least 10 bins from areas throughout the block and count the number of uninfested and scale-infested fruit. Calculate the percentage of fruit with scales. At the same time, you can estimate the percentage damaged by citrus thrips, cutworm, katydid, and peelminer.

Detailed evaluations of parasitism in *Aphytis*-release blocks. In orchards where biological control agents such as *Aphytis* and *Comperiella* wasps are used to control scale, visually monitor all stages of scales on twigs, fruit, and leaves in August, September, and October.

- Collect 10 scale-infested fruit (preferably from different areas of the block). Do not take more than one or two fruit per tree, avoiding trees in the outside rows.
- Record the number of second- and third-instar California red scales and the number of these that are parasitized. To determine whether a scale is parasitized, flip the cover over and search for *Aphytis* eggs, larvae, and pupae and *Comperiella* larvae and pupae.
- Calculate the percentage of parasitism by dividing the number parasitized by the total number of second- and third-instar scales examined. If biological control is functioning properly, you should see the percentage of parasitism increase from just a few percent in August to a high percentage in October.

Expectations of when parasitism is at sufficient levels vary by growing region, cultivar, and whether fruit are sent to a packinghouse that employs high-pressure washers to remove scales.

Male California red scales are tiny brownish specks relative to the size of the ½-inch (13-mm) counting squares on this sticky trap. The number of male flights coincides with the number of scale generations, usually two to four generations per year, depending on the growing region.

Purple Scale

Lepidosaphes beckii

Purple scale is an occasional pest in California only on citrus and only in certain coastal areas. Parasites usually provide good control.

Description and Biology

Purple scale develops through the same number of life stages as illustrated for California red scale (Figure 22): females

Purple scale is occasionally abundant in certain coastal areas. It causes yellow haloes on leaves around where it feeds. Purple scales also encrust this twig, where they are easily overlooked.

develop through three instars, males have five instars. Females and males develop the same during the first instar. Beginning with the second instar (after the first molt), males and females develop differently. Females molt once more (from second to third instar), remain immobile, and gradually extend their cover so that it becomes much wider at one end and about 1/12 to 1/8 inch (2–3 mm) long.

After the second instar, males develop through a prepupal and pupal stage before becoming adults. The male cover remains narrow overall and grows to about one-half the length of the female cover. At maturity, the adult male becomes mobile, emerges from beneath his cover, and seeks a female to mate. The adult male has 1 pair of wings and a yellowish and brown body that is about 1/50 inch (0.5 mm) long. Males live a few days or less and are easily overlooked.

Mated females lay 40 to 80 eggs under their cover. Over several weeks, eggs hatch into whitish crawlers (mobile first instars). After settling on branches, fruit, leaves, or twigs, these first instars secrete a cover of pale, cottony wax threads. The cover of male and female second-instar and older scales is variably colored, commonly brownish to purple, and resembles a mussel seashell.

Sticky Tape Traps for Timing Scale Sprays

Early-instar scales are the stages most susceptible to insecticide sprays. A reliable method of timing sprays for first instars, such as spray oil for armored and soft scales, is to monitor for crawlers using sticky tape traps.

Wrap double-sided sticky tape around 1-year-old branches that are about ½ inch (13 mm) in diameter. Choose twigs that are infested with live female scales. For California red scale, place traps on twigs that have both gray and green wood. Replace the tape traps weekly during times of year when you expect crawlers to be present. Preserve the used traps by sandwiching them between clear plastic and light blue paper and label each paper with the trap date and location. Visually compare the abundance of crawlers caught in traps for the different monitoring dates.

Sprays targeting first-instar California red scales can be timed by monitoring scale crawlers with traps of transparent tape that is sticky on both sides. Tightly encircle each of several twigs near female scales. Double over the end so you can grasp and easily unwrap the tape. Place a colored flag next to each trap so it is easily found.

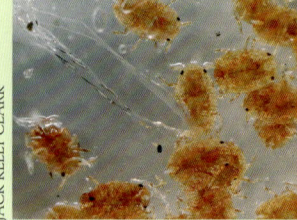

An enlargement of scale crawlers caught in a sticky tape trap. Crawlers search for a suitable place to settle and feed. When trapping crawlers to time an application, replace traps weekly and visually compare crawler abundance in traps among monitoring dates.

The purple scale cover resembles a mussel seashell. The single male cover here (center) matures at a smaller size and is narrower than female covers. Adult parasites of purple scale usually chew a single circular emergence hole, commonly near the narrow end of the cover of older scales (photo right).

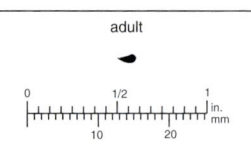

PURPLE SCALE
adult

Purple scale occurs only near the coast where the climate is mild and relatively humid. It prefers the cooler, shady (often north) part of trees. Temperatures above 80°F (27°C) and humidity below 70% greatly reduce populations. Two generations are produced between May and October; a third (overwintering) generation may be partially completed before the cold weather starts.

Damage

Purple scale attacks all parts of the tree. It causes yellow haloes on leaves around where scales feed. On young fruit, scale feeding sites remain green. High populations of purple scale can cause defoliation and kill twigs. Severe damage is usually limited to patches on the lower north side of trees.

Purple scale is usually under excellent biological control. This *Aphytis lepidosaphes* parasite is the primary natural enemy. Up to eight individuals may mature on a single scale, although most scales yield two parasites. Female wasps also puncture and kill many scales to feed on their body fluids, but this host feeding is not obvious.

Natural Enemies

A parasitic wasp, *Aphytis lepidosaphes* (family Aphelinidae), is the most effective natural enemy. *Aphytis lepidosaphes* develop as larvae externally on the body of immature scales, hidden under the scale covers (Figure 26). Purple scale predators include the *Rhyzobius lophanthae* lady beetle, twicestabbed lady beetle (*Chilocorus orbus*), and (in coastal citrus) steelblue lady beetle (*Halmus chalybeus*).

Soft Scales

Soft scale pest species in California citrus include citricola scale, brown soft scale, and black scale. Soft scales produce honeydew, and their cover is the actual body wall of the insect and cannot be removed.

Citricola Scale

Coccus pseudomagnoliarum

Citricola scale is a serious pest in the San Joaquin Valley, but it is rare in southern interior and coastal California because of more effective biological control by introduced parasites. In addition to citrus, citricola is common on hackberry (*Celtis chinensis*) in the Central Valley and is reported to occasionally occur on other tree hosts, such as pomegranate.

Description and Biology

Citricola scale occurs on leaves, interior twigs, or exterior twigs depending on the life stage and time of year (Figure

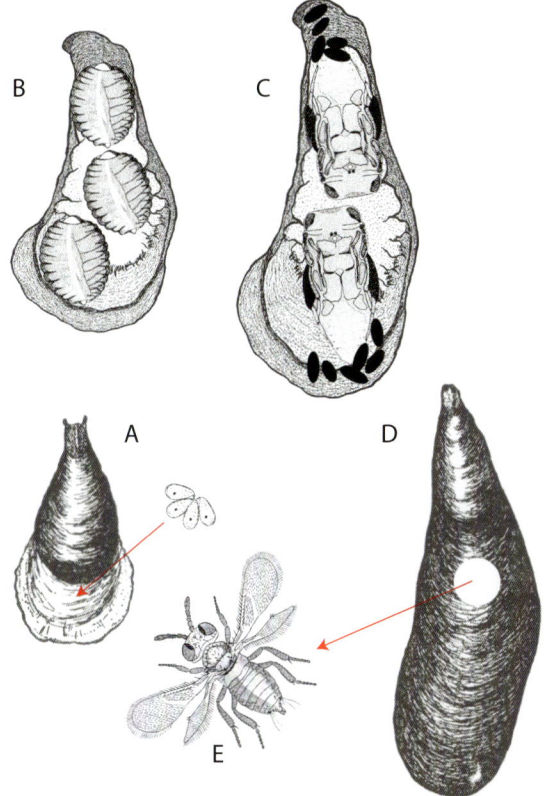

Figure 26. Life cycle and stages of *Aphytis lepidosaphes*, the primary parasite of purple scale. A. Several eggs in a cluster are laid by the *Aphytis* female underneath the cover and on the body of second- or third-instar purple scales. B. The maggotlike *Aphytis* larvae feed externally on the scale body, hidden under the cover (which has been removed here). C. As larvae mature into pupae, they excrete feces (meconium). These dark, oblong fecal pellets are often obvious when inspecting under the cover of parasitized scales. Usually only one or two *Aphytis* survive to the pupal stage on each parasitized scale. D. After *Aphytis* pupae mature into adults, the adult parasites chew a round hole in the scale cover and emerge. E. After mating, female *Aphytis* seek more purple scales to parasitize. In addition to the scales killed by parasitism, female *Aphytis* kill additional scales by puncturing their covers and sucking out the body fluids. This scale mortality from parasite host feeding is not easily observed. Adapted from DeBach and Landi 1961, Quayle 1938.

A heavy infestation of citricola scale produces copious amounts of honeydew, which covers leaves and fruit and favors the growth of blackish sooty mold. Sooty mold causes fruit to be downgraded at the packhouse, and a heavy coating of it on leaves reduces photosynthesis.

27). Females lay 1,000 to 1,500 eggs during early May through early July. Eggs hatch beneath females after 2 to 3 days, and crawlers emerge and settle mostly on the underside of leaves. In severe infestations, crawlers also settle on the upper leaf surface and on twigs, and occasionally on fruit, but scales rarely survive on fruit. On leaves, the newly settled scales are flat and translucent and barely visible with the naked eye. First instars grow slowly during the summer and become yellowish and plump by late summer.

In the fall (starting in October), nymphs molt into mottled brown second instars and start migrating to interior twigs. This migration inward from leaves to twigs peaks in February. In the late winter (March), the second-instar scales migrate down the branches toward the tips of twigs. At the ends of the twigs, the scales develop rapidly and become mottled gray. They mature into gray to mottled brown adult females by late April or early May and begin depositing eggs. There are no males.

Citricola scale may be confused with brown soft scale.

Mature citricola scale females occur on outer twigs and are mottled brown and gray or uniformly gray. Underneath, each female lays hundreds of eggs, a few at a time, that hatch into crawlers during late May to early August.

During late summer, citricola scale first instars are plump and yellowish. They often occur along a leaf vein.

When monitoring citricola scale during late summer, be sure to distinguish whether nymphs are alive. In contrast to a live yellowish first instar (photo left), the flat, brownish nymphs are dead.

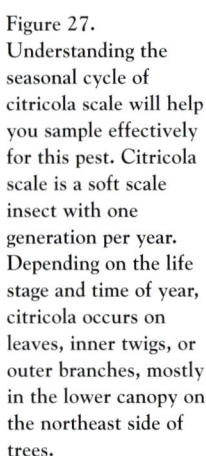

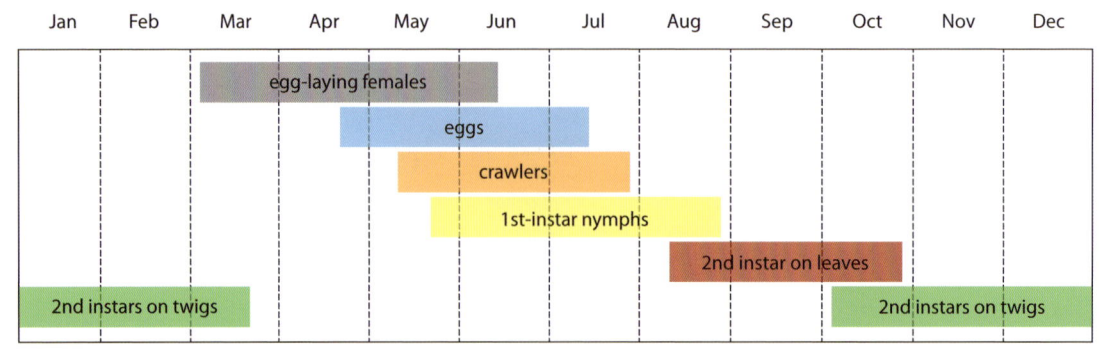

Figure 27. Understanding the seasonal cycle of citricola scale will help you sample effectively for this pest. Citricola scale is a soft scale insect with one generation per year. Depending on the life stage and time of year, citricola occurs on leaves, inner twigs, or outer branches, mostly in the lower canopy on the northeast side of trees.

In late fall and winter, the 2nd-instar scales are deep inside the tree and difficult to sample.

In April, the scales molt into females, move to the ends of the branches, and it is easy to do a twig sample.

In May to early July, females are dying, eggs are hatching, and the nymphs have not all moved out yet onto the leaves. Samples at this time could be misleading.

In late July through September, the 1st- and 2nd-instar nymphs are on the leaves on the outside of the tree and are easily sampled.

By November, citricola scale nymphs become mottled dark brownish, and some begin migrating to twigs. Migration to twigs peaks in about February and March.

Citricola scale has only one generation per year, so nymphs are uniform in size. Brown soft scale has three to five overlapping generations per year, so several different-sized life stages occur at the same time. Citricola scale occurs on leaves only during the summer and fall and on twigs during late fall, winter, and spring. Brown soft scales are present on both leaves and twigs throughout the year. In comparison with the female citricola scale, the mature brown soft scale is smaller and usually yellowish or dark brown, rarely gray.

Damage

Citricola scales suck phloem juices from leaves and twigs. A severe infestation may kill twigs and reduce tree vigor, flowering, and fruit set. This soft scale excretes honeydew on fruit and leaves. Dark sooty mold grows on the honeydew, downgrading the fruit and interfering with photosynthesis in extreme cases.

Natural Enemies

Parasitic wasps that attack citricola and other soft scales include *Metaphycus angustifrons*, *M. luteolus*, *M. stanleyi*, *M. nietneri*, *M. helvolus* (family Encyrtidae), and *Coccophagus lycimnia* and *C. scutellaris* (Aphelinidae). In the San Joaquin Valley, parasites sometimes control citricola scale in orchards near urban areas and in orchards with high populations of brown soft scale. This localized biological control occurs because brown soft scale and various other soft scales on ornamentals serve as alternate hosts for the parasites that also attack citricola scale. These other scales allow parasites to remain abundant locally when citricola scale is not in the second-instar nymphal stage that parasites can successfully attack.

Various predators discussed earlier in the "General Predators" section feed on citricola and other soft scale species. The twicestabbed lady beetle (*Chilocorus orbus*) is a common predator that specializes on scales. The adult is black with two red spots. Its young larvae are easily overlooked as they often feed hidden underneath the scales (Figure 28).

Monitoring

Check for citricola scale at all times of the year when monitoring for other scales. Monitor the number of live females per twig in April to May, counting scales per twig. Observe the percentage of second-instar scales that are parasitized in late spring. Monitor the nymphs in July to September using either presence-absence sampling or counts of the number of nymphs per leaf. The emergence of crawlers (mobile first-instar nymphs) can be monitored as described in the "Sticky Tape Traps for Timing Scale Sprays" sidebar (p. 106).

Be sure to distinguish whether nymphs are alive. Live nymphs are plump and yellowish during late summer. Dead nymphs, such as those killed by hot weather, are flat and brownish. Nymphs that are dying sometimes balloon up, especially after they have been treated with oil. Also be sure to distinguish citricola scale from the brown soft scale.

Honeydew and sooty mold on leaves or fruit may indicate an infestation by the citricola scale. However, heavy populations of aphids or other soft scales also produce honeydew and sooty mold. Citricola scale should be managed before its honeydew and sooty mold become obvious.

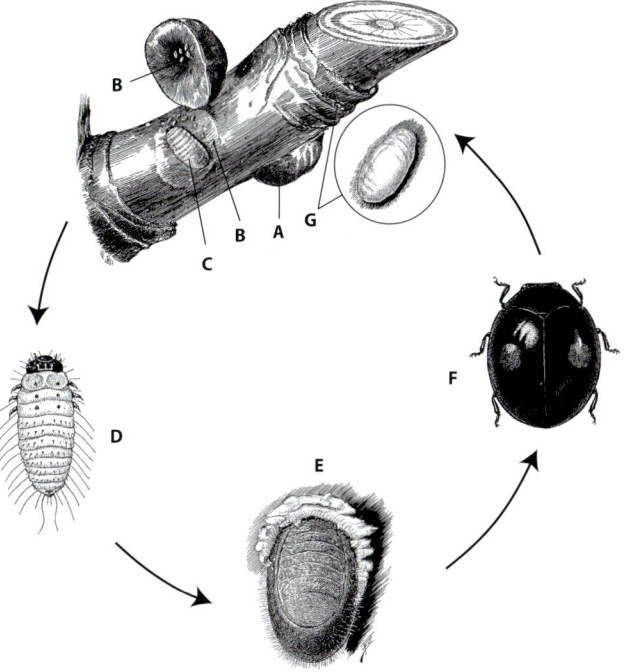

Figure 28. Lady beetle life cycle and stages, illustrated here with a predator of scales. A. Female soft scales such as this are immobile and dome shaped. B. Scale eggs and crawlers occur underneath, as revealed by lifting the mature female from bark. C. Young beetle larvae and certain other small predators feed on eggs and crawlers beneath the female. D. Beetle larvae develop through four instars, and late-instar larvae feed openly on scales. E. Beetle larvae pupate on or near the host plant. F. The lady beetle adult emerges from the pupal case and locates and feeds on scales. G. After mating, the female beetle lays eggs from which larvae hatch and seek prey. Adapted from Simanton 1916.

Brown Soft Scale

Coccus hesperidum

Brown soft scale occurs on citrus in all growing areas. It also feeds on avocado, other subtropical fruit trees, and many ornamentals, especially broadleaf evergreens. Effective parasites usually prevent it from becoming an economic pest. High populations can occur when biological control is disrupted by ants or broad-spectrum pesticides.

Description and Biology

Brown soft scales are oval, somewhat flattened, and yellowish to brown. They occur on leaves or twigs and rarely on fruit. Female brown soft scales lay a few eggs at a time over an extended period, mostly throughout the summer. Eggs hatch almost immediately, and the crawlers soon settle and start to feed. Nymphs molt twice and mature into females. No males have been observed in California.

Brown soft scale populations are usually highest from midsummer to early fall. There are three to five overlapping generations per year, and several different-sized life stages usually occur together at the same time. Brown soft scale resembles citricola scale, but citricola scale has only one generation per year, so its nymphs are uniform in size. Citricola scale occurs on leaves only during the summer and fall and on twigs in the late fall, winter, and spring. Brown soft scale is present on both leaves and twigs throughout the year.

Damage

Feeding by high populations of brown soft scale kills twigs and reduces tree vigor and fruit yield. Blackish sooty mold grows on the scale's honeydew, which may cause fruit to be downgraded at the packinghouse. Honeydew attracts ants, which interfere with the biological control of scales and other pests.

Parasite Encapsulation

Parasites sometimes fail to develop because they are killed by the host's defenses, called encapsulation. Encapsulation of *Metaphycus* species parasites by brown soft scale and citricola scale can sometimes be recognized in the field. When a soft scale encapsulates parasites inside its body, the scale body often darkens overall and develops dark spots where the parasite laid its eggs. Although these soft scales may not die from parasitization, once they mature, the number of eggs the scale can lay may be significantly reduced because it diverted resources to encapsulation.

Parasitized soft scales sometimes kill the immature parasites by encapsulation. When a soft scale encapsulates parasites inside its body, as with this brown soft scale nymph, the soft scale body develops dark spots where the parasite laid its eggs.

Psocids and Sooty Mold

Psocids are not pests in citrus, but they are commonly present when trees are infested with soft scales or other honeydew-producing insects. Psocids feed on fungi, including the blackish sooty mold that grows on honeydew excreted by Homoptera.

Psocids are small soft-bodied insects with long, slender antennae. The adults are usually brown or gray and have four wings or no wings. Winged adults resemble the adults of psyllids, such as the Asian citrus psyllid pictured later in this chapter. Nymphs are wingless and commonly pale colored.

Females lay eggs in clusters and cover them with silken threads. Psocid egg clusters may superficially resemble the cocoon of brown lacewings or the egg sac of certain spiders as pictured earlier in the section "General Predators."

Psocids feed on honeydew and sooty mold resulting from citricola and other pests that suck phloem sap. An adult, about 1/6 inch (4 mm) long, and a pale nymph are shown here.

Some psocid species lay their eggs in clusters and cover them with silken threads. Psocids become common in citrus orchards during spring if honeydew-producing aphids, mealybugs, soft scales, or whiteflies are present.

Young brown soft scales are yellowish, mottled, and rounded. Unlike the similar-looking citricola scale, brown soft scales can occur on both leaves and twigs throughout the year.

Brown soft scale is the only soft scale on citrus with multiple generations per year. Different-sized scales occur together, which helps distinguish this species.

BROWN SOFT SCALE

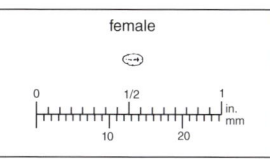

Natural Enemies

Parasitic wasps attacking brown soft scale include *Metaphycus angustifrons*, *M. luteolus*, *M. stanleyi*, and *Coccophagus* spp. The female parasites primarily lay their eggs in the early instars, before scales mature and can produce offspring. The Australian lady beetle (*Rhyzobius lophanthae*), brown and green lacewings, and twicestabbed lady beetle (*Chilocorus orbus*) prey on all stages of brown soft scale.

Because brown soft scale generations overlap, it provides parasites with the nymphal stages susceptible to parasitism throughout the year. When brown soft scale occurs on citrus or other host species nearby, its presence allows parasites and predators to remain abundant locally and move to other soft scale species when those scales produce life stages susceptible to parasitism. Parasitism of brown soft scale is very successful unless disrupted by ants, dust, or broad-spectrum pesticides.

Monitoring

Look for brown soft scale from June through October, when disruption of biological control can be a problem. Check the level of parasitism by examining scales for parasite exit holes and for parasites developing within the scale body. Parasitized scales often change color during the later stages of parasite development. Sometimes the parasite body itself is visible through the scale's surface.

This colony of brown soft scales is heavily tended by Argentine ants. If brown soft scale is abundant, the cause is usually disruption of its natural enemies, such as by ants, dust, pesticides, or weather.

Metaphycus species adult parasites are tiny, yellowish orange wasps, as with this female laying eggs in a soft scale nymph. Adult female parasites also host feed on the body fluids of young scales, although this host-feeding mortality is difficult to recognize.

Metaphycus are internal parasites, as with the five wasp larvae visible here through the surface of this brown soft scale.

TWICESTABBED LADY BEETLE

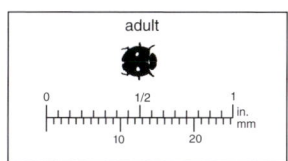

The adult twicestabbed lady beetle has two reddish spots on its shiny black wing covers. Both adults and larvae prey on armored and soft scale adults and nymphs.

This larva of the steelblue lady beetle (*Halmus chalybeus*) is pale gray or yellowish with dark spots. Rows of spines cover its body. *Halmus* adults and larvae feed on scales in coastal citrus and avocado. The adult is a dark metallic blue, hence the common name steelblue lady beetle.

Within any colony, many individual brown soft scales are usually discolored or dark because they are parasitized. A mature parasite has emerged from the scale at the right, as evidenced by the round hole.

Black Scale

Saissetia oleae

Black scale is a major citrus pest in southern California. This soft scale is an uncommon problem on citrus in the San Joaquin Valley, where it is mostly found on grapefruit or on citrus trees near olives. Black scale feeds on many other crops and ornamentals, including cherry, oleander, olive, and pistachio. Introduced parasites provide substantial control in some districts.

Description and Biology

In the interior areas of southern California, black scale produces one generation each year. In coastal districts, it usually has two overlapping generations. The females can reproduce without mating and lay 1,000 to 2,000 eggs over a period of 2 to 3 months, mainly during May and June. In cooler coastal areas with two broods per year, egg laying and crawlers also occur from October through November.

First instars settle and feed on leaves. In the late second instar, a ridge develops on the scale's back and later expands into an H shape. After the second molt, the young scales migrate to twigs, where they grow rapidly and become nearly circular. These older nymphs become mottled dark and gray and develop a leathery texture, called the "rubber stage." Once egg laying starts, the scales become darker, harder, and distinctly hump shaped, and the H-shaped ridge on top often disappears. Male black scales are rarely seen.

Damage

Feeding by high populations of black scale reduces tree vigor, eventually leading to leaf or fruit drop and twig dieback. Fruit may be covered with honeydew, which attracts ants and supports the growth of blackish sooty mold. Noticeable sooty mold causes fruit to be downgraded at the packinghouse.

Natural Enemies

Predators of black scale include brown and green lacewings, twicestabbed lady beetle, and (in coastal areas) steelblue lady beetle. Parasitic wasps are especially important and provide substantial biological control in southern California, including *Metaphycus flavus*, *M. helvolus*, and *M. luteolus*, which also parasitize brown soft scale and citricola scale. However, black scale biological control is especially susceptible to disruption by ants. *Metaphycus helvolus*, the most important parasite of black scale, takes a relatively long time to deposit its eggs. This parasite's long presence on scales, while it slowly deposits its eggs, greatly increases the likelihood that ants will locate and attack it in comparison with other parasite species that take only a few seconds to complete their egg laying.

Metaphycus helvolus. *Metaphycus helvolus* is a tiny, yellowish orange wasp. Females lay eggs primarily inside second-

This crawler, or mobile first instar, is one of hundreds that will eventually emerge from eggs (photo left), exposed by removing the female black scale.

Second-instar and immature female black scales occur on leaves and have a raised ridge or H-shape on their top. The H-shaped ridges are obvious here, but they often disappear once females mature.

Mature female black scales occur on branches. They are dark, hard, and distinctly hump shaped.

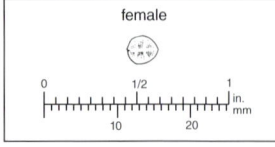

and early third-instar scales. Eggs hatch into larvae that feed within the scales and kill them. Usually one parasite develops inside each scale, although occasionally two or three wasps emerge as adults from a single scale. Adult female parasites also host feed on the body fluids of young scales, often causing higher mortality through host feeding than through parasitism. This host-feeding mortality is difficult to recognize.

The effectiveness of Metaphycus helvolus depends on the climatic region and conditions in the orchard, as well as on ant control. The parasite survives better in the coastal areas, where two overlapping scale generations provide susceptible nymphal stages for a longer time than in the interior regions. Warm winter weather also helps the parasite because it causes the scale's development to be less synchronized. Cold weather and other adverse conditions cause black scale populations to consist more of a single life stage, thereby often drastically reducing Metaphycus populations.

Metaphycus helvolus may be commercially available for purchase and release. Alternatively, it can be collected in the field and relocated to infestations to supplement existing parasite populations.

Monitoring

In southern California where it is a problem, look for black scale during summer and fall. Also look for ants, which are a common cause of black scale outbreaks. Especially watch for newly settled (first-instar) scales on the underside of leaves in late June or early July, as the end of egg hatch is the most effective time to apply spray oil.

If scales are abundant and parasite introductions are planned during late summer and early fall, also look for black scales in other citrus orchards with high parasite populations. Parasites can be collected by cutting branches with black scales (before emergence holes are present) and putting them in orchards where parasite activity is low. If ants are present, control them before releasing any parasites.

Cottony Cushion Scale

Icerya purchasi

Cottony cushion scale is under complete biological control, except where natural enemies are disrupted by insecticide treatments and for certain varieties of citrus (grapefruit and young mandarins) with dense foliage that promote this pest. Insecticides are usually not necessary to control this species.

The cottony cushion scale feeds on a wide variety of fruit and nut trees and ornamentals such as *Nandina* and *Pittosporum* spp. It was a major pest on citrus in the 1880s. Foreign exploration for natural enemies in this pest's native country (Australia) and the introduction of a parasite and predator from there into California resulted in one of the earliest and most impressive examples of biological control.

Description and Biology

This scale's most distinguishing feature is the adult female's fluted, cottony egg sac. The white, elongated egg sac erupts from beneath and to one side of the cover as the female secretes the sac and lays from 600 to 800 eggs inside. The eggs hatch in a few days in the summer but take up to 2 months in the winter.

The newly hatched nymphs are red with dark legs and antennae. First and second instars settle on leaves, usually along the veins, or on small twigs. As they age, the first and second instars change from reddish to cottony whitish.

Parasitic wasps, especially *Metaphycus helvolus*, provide substantial biological control of the black scale in southern California. Look for the exit holes of adult parasites, as with the left- and right-most scales shown here.

The adult female cottony cushion scale is easily recognized by its white, fluted egg sac. The black and red vedalia beetle is an effective predator of this scale. A red vedalia first-instar larva and oblong eggs of this lady beetle are on top of the scale egg sac.

COTTONY CUSHION SCALE

The covers of third-instar and adult females are variably brown, orange, red, or yellow. White powder may form on top the adult female's cover. Female third instars move to the heavier scaffolding, and adults occur mainly on heavy branches and the trunk. Cottony cushion scales are rarely found on fruit.

In the San Joaquin Valley, cottony cushion scale develops about three generations per year and overwinters as mixed stages of first-instar through adult female scales (Figure 29). Eggs are produced during the winter (generation 1). During April and May, the population becomes synchronized and consists primarily of adult females. The eggs from these females hatch in June (generation 2). During the summer, the populations are generally very low due to heat. Another period of egg production occurs in the fall (generation 3).

Males are rarely seen, and females can reproduce without mating. Males develop differently from the females beginning in the third instar. Male nymphs seek a secluded place on the tree or ground, where they form a loose, cottony cocoon. A tiny, delicate, winged adult male develops inside.

Damage

Cottony cushion scales suck phloem sap from leaves, twigs, and branches, reducing tree vigor. A heavy infestation may result in leaf and fruit drop, and twigs and small branches may die back. Yield may be reduced if scale densities are heavy. The secreted honeydew and subsequent growth of sooty mold cover the fruit, causing downgrading at the packinghouse.

Natural Enemies

In most orchards, vedalia and *Cryptochaetum* effectively control the cottony cushion scale if these natural enemies are conserved as discussed earlier in the section "Scale Insect Management." Consult *Stages of the Cottony Cushion Scale* (Icerya purchasi) *and its Natural Enemy, the Vedalia Beetle* (Rodolia cardinalis) (Grafton-Cardwell 2002) for more discussion and photographs.

Vedalia. The vedalia beetle *(Rodolia cardinalis)* was introduced from Australia in the early 1890s. If vedalia are present, they will probably control cottony cushion scale, as long as you do not disrupt them with pesticides. Vedalia beetles have a number of qualities that make them an excellent predator. Vedalia beetles are very specific predators, feeding only on cottony cushion scales. They grow very rapidly and can complete two generations in the time it takes cottony cushion scale to complete one generation. They attack all stages of the scales and have a voracious appetite. When vedalia beetles arrive in an orchard, they can control a

Opening this egg mass reveals many cottony cushion scale eggs, two recently hatched crawlers, and a first-instar vedalia larva. Note that the cottony cushion scale crawler (left arrow) has black legs and black antenna, whereas young vedalia have red legs and no obvious antennae.

Cottony cushion scale first and second instars secrete thick, whitish material as they age. Immediately after molting, the covering disappears and nymphs are mostly reddish as with the one at the left. These early instars are the stages most susceptible to insecticides.

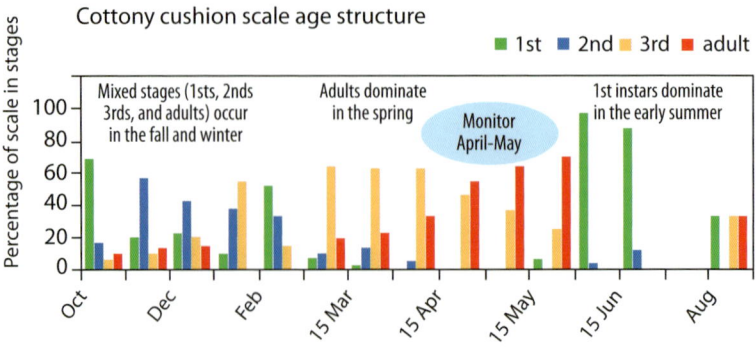

Figure 29. Cottony cushion scale seasonal development. Cottony cushion scale populations are sometimes composed of only one life stage (synchronous), and sometimes all stages are present (mixed). April to May is the best time to sample for cottony cushion scale because the population consists mostly of females and they are large and easy to see. April and May is also the time to look for vedalia beetle, which when present usually provides good control.

serious cottony cushion scale problem in 4 to 6 weeks. However, vedalia beetles are very sensitive to heat, and this can reduce their control of scales. They halt egg production and larval development when San Joaquin Valley daily temperatures exceed 90°F (33°C), usually in June. Thus, vedalia must arrive in March or April to have sufficient time to control the scale population before temperatures increase.

Adults are red and black lady beetles, about 1/12 to 1/6 inch (2–4 mm) long. Black predominates on some adults, while others have more red. A covering of fine hairs often gives adult vedalia a grayish appearance.

The female beetle lays red eggs, sometimes singly but usually in groups of three or more, attached to the cottony cushion egg sac. The first-instar larvae dig their way into the egg mass and feed on the eggs. First-instar vedalia larvae are reddish and resemble the crawlers of cottony cushion scale. However, cottony cushion scale crawlers have black legs and black antenna, and young vedalia larvae have red legs and no obvious antennae. Second-instar larvae look similar to first-instar vedalia larvae, only larger. Third-instar vedalia larvae take on a gray color and are elongate and tapered at both ends. Second- and third-instar vedalia larvae move freely up and down branches and feed on all scale stages. When ready to pupate, the third-instar larva molts into a fourth instar that crawls out to a citrus leaf, stops feeding, glues itself to the leaf, and changes into a pupa.

The skin of the fourth instar splits over the back of the pupa as the pupa changes shape. A healthy pupa will be reddish and move slightly when touched.

Cryptochaetum. The parasitic fly *Cryptochaetum iceryae* (family Cryptochaetidae) is very effective in the coastal areas and (during late fall through early spring) in interior areas. *Cryptochaetum* can occur at lower levels wherever cottony cushion scale occurs. The adult fly is dark blue or green to black and about 1/12 inch (2 mm) long. The female deposits its eggs inside the scale body. The maggots feed and usually pupate inside the scale. Pupae are black with two tiny protruding breathing tubes (spiracles). The fly produces about five or six generations per year.

Monitoring

April-May is the best time to sample for damaging populations of cottony cushion scale because the population consists of females and they are large and easy to see. Monitor by opening the canopy and searching interior limbs and the trunk for adult female scales. Be sure to determine whether the scales are live or dead. Live female scales have yellow gut contents and red eggs or crawlers in their egg sac. Also search for vedalia's red eggs on the egg sacs, larvae and adults in and among the scales, and pupae on the leaves. If live stages of vedalia are present, the cottony cushion

Older vedalia larvae are reddish to gray. They are elongate and tapered at both ends, as with this vedalia eating a scale nymph.

The skin of the last-instar larva has split open to show the red coloration of the vedalia pupa developing into an adult. Often only the empty pupal case can be seen attached to leaves on the outside canopy.

VEDALIA

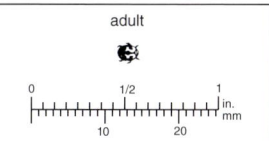

The *Cryptochaetum iceryae* adult is a dark blue or green to black fly. After feeding inside as a larva, *Cryptochaetum* often pupates inside its dead host. The emerging adult leaves an exit hole in the dead, mummified scale.

CRYPTOCHAETUM

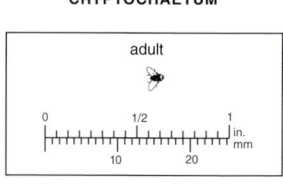

The dark pupal case of the *Cryptochaetum* parasite sometimes occurs outside its host, as with this pupal case attached to the remains of a dead cottony cushion scale. *Cryptochaetum* is very effective in coastal areas and (during cooler times of the year) in interior areas.

scale population will likely be controlled in 6 weeks. If no vedalia are present in April, you can collect vedalia from other orchards and release them in the problem orchard. Relocating as few as 20 vedalia adults or larvae can establish a population. If no vedalia stages are present by the end of May, there is not sufficient time for control by vedalia (6 weeks) before the summer heat, and a pesticide treatment may be warranted.

THRIPS

Thrips (spelled the same for singular or plural) are sucking insects that, with some species, discolor, distort, and scar fruit and leaves. Thrips (order Thysanoptera) develop from an egg, through two larval stages, one or two pupal stages, then into adults (Figure 30). Adults and larvae are tiny, slender insects and as adults have fringed-tipped wings.

Citrus thrips, flower thrips, bean thrips, and greenhouse thrips, in approximate order of commonness, are plant-feeding thrips found in California citrus (Table 9). It is important for you to distinguish between these species because their distribution, monitoring methods, and management tactics differ. For example, greenhouse thrips are common only in the coast. Citrus thrips are damaging pests that look similar to nonpest flower thrips. Both species infest trees at petal fall, thus they can be easily confused at that time. Bean thrips do not directly damage citrus, but they are a quarantine concern because they can hide in the navel of the fruit. You may also find other plant-feeding thrips species that are not pests of citrus as well as predatory thrips that are important natural enemies of mites, plant-feeding thrips, and other pests.

Because they often feed on citrus flowers but are not pests in citrus, western flower thrips (at tips of arrows) must be distinguished from the citrus thrips.

Citrus thrips (at tip of arrow) is the primary pest thrips because larvae feed under the sepals of young fruit, causing scars that later become obvious on rinds as fruit grow.

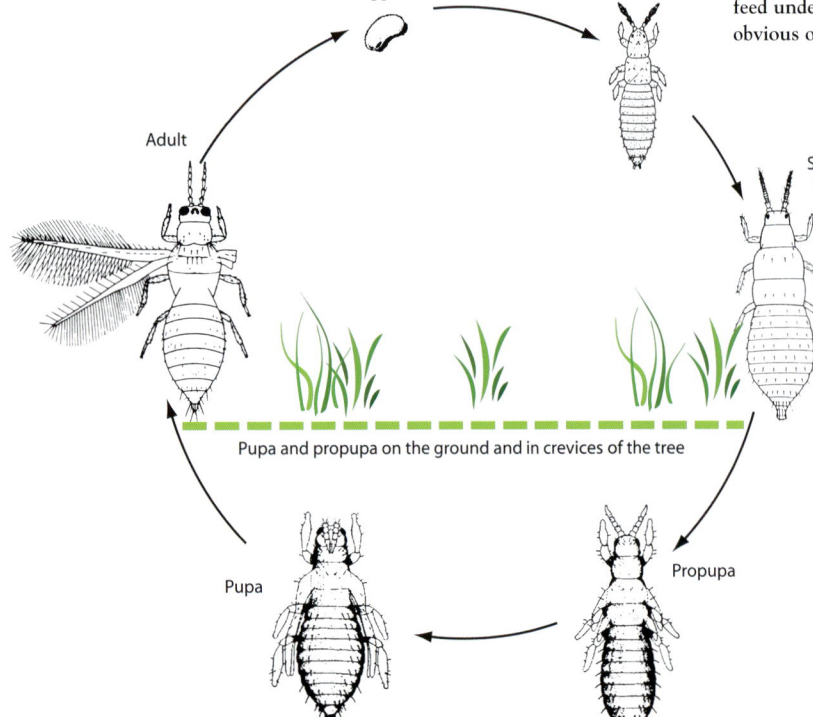

Figure 30. Citrus thrips life cycle and stages. Citrus thrips are present throughout the year, but economic damage occurs only during late April through June when larvae feed on young fruit. *Sources*: Thrips by Ebeling in Quayle 1938, McKenzie 1935.

Table 9. Selected Thrips in Citrus and How to Distinguish Them.

Name			Appearance	
Common	Scientific	Where most occur	Adults	Larvae
Pest thrips				
citrus	Scirtothrips citri[1]	under sepals of young fruit when weather is cool or overcast; on the fruit surface when it is sunny; on succulent leaves anytime	body light orangish yellow to white, no bands on abdomen	light orangish, yellow, or white body
greenhouse	Heliothrips haemorrhoidalis[1]	where fruit touch in clusters, upper leaf surface	black body with pale wings	white to yellowish body
bean	Caliothrips fasciatus[1]	in the navel of oranges	in fruit navels, they often appear blackish without banding or distinguishing characteristics; otherwise, uniformly dark or grayish black body, two dark and two pale bands on forewings, and legs and antennae banded light and dark	do not occur on citrus
Importance unknown				
—	Neohydatothrips burungae[1]	on coastal lemon	three red spots atop head, banded antennae, brown lines separating segments ("tiger stripes") only on the upper side of abdomen; can be distinguished from similar-looking avocado thrips as shown in Figure 34	—
Innocuous thrips				
western flower	Frankliniella occidentalis[1]	on or near flowers	thick, bristlelike hairs at the tip of the abdomen, which other species lack; body black, brownish, yellow, white, or orange; some individuals have brown abdominal bands; abdomen extends beyond wing tips at rest	yellow to orangish body
onion	Caliothrips faciatus[1]	on citrus growing near preferred hosts that dry up or are cut	pale yellow to dark brown body	do not occur on citrus
avocado[2]	Scirtothrips perseae[1]	in traps in citrus growing near avocado	three red spots atop head, banded antennae, brown lines separating segments on upper side and underside of pale yellow abdomen, wing tips at rest extend beyond abdomen	do not occur on citrus
Beneficial predatory thrips				
banded	Aeolothrips spp.[3]	among pest mites and thrips	black body; white wings have two distinguishing black bands	yellow body
black hunter	Leptothrips mali[4]	among mites, scales, and pest thrips	dark brown or entirely black body, white wings, much more active than similar-looking greenhouse thrips	dark reddish brown or blackish body
Franklinothrips or vespiform	Franklinothrips orizabensis, F. vespiformis[3]	among mites and pest thrips	mostly black body, with pale or white areas; distinctly narrow where abdomen meets thorax	yellow to orange body, swollen abdomen with red or dark orange band, body stout or oval shaped
sixspotted[5]	Scolothrips sexmaculatus[1]	in colonies of mites	three dark blotches on each forewing, body pale to yellowish	yellow to whitish body

1. Thripidae family.
2. Avocado thrips rarely occur on citrus fruit or trees and do not reproduce in citrus, but adult avocado thrips can be found in traps in citrus growing near avocado.
3. Aeolothripidae family.
4. Phlaeothripidae family.
5. Sixspotted thrips primarily prey on mites and are not an important predator of pest thrips.

Identifying Thrips Species

Thrips are easily overlooked because they are tiny and prefer hidden locations. Their presence may not be recognized until fruit or leaves exhibit characteristic damage, as pictured later. However, there are many other causes of damage to the rind of citrus, including wind rubbing, certain other invertebrates, and some pathogens and disorders, as summarized in Table 3 in the chapter "Managing Pests in Citrus" and in A *Photographic Guide to Citrus Fruit Scarring* (Grafton-Cardwell et al. 2003). For instance, rind injury resembling that of citrus thrips can also be caused by amorbia caterpillars, but the latter is deeper into the rind of the fruit. Correctly identify the thrips in your orchard so you can choose the best methods for monitoring that species and preventing its damage.

Identify thrips based on insect appearance (such as body shape and color) and behavior (whether they occur singly or in groups and are sluggish versus fast moving), as well as their occurrence together with characteristic damage. Also helpful for identification are the location of thrips on the plant (on fruit versus leaves), the time of year when they are observed, and whether they occur together with other invertebrates (such as prey species, indicating they may be predaceous thrips).

To reliably identify thrips, carefully prepare specimens and use a binocular dissecting microscope to examine features such as body sculpturing and the number, length, and location of hairs on their body and wings, as illustrated later in the pest species sections. Additional online tools for distinguishing among thrips species include *Thrips of California* (Hoddle, Mound, and Paris 2008) and *Pest Thrips of North America* (O'Donnell, Moritz, and Mound, undated). Because new species are periodically introduced, if thrips appear to be a new species or are causing unusual damage, collect and submit adult and immature specimens to your county agricultural commissioner for identification.

Adults of plant-feeding thrips that may be found in citrus but are typically not citrus pests include avocado thrips (which occurs in traps in citrus growing near avocado) and onion thrips (a migrant from many other crops, which can occur in traps and occasionally on citrus fruit). *Neohydatothrips burungae* (reported on coastal lemon) is of unknown importance. Beneficial predatory thrips in citrus include banded thrips (*Aeolothrips* species), black hunter thrips (*Leptothrips mali*), *Franklinothrips* species, and sixspotted thrips (*Scolothrips sexmaculatus*), as described in their photo captions.

Management

The following three species of pest thrips are managed differently because of differences in location and the type of

The adult citrus thrips has a mostly yellowish or light orange body without obvious dark lines or bands. See Figures 32–35 for tips on distinguishing citrus thrips from other thrips species.

Adult western flower thrips vary greatly in color. Their body can be all black, all yellow, or a mix of light and dark colors. In comparison with citrus thrips, western flower thrips has hairs on the tip of its abdomen that are longer and stouter. At rest, its abdomen extends beyond the tips of its wings, while citrus thrips wings extend beyond the rear of the abdomen.

Onion thrips adults have a pale yellow to light brown body, pale wings, and gray eyes. Onion thrips are not a pest in citrus, but they are often caught in sticky traps and occasionally are found on fruit as they have many broadleaved and grain crop hosts.

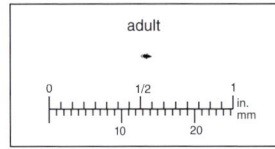

Neohydatothrips burungae occurs at least on lemons and avocados in southern California, but it is of unknown importance. *Neohydatothrips* has brown bands ("tiger stripes") only on the upper side of its abdomen. Avocado thrips is lighter colored and has brown lines separating segments on both the upper side and underside of its body. Citrus thrips lacks these abdominal bands. See Figure 34 to accurately distinguish these species.

This predaceous sixspotted thrips occurs on citrus, although normally not around petal fall, when most citrus thrips monitoring is done. Sixspotted thrips is usually abundant only where spider mites (its primary prey) are common. Three dark spots on each forewing distinguish the adults of the sixspotted thrips from other thrips.

damage they cause. Citrus thrips are managed using a combination of cultural, biological, and insecticidal controls. Bean thrips management relies on cultural practices because their migratory behavior makes them difficult to control with pesticides or natural enemies. Greenhouse thrips are managed primarily by cultural and biological methods, along with an occasional spot application of an insecticide that has minimal impact on natural enemies.

Cultural practices such as fertilization and pruning influence the extent and timing of leaf flush, thereby affecting citrus thrips damage. Citrus thrips prefer to feed on succulent leaves and are more likely to feed on and scar young fruit when young leaves are not available. For bean thrips, keep orchards and bordering areas free of weed hosts to help reduce bean thrips movement from alternate hosts to citrus fruit. Where greenhouse thrips populations are high, harvest fruit early in affected areas of the orchard to minimize the amount of damage. Because much of the greenhouse thrips population resides on the fruit, greenhouse thrips are removed at harvest. Early harvest also reduces the crop-to-crop overlap time, so it also minimizes the greenhouse thrips' movement to and damage of the following year's crop.

Natural enemies of thrips include predatory mites, predatory thrips, and (for greenhouse thrips) tiny parasitic wasps such as *Thripobius*. Natural enemies may not be specific enough or arrive early enough to reduce citrus thrips densities below economic thresholds or prevent bean thrips from contaminating fruit navels. However, natural enemies can be helpful for citrus thrips and are believed to be very important for greenhouse thrips.

Citrus and greenhouse thrips are less of a problem in orchards that receive minimal pesticide applications. Thrips population levels increase after broad-spectrum, persistent insecticides are used because the insecticides both reduce natural enemy abundance and increase pest thrips populations by stimulating females to lay more eggs (pesticide-induced hormoligosis).

Citrus Thrips

Scirtothrips citri

Feeding by citrus thrips scars fruit rinds, typically with a partial or complete circular ring scar around the calyx. This

The adult greenhouse thrips is black with pale wings. The larva (lower right) is whitish to pale yellow. Compared with other thrips, this species is readily distinguished because it moves slowly and feeds in colonies that produce dark excrement.

Black hunter thrips are dark reddish brown or blackish, as with the larva shown here. Adults have white wings, resembling greenhouse thrips. However, unlike the sluggish greenhouse thrips that occur in groups, black hunter thrips are fast-moving, solitary predators of mites and small insects.

Bean thrips do not directly damage fruit but can be a pest because adults overwinter in the navel of citrus, causing exported fruit to be rejected by some foreign countries. Adult bean thrips have a uniformly dark body, with two dark and two pale bands on their forewings.

Franklinothrips are predators of greenhouse thrips and other small insects in south coastal growing areas. *Franklinothrips vespiformis* (shown here) and *F. orizabensis* are virtually indistinguishable. Adults of both species are mostly black with pale to white areas, including at their thin waist.

Banded thrips, also called banded-wing thrips, are most often observed when their pest mite and thrips prey are abundant. In this species, wing bands alternate crosswise to the body. Other *Aeolothrips* species have wing bands oriented lengthwise or a combination of crosswise and lengthwise bands.

The *Franklinothrips* second-instar larva has a swollen abdomen with a distinct red or dark orange band or dot. To the naked eye, it may appear to be a fast-moving reddish dot.

damage is of greatest economic importance on San Joaquin Valley mandarin and navel oranges, citrus grown in the desert, and sometimes on coastal lemons.

Description and Biology

Citrus thrips are tiny, elongate insects that vary in body color from whitish or pale yellow to orange. Adults and second through fourth instars (pupae) are about 1/25 inch (1 mm) long. Eggs and first instars are even smaller.

In the fall, overwintering eggs are laid mostly in the last growth flush of the season. Adults and immature stages (except eggs) are uncommon from October through February. The hatch of overwintering eggs typically coincides with the new spring growth in about March. The active, spindle-shaped first and second instars feed on tender leaves and fruit, especially under the sepals of young fruit if present. Third and fourth instars (propupa and pupa) occur in debris on the ground or in the crevices of trees (e.g., under bark). They have wing pads and do not feed but can move slowly if disturbed. After pupation, the adults move to foliage or fruit (Figure 30). During the spring and summer, each female lays about 25 eggs in new leaf tissue, young fruit, or green twigs.

Citrus thrips development from egg to peak egg-laying adult requires about 430°D above 58.3°F (14.6°C). At typical early spring temperatures in the San Joaquin Valley, development time from egg to the adult egg-laying stage is about 30 to 40 days. Citrus thrips can produce up to eight generations annually (Figure 31). The most important generations are the second and third because these coincide with young fruit that are most sensitive to feeding damage.

Distinguishing Citrus Thrips from Western Flower Thrips. Be sure to distinguish citrus thrips from western flower thrips. Flower thrips feed on flower parts but do not damage citrus fruit, and so they should not be included in thrips counts. Shortly after petal fall, you may see both immature citrus thrips and flower thrips moving about young fruit. But when the petals completely disappear, the flower thrips adults disperse to other plants. Appearance and behavior help distinguish these species (Figures 32, 33). Flower thrips at maturity are larger, straw colored to dark, have a more elongate or cigar-shaped body, and are slower moving than citrus thrips. Citrus thrips are shorter relative to their width, bright yellow, and have a more oval-shaped abdomen. Immature stages also differ, with the western

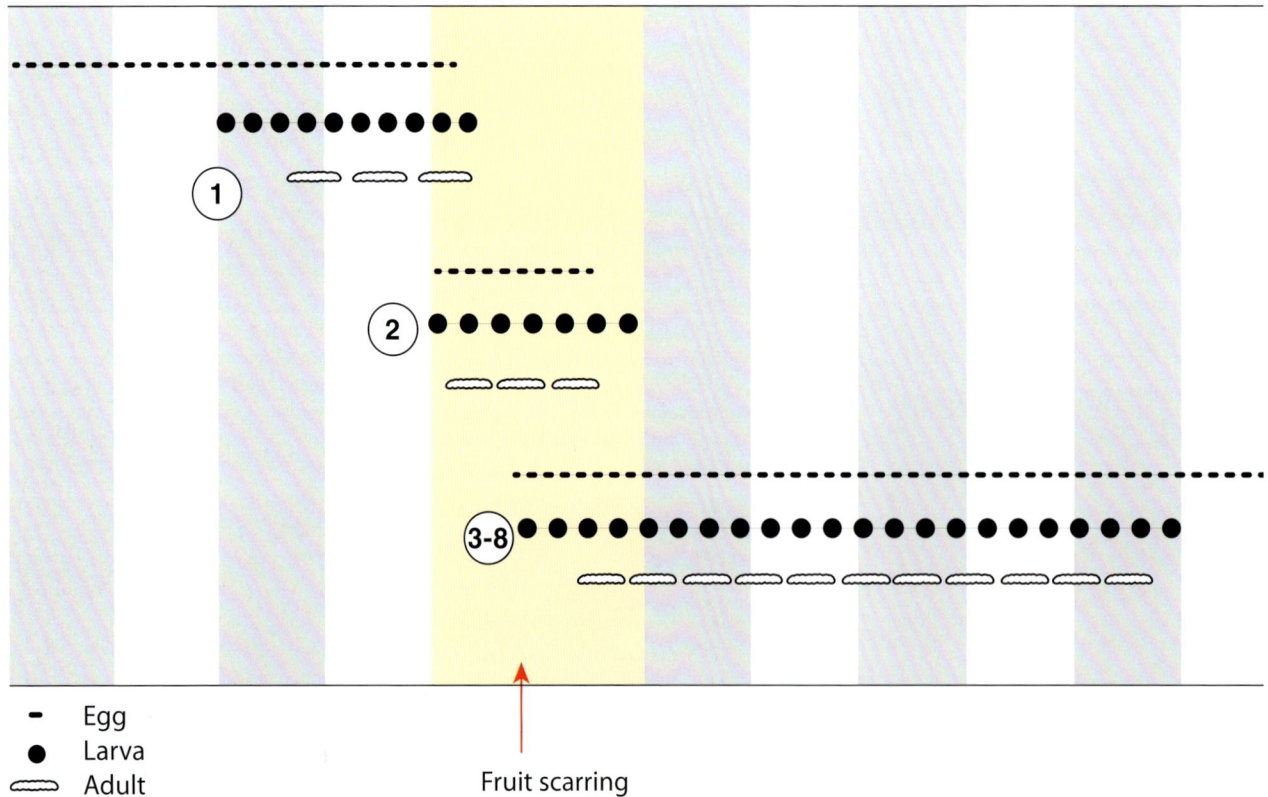

Figure 31. Seasonal development of citrus thrips on navel orange in the San Joaquin Valley. All stages of citrus thrips occur throughout the year except during winter, when only eggs may be present. Citrus thrips produce up to 8 generations if the weather is favorable, but the most important are the second (2) and third (3) generations because they develop during early fruiting (yellow-shaded area). Larvae feeding beneath the sepals of small fruit until fruit are about 1½ inches (4 cm) in diameter cause the economic damage.

flower thrips being longer, having distinct spines, and moving in a more serpentine manner (sinuous or winding).

Distinguishing Citrus Thrips from *Neohydatothrips*. *Neohydatothrips burungae* occurs at least on lemons and avocados in southern California, but it is of unknown importance. *Neohydatothrips* have dark abdominal stripes (bands) on top of the abdomen and three red head spots between their eyes. Citrus thrips lacks these features, but thrips coloration can vary. See the *"Neohydatothrips"* section for discussion and illustrations on how to reliably separate these species.

Damage

Citrus thrips puncture epidermal cells with their mouthparts, causing scabby, grayish or silvery scars. On young leaves, citrus thrips feeding causes thick, gray streaks on

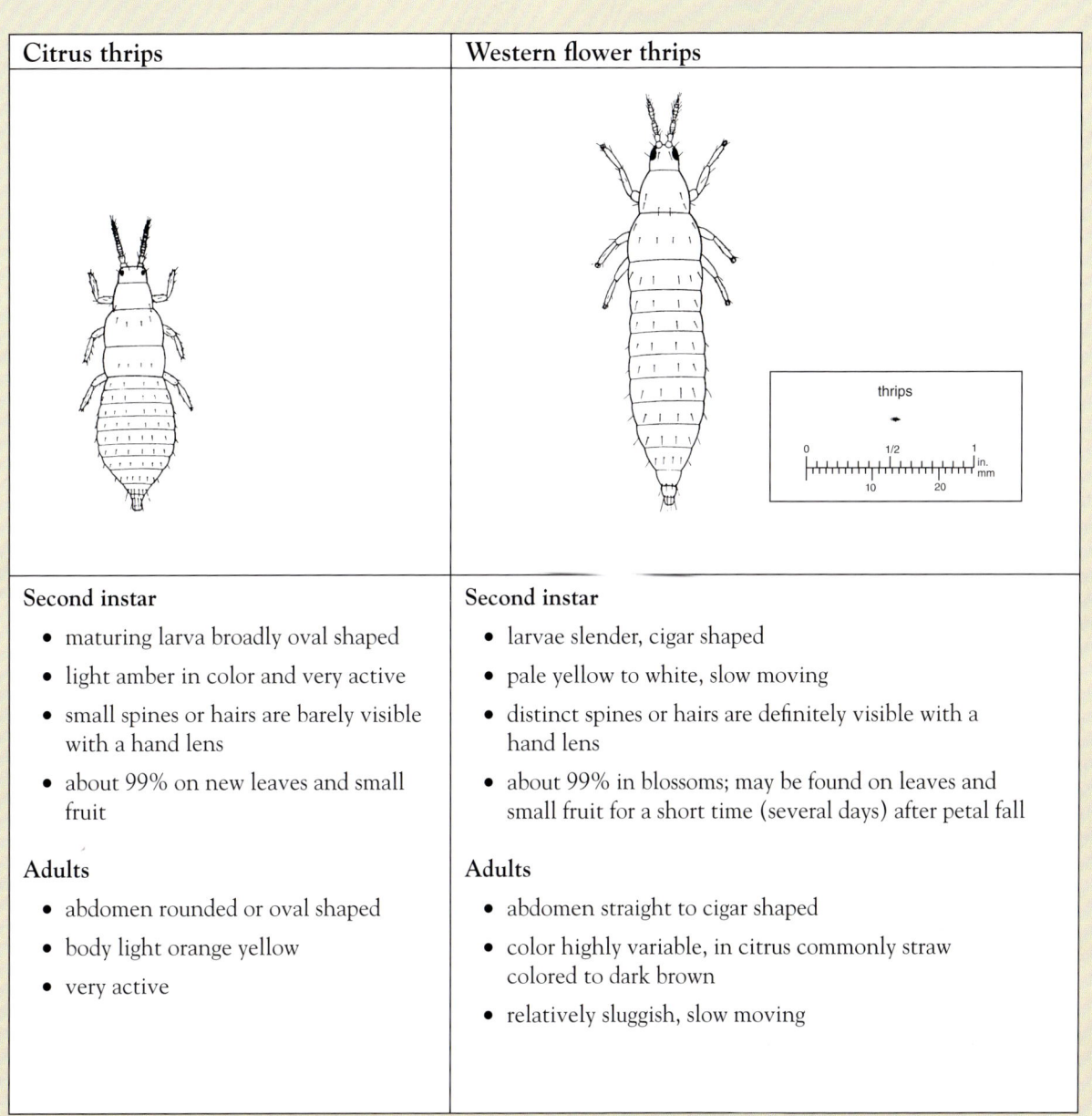

Citrus thrips	Western flower thrips
Second instar • maturing larva broadly oval shaped • light amber in color and very active • small spines or hairs are barely visible with a hand lens • about 99% on new leaves and small fruit	**Second instar** • larvae slender, cigar shaped • pale yellow to white, slow moving • distinct spines or hairs are definitely visible with a hand lens • about 99% in blossoms; may be found on leaves and small fruit for a short time (several days) after petal fall
Adults • abdomen rounded or oval shaped • body light orange yellow • very active	**Adults** • abdomen straight to cigar shaped • color highly variable, in citrus commonly straw colored to dark brown • relatively sluggish, slow moving

Figure 32. How to distinguish citrus thrips from flower thrips, illustrated for second instars. It is important to distinguish these species, as citrus thrips is an important pest that scars fruit, while flower thrips does not damage citrus and should not be treated. *Source:* Larvae by Ebeling in Quayle 1938.

Figure 33. Citrus thrips and western flower thrips compared.

Natural Enemies

Predatory mites in the genus *Euseius* and the whirligig mite (*Anystis agilis*, family Anystidae) commonly feed on thrips. Brown and green lacewings, dustywings, minute pirate bugs, and spiders discussed earlier under "General Predators" feed in the tree on thrips and other pests. Various natural enemies also feed on thrips pupating on the ground in litter, but their impact is unknown. Banded thrips (*Aeolothrips* spp.), black hunter thrips (*Leptothrips mali*), and *Franklinothrips* species, all pictured earlier in this section, are beneficial predators on plant-feeding thrips and other pests.

Young leaves damaged by thrips feeding are often thickened, deformed, and show grayish streaks on both sides of the midrib. However, this injury is not economically important, so preventing damage to leaves is not a goal of citrus thrips management.

On young fruit, citrus thrips usually feed under or near the sepals (calyx). If citrus thrips were numerous, the feeding results in extensive scarring, which does not become apparent until the fruit grows.

Scabby, gray, or silvery scars from citrus thrips feeding become apparent as damaged rind tissue moves outward from beneath the sepals (calyx). This conspicuous full or partial ring of scarred tissue of variable width around the stem end is characteristic of citrus thrips scarring. Although amorbia causes similar injury, caterpillar chewing occurs deeper into the rind.

both sides of the midrib, and the leaves often become distorted as they expand. However, extensive leaf damage has shown no effect on tree growth, even on young trees. Thus, treatments of nonbearing trees for citrus thrips are not recommended.

Second-instar citrus thrips feeding under the sepals of young fruit cause most of the economic damage. Developing fruit are susceptible to citrus thrips injury from petal fall until fruit are about 1½ inches (3.7 cm) in diameter. This injury does not become fully apparent until fruit grow. Damaged rind tissue moves outward from beneath the sepals (calyx) as the fruit grows, producing a conspicuous irregular ring or partial ring of scarred tissue of variable width around the stem end. Citrus thrips feeding does not harm internal fruit quality or reduce yields. However, this cosmetic damage can cause fruit to be downgraded in the packhouse.

The second and third generations of thrips cause economic damage because their development coincides with the presence of young fruit susceptible to rind scarring (Figure 31). Damage occurs mostly to fruit on the outside of the canopy, where injury from equipment and wind also occur. Rind scarring from citrus thrips can resemble chewing damage from young amorbia, as pictured later in the "Caterpillars" section.

There are many other causes of damage to fruit rinds, as listed in Table 3 in the chapter "Managing Pests in Citrus." Correctly diagnose the cause(s) of injury so you can identify the effective management actions. Also consult *A Photographic Guide to Citrus Fruit Scarring* (Grafton-Cardwell et al. 2003) for more information.

***Euseius* species.** *Euseius* species are predaceous mites that help control citrus red mite and also feeds on citrus thrips, especially in orchards where broad-spectrum pesticide use is minimized. *Euseius tularensis* is common in the San Joaquin Valley. *Euseius stipulatus* and *E. hibisci* are common in southern and coastal California.

Euseius hatch from an egg and develop through a six-legged larval stage and 2 eight-legged immature nymphal stages. Adults are pear shaped and shiny. Nymphs look like small adults. Except when feeding or molting, predator mites move quickly and avoid direct sunlight. When held on a leaf in the sun, the tiny, shiny body can be observed running rapidly along the main vein or across the leaf. See the "Spider Mites" section for more description and photographs of *Euseius*.

Euseius species are generalist feeders and include citrus red mites, citrus thrips, pollen, and leaf sap in their diet. Because they are not specific to citrus thrips, they assist with control but do not always reduce thrips populations below an economic threshold. Densities of at least 0.5 *Euseius tularensis* per leaf are needed to provide aid in citrus thrips control in the San Joaquin Valley. *Euseius* are also good indicator species of the abundance of other natural enemies. When you can see that *Euseius* are common, the orchard will also have more of the other natural enemies. Unproven efforts to increase the number of *Euseius* in citrus orchards have included growing cover crops, releasing pollen, and augmenting their numbers (releasing *Euseius*). However, reducing broad-spectrum pesticide use and pruning the trees in fall to stimulate good flush in spring had a greater effect on increasing predatory mites. Be aware that fall pruning can increase the risk of freeze damage because growth stimulated in the fall may still be immature when cold temperatures arrive.

Monitoring

For cultivars and locations where citrus thrips are a pest, check young fruit for thrips larvae at least twice weekly from petal fall until fruit are greater than 1½ inches (3.7 cm) in diameter. For navel oranges and mandarins, monitor for about 6 to 8 weeks in spring. For lemons, monitor June through October. Monitor during midday (10 a.m. to 2 p.m.), when thrips are most active. Examine the surface of young fruit for citrus thrips larvae, pick fruit, and, through a magnifier or hand lens, look closely under the sepals (calyx) for citrus thrips larvae. Also inspect for predaceous mites on the underside of inner canopy foliage.

On young fruit, thrips occur mostly under the sepals when weather is overcast or cool, such as in early morning. Larvae occur more openly on the fruit surface when it is sunny, as it usually is at midday in California citrus orchards. Because inspecting under the sepals is time consuming and destroys potentially harvestable (marketable) fruit, monitoring is recommended during midday, as some growers and PCAs look for thrips only on the exposed surfaces of young fruit on the tree.

Monitoring Thrips. Select trees that are three or four rows in from the outside edge of the block. Sample 25 young fruit from each corner of the block for a total of 100 fruit. Take only one or two healthy, dark green fruit from outside, sunny branches of each tree. Look for thrips on the exposed fruit surface and stem end of the fruit under the sepals (calyx). Count fruit as infested only if it has one or more wingless first- or second-instar larvae (ignore adults and any other stages). Record the total number of fruit infested with imma-

Anystis agilis is a relatively large, long-legged, reddish mite, as with this adult feeding on a thrips larva. *Anystis* feeds on various small insects, including aphids, psocids, and mites, but in citrus it prefers to feed on citrus thrips. It is called whirligig mite because of its fast circular motions.

Euseius tularensis (at left) is white or yellow when it feeds on citrus thrips, leaf sap, or pollen. *Euseius* is mostly reddish when feeding on citrus red mite.

To monitor citrus thrips, examine the exposed surface of young fruit during midday (about 10 a.m. to 2 p.m.). Then pinch or clip off the stem of young fruit and twist off the calyx. Look for thrips larvae on the underside of the calyx (sepals) and on the fruit where it was covered by the calyx and button (the doughnut-shaped structure where the stem connects to the fruit).

ture citrus thrips and calculate the percentage of infested fruit. On very susceptible varieties, such as San Joaquin Valley navels, monitor fruit at least twice a week after petal fall, and continue monitoring as long as susceptible fruit is on the tree.

Monitoring Predatory Mites. Examine the underside of twenty 5-leaf terminals with fully expanded leaves from shady areas of the canopy (a total of 100 leaves) and count the number of adult predatory mites. Calculate and record the average number of predatory mites per leaf.

Monitoring Results. A combination of factors, such as the percentage of infested fruit, the number of predatory mites per leaf, and the cultivar, should be considered when making treatment decisions. For example, Valencias can tolerate twice the percentage of citrus thrips infestation than navels or mandarins can because the Valencia orange has a tougher, less-susceptible rind. As citrus fruit size increases, its susceptibility decreases. If predatory mite abundance is greater than 0.5 per leaf, the treatment threshold also doubles.

If monitoring indicates that treatment may be needed, proper application timing and method are critical to provide effective control. Citrus thrips infestations of fruit can increase very quickly, in a matter of a few days, and damage must be controlled quickly. A limiting factor for many growers is the availability of spray application equipment. Because only the outside fruit of a citrus tree is affected by citrus thrips, a weed sprayer can be modified to apply the insecticide, providing more control over the timing of treatment. Good timing will reduce the likely need for a second treatment, thus reducing the development of insecticide resistance. Consult the online *UC IPM Pest Management Guidelines: Citrus* for current sampling and treatment thresholds for your situation.

Weather Monitoring and Damage Prediction. Weather during March and April appears to greatly influence the extent of thrips fruit scarring. At least in the southern San Joaquin Valley, cool temperatures in early March and warm temperatures during bloom are associated with higher levels of thrips scarring. Relatively warm temperatures during early March and relatively cool temperatures during bloom are associated with less scarring. The reason for the difference is that the rate of development of the citrus thrips and the citrus tree are both influenced by temperature, but in different ways. In some years, citrus thrips emergence coincides with young fruit development, and in other years it does not. In addition, wet weather in the spring seems to reduce successful emergence of the adults. Thus, citrus thrips tend to be a more severe problem in the absence of spring rains during bloom.

To help you plan whether citrus thrips management is warranted in a particular season, consider using the citrus thrips damage estimator model available online at www.ipm.ucdavis.edu. For example, if spring weather leads the model to predict more-extensive thrips scarring, increase the frequency of fruit inspections and be prepared to take immediate action if fruit sampling reveals that thresholds are exceeded.

Neohydatothrips

Neohydatothrips burungae

Neohydatothrips burungae was discovered in San Diego County in 2004 and has been found on avocado and lemon in coastal California. This species has previously been reported throughout Central America and in Mexico, where it is relatively common on avocado and mango. *Neohydatothrips burungae* closely resemble avocado thrips (*Scirtothrips perseae*) and citrus thrips. Avocado thrips are not a pest in citrus, but the adults are sometimes found in traps in citrus growing near avocados. The importance of *N. burungae* in California is unknown.

Citrus thrips do not have dark bands or lines on their abdomen and do not have red eye spots on their head. Both avocado thrips and *Neohydatothrips* have brown bands on their abdomen and 3 red eye spots arranged in a triangle on top of their head. In comparison with avocado thrips, *N. burungae* have darker brown shading on the thorax and darker abdominal stripes (brownish rings around the top front of each abdominal segment) that occur only on top of their abdomen, not underneath. Citrus thrips are usually pale overall (yellowish), *Neohydatothrips* are distinctly darker (brownish) with obvious dark banding, and avocado thrips

	Citrus thrips	Neohydatothrips	Avocado thrips
Distinguishing characters			
Lines (bands) separating abdominal segments	No dark lines or bands.	Dark bands ("tiger stripes") only on the upper side of its abdomen.	Bands on both the upper side and underside of the abdomen, but lighter colored than in *Neohydatothrips*.
Three red eye spots on top of head between the true eyes	No	Yes	Yes (beneath arrow)
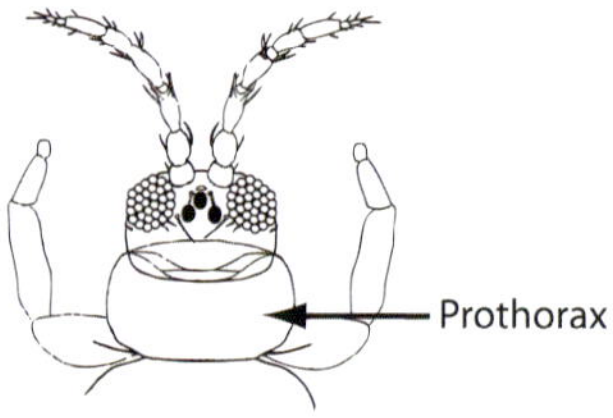 Setae (hairs) on rear corners of prothorax (left corner circled in red)	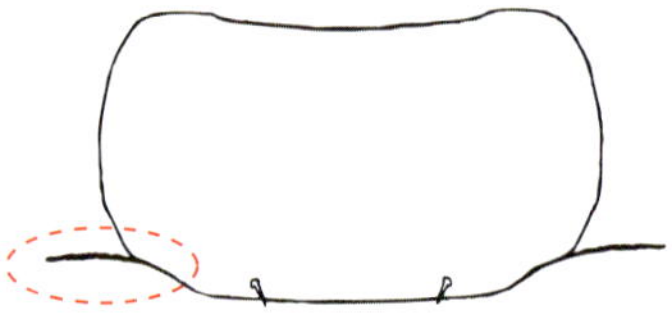 1 long setae on each rear corner (both citrus thrips and *Neohydatothrips*)		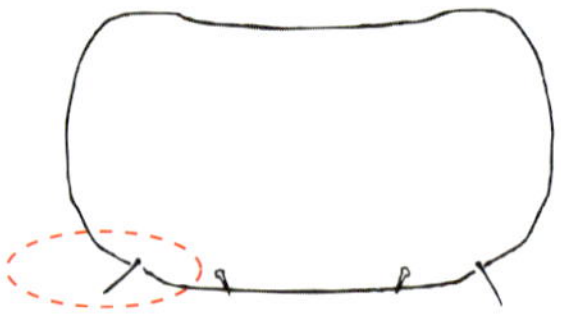 Short setae at corner
Setae on forewing (front wing) All three spp. have relatively long hairs projecting from the front margin (the main vein) of the forewing.	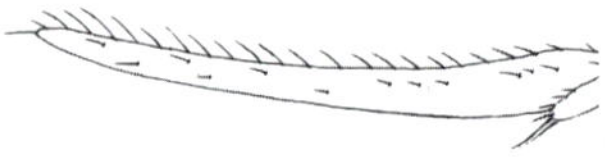 Short hairs behind forewing main vein are not in a continuous row; there are gaps between hairs.	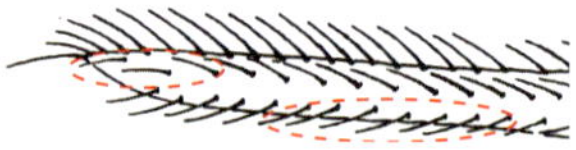 Continuous rows of hairs along both the middle and back edge (rear vein, circled).	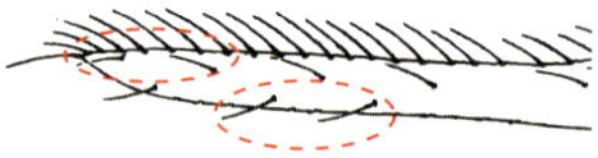 Relatively few hairs in both the middle and back edge (circled) of wing.

Figure 34. How to distinguish adults of citrus thrips, *Neohydatothrips burungae*, and avocado thrips. *Neohydatothrips* occur at least on lemons and avocados in southern California but are of unknown importance. Avocado thrips (*Scirtothrips perseae*) are not a pest of citrus, but avocado thrips adults can be caught in traps in citrus growing near avocado. These species can generally be distinguished based on dark abdominal stripes (bands) and whether they have three red head spots between their eyes (citrus thrips lack both). However, coloration can vary, and thrips species are more reliably separated by carefully preparing specimens and using a binocular dissecting microscope to examine tiny characters such as the arrangement and length of hairs on their thorax and wings, as illustrated here. Illustrations adapted from Bailey 1941, Gill 1997b, Kono and Papp 1977, Watson 2005.

are intermediate in color. However, coloration is variable and may not be a reliable way to distinguish these species.

The reliable way to distinguish these thrips is according to differences in the position and size of setae (stout hairs) on the thorax and wings (Figure 34). For example, *Neohydatothrips burungae* have a continuous or complete row of short, stout hairs on both midveins within their forewings. Avocado and citrus thrips have sizable gaps in both these rows of hairs due to the relatively few hairs along these midveins on their front wings. Careful preparation of several specimens and a good microscope are necessary if you want to recognize these characters.

Greenhouse Thrips

Heliothrips haemorrhoidalis

Greenhouse thrips are occasional pests on Valencia oranges and lemons growing near the coast. They rarely damage citrus in interior areas. They are more likely to cause damage when prolonged mild weather occurs during winter through summer. Greenhouse thrips are primarily a pest of coastal avocados and various ornamentals, especially broadleaf evergreens.

Description and Biology

Adult greenhouse thrips are tiny black insects with white legs. Adults have white to translucent wings that at rest fold back over their thorax and abdomen. Adult bean thrips also have a dark body, but their antennae, legs, and wings have dark and pale bands (Figure 35). Adult female greenhouse thrips reproduce without mating and insert their eggs just under the surface of leaves and fruit. Where eggs occur, blisters develop in leaf tissue just before eggs hatch. These tiny egg blisters can be seen with a hand lens.

Larvae, propupae, and pupae are white to pale yellowish. As they feed, larvae secrete blackish globules of fecal fluid from the tip of their abdomen. Propupae and pupae do not feed but remain on the plant among the feeding adults and larvae. All stages of greenhouse thrips are sluggish, and the adults rarely fly. There are approximately five to six generations annually in coastal California.

Greenhouse thrips prefer moderate temperatures and humidity. Cold winters or hot, dry "Santa Ana" winds cause high mortality of adults and immatures. Temperatures below freezing or above 100°F (38°C) cause significant mortality, particularly if they occur over several days.

Damage

Greenhouse thrips suck chlorophyll and other contents out of epidermal cells of leaves and fruit, causing cells to become pale. The injury is most likely to occur where thrips congregate, primarily where fruit touch other fruit or leaves. Heavier than normal crops contain more touching fruit, so heavy crops are more likely to become damaged by greenhouse thrips.

In addition to bleaching, affected areas on fruit and leaves appear to be dirty or dark spotted because of thrips excrement. Greenhouse thrips feed in groups, gradually expanding their discolored feeding area out from the initial infestation point. Unlike with citrus thrips, no actual scars or leaf deformities develop because of greenhouse thrips feeding, but fruit discolored by greenhouse thrips may be downgraded.

Natural Enemies

Thrips predators include lacewings, pirate bugs (*Orius* spp.), and mites (*Euseius* spp.). At least three species of predatory thrips (pictured earlier) attack greenhouse thrips: *Franklinothrips orizabensis*, *Franklinothrips vespiformis*, and black hunter thrips (*Leptothrips mali*). In coastal growing areas, greenhouse thrips are parasitized by two species of tiny wasps: *Thripobius semiluteus* and *Megaphragma mymaripenne*.

Thripobius. *Thripobius semiluteus* (family Eulophidae) is the most effective natural enemy of greenhouse thrips. This tiny wasp lays its eggs in young (mostly second-instar) thrips larvae. Parasitized larvae swell, especially around their head, and the sides of their bodies are more parallel than the tapered, unparasitized thrips larvae. About 2 weeks before the wasps' emergence, parasitized larvae turn black and are immobile, in contrast to the pale yellow to whitish, slow-moving unparasitized larvae. Unlike healthy black adult thrips, the black parasitized thrips are smaller, do not move, and lack wings. Egg-to-adult development time for *Thripobius* is about 3 weeks when temperatures average 70°F (21°C).

Greenhouse thrips populations periodically become scarce in coastal citrus, making it difficult for *Thripobius* to maintain their populations in citrus. However, *Thripobius* is

Greenhouse thrips feeding causes pale or whitish to brown discoloration of leaves and fruit. Damage initially occurs in a patch (as on this avocado leaf), then spreads outward. Discolored patches contain specks of black thrips excrement.

Greenhouse thrips larvae occur in groups and are white to pale yellow and slow moving. Dark gut contents may be visible through their integument. At the tip of their abdomen, they often carry a droplet of dark excrement. The greenhouse thrips adult (center) has a black body with white legs and white to translucent wings.

The *Thripobius semiluteus* adult is a tiny yellow and black parasitic wasp. *Thripobius* prefers to lay eggs in second-instar greenhouse thrips. It may not be apparent that thrips are parasitized until thrips darken about 2 weeks after *Thripobius* has laid its eggs.

The normally yellow to whitish greenhouse thrips larvae (right) turn black and swell around the head when parasitized. Three parasitized thrips have become *Thripobius semiluteus* pupae (photo left). Unlike the immobile parasite pupae, the adult greenhouse thrips (bottom) moves.

well established in coastal avocado. Parasites either move into the orchard on their own or parasitized larvae can be collected in nearby avocado orchards and relocated to shady spots in citrus near greenhouse thrips.

Megaphragma. The adult *Megaphragma mymaripenne* (Trichogrammatidae) is a yellowish wasp about $1/50$ inch (0.5 mm) long. Females oviposit inside greenhouse thrips eggs, and one wasp larva feeds inside each thrips egg, hidden from view embedded in plant tissue. Eggs killed by *Megaphragma* develop a round hole where the *Megaphragma* adult emerged, usually in the raised middle of the egg blister. In contrast, when a greenhouse thrips emerges, part of the egg shell is often visible at the side of the egg blister.

Monitoring

Mark locations of previous years' greenhouse thrips infestations. Check these areas in late March or April through mid-May to determine whether thrips may become a problem in the current year. Greenhouse thrips infestations tend to reoccur in orchard areas with a moderate microclimate. Carefully inspect fruit where it touches other fruit or foliage, looking for thrips and their feeding injury. If greenhouse thrips appear to be a problem, consult the online *UC IPM Pest Management Guidelines: Citrus*.

Bean Thrips

Caliothrips fasciatus

Bean thrips are a problem when they migrate into orchards and contaminate the navels of oranges shipped overseas. Bean thrips do not directly damage fruit or reproduce on citrus, but, because of quarantines, they may need to be managed to allow navel oranges to be exported to countries that consider bean thrips an exotic pest threat.

Adult bean thrips (pictured earlier under "Identifying Thrips Species") have a uniformly dark, grayish black body and are about $1/25$ inch (1 mm) long. Their forewings have two dark and two pale bands (Figure 35), and (visible under magnification) the legs and antennae are also banded light and dark. When fruit are cut in cross section to inspect whether their navels are contaminated and when examining thrips in traps, adult bean thrips often appear blackish without banding or distinguishing characteristics. Bean thrips' pale larvae do not occur on citrus.

Adult bean thrips migrate into orchards in fall when their host weeds die or dry up or when field crops they infest are harvested. Bean thrips enter the navel of oranges, where they overwinter and contaminate harvested fruit, making it unsuitable for export. Although various natural enemies feed on bean thrips, biological control is of little importance where bean thrips are a quarantined pest.

Monitoring

Growers planning to ship navel oranges to certain markets may be required to comply with an export certification plan. Be sure to read and understand all of the conditions of the current protocol, which is periodically revised. The methods summarized below are not the current export protocol.

Monitoring methods include cutting through fruit navels and looking for thrips and trapping orchard borders for migrating adults using green sticky cards. Where bean thrips are a problem, also inspect around orchard borders for the weeds on which bean thrips feed so these alternate hosts can be controlled before thrips mature and migrate to fruit. For

Species	General appearance visible to the naked eye or through a hand lens	Distinguishing characters visible with a dissecting binocular microscope	
		Head and thorax	Forewing
Citrus thrips, *Scirtothrips citri*	Body uniformly yellow or light orangish. Wings and legs yellow to whitish.	Hairs on head all relatively short. Long hairs on the rear margin of the pronotum but not on the front margin. Body relatively smooth.	Long hairs projecting from margin of forewing. Short hairs behind forewing main vein are not in a continuous row.
Western flower thrips, *Frankliniella occidentalis*	Body variable: entirely yellow, uniformly dark, or bicolored dark and yellow. Wings pale to translucent.	Two pairs of relatively long hairs on head, plus short hairs. Long hairs on the front and rear margins of the pronotum. Body relatively smooth.	Long hairs projecting from margin of forewing. Short hairs behind forewing main vein are in a continuous row.
Greenhouse thrips, *Heliothrips haemorrhoidalis*	Body uniformly blackish. Forewings and legs whitish to translucent, without banding.	No obvious hairs on pronotum. Body with heavy reticulate sculpture, but without numerous distinct marks within outline of each polygon.	No long hairs projecting from margin of forewing. Wing clear to pale colored.
Bean thrips, *Caliothrips fasciatus*	Body uniformly dark, grayish black. Antennae, forewings, and legs banded light and dark.	Hairs on pronotum, medium to short in length. Body with reticulate sculpture, with numerous markings within each polygonal shape.	Long hairs projecting from forewing margins. Forewing clear with two dark and two pale bands.

Figure 35. Distinguishing characters for adults of four species of thrips found in California citrus. Under good magnification (using a binocular dissecting microscope), carefully prepared specimens of thrips can be distinguished by the arrangement and length of hairs on their head, thorax, and wings and by their smooth versus sculptured body surface. Note that pale to translucent wings may appear dark if the wings are folded over a dark-bodied insect. Adapted from Gill 1997b; Mound and Kibby 1998; O'Donnell, Mound, and Parrella, undated; Kono and Papp 1977.

A bean thrips adult appears to be mostly black when caught in a sticky trap. Dark and pale bands on the forewings, legs, or antennae may also be visible, depending on its orientation in the trap.

When monitoring fruit on trees, it is easier to spot the blackish bean thrips if thin slices of orange are placed on a white or blue background. Using magnification, look for thrips around the navel of each slice.

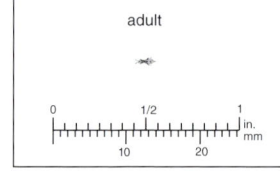

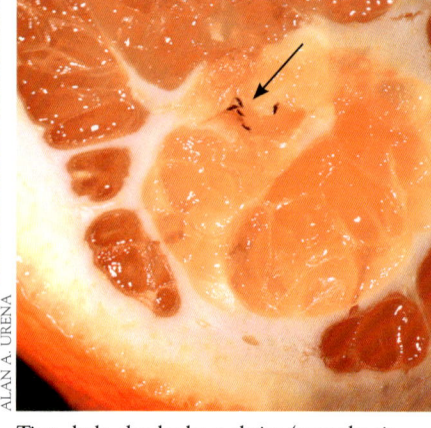

Tiny, dark, slender bean thrips (near the tip of the arrow) are revealed here by cutting into a navel. To the naked eye, bean thrips usually appear blackish without banding or distinguishing marks. If cutting fruit to look for thrips, make multiple thin slices through the entire navel and examine navels through a magnifier to avoid missing these tiny insects.

a list and photographs of weed hosts of bean thrips, consult the online UC IPM *Pest Management Guidelines: Citrus*.

Cutting Fruit. When monitoring fruit collected from trees in the field, collect oranges from the lower canopy of outer rows. To detect bean thrips, cut the fruit into thin slices: start at the navel end and continue to cut until the fruit has been sliced up to the bottom of the navel. It is easier to spot bean thrips if the orange slices are placed on a white or blue background. Examine the slices using magnification, such as with a hand lens or a hands-free magnifier. Bean thrips are about 1/25 inch long (1 mm), and to the naked eye they appear blackish without banding.

Be sure to differentiate the blackish bean thrips from small, dark pieces of the fruit stamen. Citrus and western flower thrips may also occur in the navel, but they are usually yellowish except for the dark phase of western flower thrips. In comparison with bean thrips, western flower thrips are larger and have more hairs, especially longer and darker hairs around their posterior (rear) end. In some countries, live western flower thrips are also of quarantine concern, but adults inside the navel do not survive the winter for as long or as well as adult bean thrips.

Trapping Adults. Green sticky traps are effective in detecting migrating adults. Hang the 3-by-4.5 inch (7.5-by-11.2 cm) traps in trees on one of the two outside rows of the block on the side of the tree facing the outside of the block. If the outside row borders a dusty road, use the second row from the outside of the block. Hang traps about 4 to 5 feet (1.2–1.5 m) above ground to minimize contaminating splash from irrigation and rainwater. Use 1 trap for every 5 acres, with a minimum of 4 traps per block, one each on the north, south, east, and west sides of the block.

Adult bean thrips appear mostly black when they are caught in traps. Some pale banding may be visible on appendages, depending on their orientation in the trap. For photographs comparing bean thrips with other insects caught in sticky traps, see the online *Citrus Year-Round IPM Program*.

CATERPILLARS (ORANGEWORMS)

Caterpillars (sometimes called orangeworms) are the larvae of moths and butterflies (order Lepidoptera). Most species feed externally on plants, chewing blossoms, fruit, and leaves. Citrus leafminer and citrus peelminer feed just under the surface of plant tissue and are discussed separately later in the section "Miners." Trees can tolerate moderate caterpillar chewing of foliage and blossoms. Caterpillars can cause serious economic damage by feeding on fruit.

Identifying which caterpillars are present is important because damage potential, monitoring, and thresholds vary with the species. The major pest caterpillars that chew citrus fruit can be difficult to distinguish to species, particularly when they are very young. Individuals of the same species often vary in color, so color is not always reliable for identification. Close observation with a hand lens or microscope is necessary for positive identification of some species.

The key in Figure 36 is designed to assist you in distinguishing the common caterpillars in California citrus. Characters important in identifying the species include the number and location of prolegs (leglike appendages underneath the abdomen), the color of tubercles (knoblike or rounded projections, sometimes with projecting hairs), and the presence and color of any prothoracic shield (a hardened plate on top of the first segment behind the head).

In addition to differences in how they look, the behavior, biology, and geographical distribution as described in the individual pest sections can help you to distinguish the species. For example, larvae of the beet armyworm, cutworms, and loopers (the families Noctuidae and Geometridae) do not produce substantial webbing. Amorbia, fruittree leafroller, omnivorous leafroller, and orange tortrix larvae (Tortricidae) produce webbing as they tie foliage and fruit together with silken threads. Tortricid larvae back up rapidly if they are touched on the head, while cutworm larvae wriggle back and forth around their center. First-instar cutworms and all tortricid larvae may drop on a silken thread when disturbed, while other species do not respond this way. The citrus cutworm and the fruittree leafroller have one generation per year, and larvae are present only in the spring. Amorbia, omnivorous leafroller, and orange tortrix have several generations, and their larvae can be found from late spring into late fall.

Management

Key management tactics include conserving natural enemies, monitoring carefully, and (if thresholds are exceeded) applying insecticides (preferably selective materials) that allow natural enemies to survive and assist with control.

Parasitic wasps and tachinid flies attack eggs or larvae and are especially important, as discussed in the individual pest sections. Assassin bugs, lacewings, minute pirate bugs, and spiders discussed earlier in the section "General Predators" also prey on caterpillars. The effect of natural enemies on caterpillar populations is not well documented, but biological control is believed to be very important because when selective insecticides are used to control citrus pests and preserve natural enemies, caterpillars are much less of a problem.

Monitoring. Monitoring methods include searching for eggs, timed counts or sweep net shake samples for larvae, and using pheromone-baited traps for adults. Male moths of several citrus pest species can be monitored using traps baited with a synthetic female pheromone (attractant) specific to each species. When males are caught, females are also present and laying eggs, allowing control tactics to be timed to target susceptible stages. The moth flight biofix combined with degree-day monitoring (temperature accumulation) can be used to predict when it is the right time to sample and treat larval stages.

The specific method and time for monitoring vary among species. For example, with cutworms, monitor for larvae during late bloom to post–petal fall, when cutworms are attracted to feed on the small developing fruit. For species such as orange tortrix, monitoring needs to continue later in the season (through July) because later generations of orange tortrix feed on larger fruit. For current, species-specific monitoring and management recommendations, consult the online *UC IPM Pest Management Guidelines: Citrus*.

Caterpillars occasionally chew fruit, but so do certain other pests such as earwigs, katydids, and snails. Identifying the cause of injury is important because damage potential, monitoring, and thresholds vary with the species. Fruittree leafroller, a minor pest in citrus, chewed these holes, which were revealed by pulling back a leaf that was tied to the fruit with webbing.

Trichogramma spp. are parasites of many Lepidoptera species. The adult *Trichogramma* wasp, 1/25 inch (1 mm) long, lays one or several of its eggs into each caterpillar egg. When caterpillar eggs are abnormally dark, they likely have parasites maturing inside.

This key will help you identify the larvae of butterflies and moths that feed openly on citrus blossoms, fruit, or leaves in California. Leafminer, peelminer, and Lepidoptera species that are found only occasionally are not included. Ask your farm advisor or agricultural commissioner for help in identifying specimens that do not seem to fit the key.

Read all (2 or 3) descriptions and compare the specimen with the drawings provided before proceeding. After you arrive at a name, compare the specimen with the appropriate photos and descriptions. Take several specimens through the key individually; your monitoring sample may include more than one species and different stages of one species.

Parts of a caterpillar useful in identifying the species

The features used in the key can be seen with a good hand lens, but you may need a low-power microscope to see certain features on small larvae. The key works best for specimens in the third instar or larger; certain characters, such as the number of prolegs and color markings, can be different in very young larvae.

1a.	Scent glands protrude from top of prothorax when disturbed, glands often retracted in slit when caterpillar is at rest (family Papilionidae). See 2.	
1b.	Scent glands absent. See 3.	

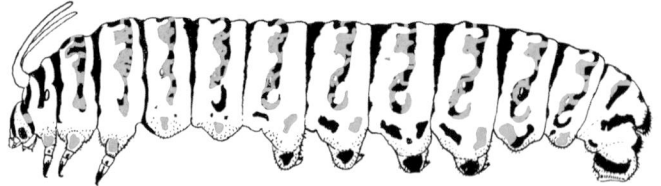

2a.	Scent glands orangish. Body color variable, in late instars commonly greenish with bands of alternating black and yellow or orange spots and bands. Larvae up to 2 inches long. California orangedog	
2b.	Scent glands usually red. Body brown or olive with light patches, resembling bird droppings. Larvae up to about 2¼ inches long Giant orangedog	

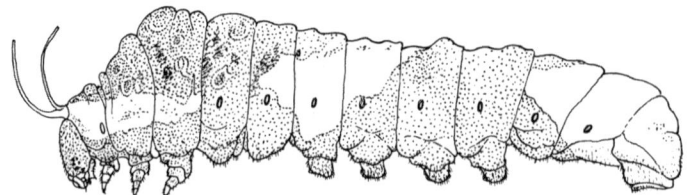

3a.	Tufts of hair on top of prothorax and on abdominal segments 1 to 4 and 8 (Lymantriidae). Western tussock moth	
3b.	Prominent tufts of hairs absent on prothoracic and abdominal segments. See 4.	

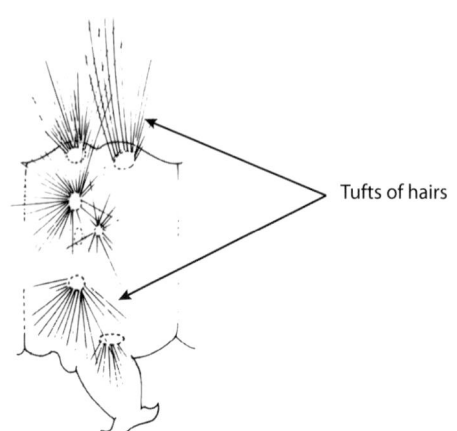

Figure 36. Key to caterpillars.

4a.	One pair of prolegs on abdominal segment 6, no prolegs on middle of abdomen (Geometridae).	
	Citrus looper	
4b.	Two pairs of prolegs near middle of abdomen, on abdominal segments 5 and 6 (Noctuidae subfamily Plusiinae).	
	Cabbage looper	
4c.	Four pairs of prolegs in middle of abdomen, on segments 3, 4, 5, and 6.	
	See 5.	
5a.	Hooks (crochets) arranged like a fan along inner edge of prolegs. Larvae up to 2 inches (5 cm) long. Larvae do not tie leaves together and do not produce obvious silk, except when pupating (Noctuidae).	
	See 6.	
5b.	Crochets arranged in circle or semicircle. Larvae not more than 3/4 inch (2 cm) long. Larvae produce silk and roll leaves.	
	See 8.	
6a.	Center of spiracles brown or black. White to yellowish dots on top of abdominal segments.	
	Variegated cutworm	
6b.	Center of spiracle white, bordered with black.	
	See 7.	

7a.	Black spot usually present on thoracic segment 2. Body usually dull green with dotted or thin, pale, lengthwise stripe. Rarely in southern California. Beet armyworm	
7b.	No black spot on thoracic segment 2. Broad white line on each side of body. Citrus cutworm	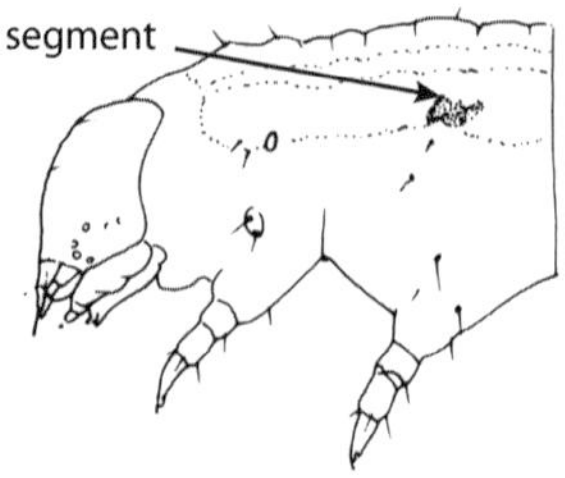
8a.	Anal comb present (Tortricidae) See 10.	
8b.	Anal comb absent. Larvae ²/₅ inch (10 mm) at maturity, much smaller than other orangeworms. Scavenge on injured fruit. Never roll leaves. See 9.	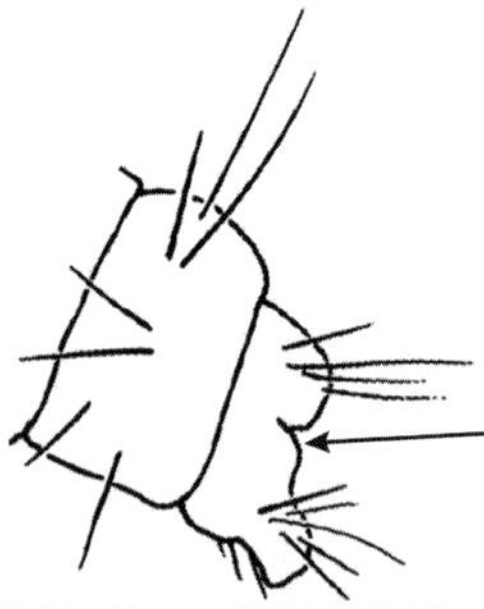
9a.	Pink body color. Pink scavenger caterpillar (Cosmopterigidae)	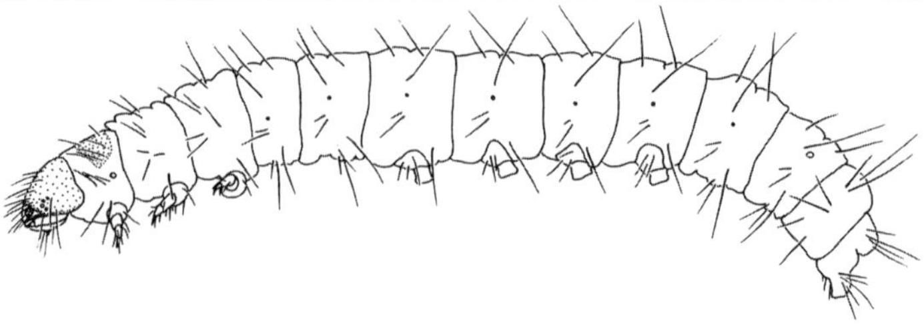
9b.	Black to dark reddish body color, less common than pink scavenger caterpillar. Black scavenger caterpillar (Blastobasidae)	

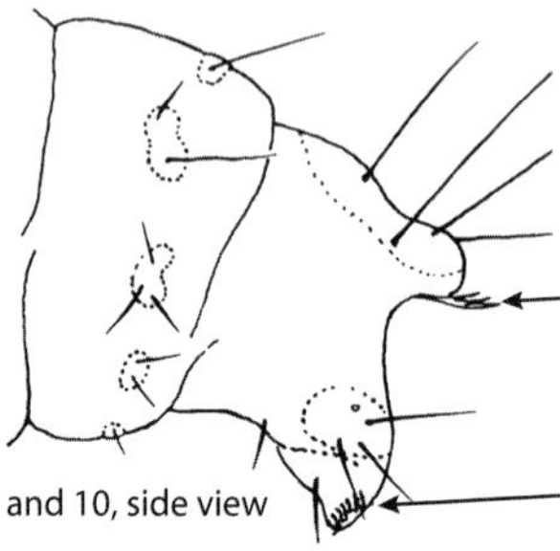

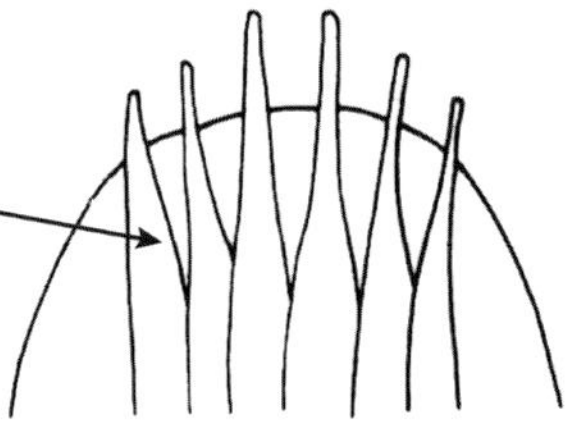

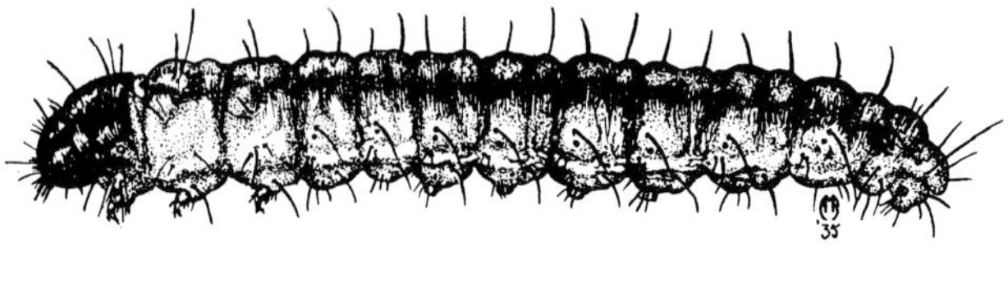

Figure 36. Key to caterpillars (continued).

10a.	Black line on thoracic segment 1 above first pair of true legs.	
	Amorbia	
10b.	Black line absent on thoracic segment 1.	
	See 11.	

11a.	Base of hairs (tubercles) chalky white. Mature larvae with light brown head and prothoracic shield. Early instars with black head and prothoracic shield. Median dark stripe along the top of the body. Occurs throughout California, except rare along coast.	
	Omnivorous leafroller	
11b.	Tubercle color not chalky white, typically colored brown, green, or yellowish.	
	See 12.	

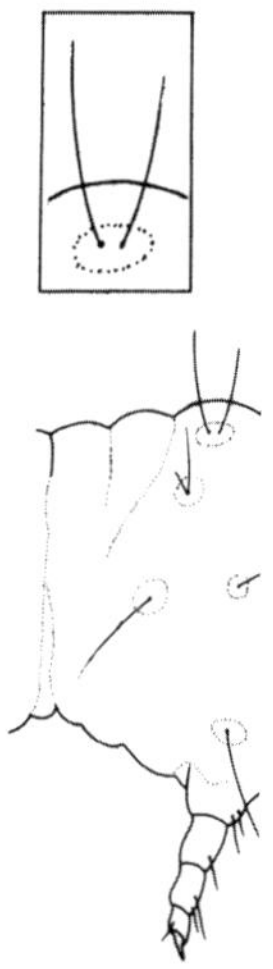

12a.	All instars have a golden brown head and brown prothoracic shield. Body brown to orange.	
	Orange tortrix	
12b.	Brown to shiny black head and a brown to black prothoracic shield, varying with the instar. Body light to dark green.	
	Fruittree leafroller	
12c.	Head is light yellow-brown to dark brown, and the prothoracic shield is light greenish brown to indistinct with no dark markings. Body pale yellow-green to darker green or light greenish brown with a darker green central stripe that may continue to the prothoracic shield.	
	Light brown apple moth	
	Larvae of light brown apple moth cannot be reliably separated in the field from other leafroller (tortricid) species. Submit suspected finds to the county agricultural commissioner.	

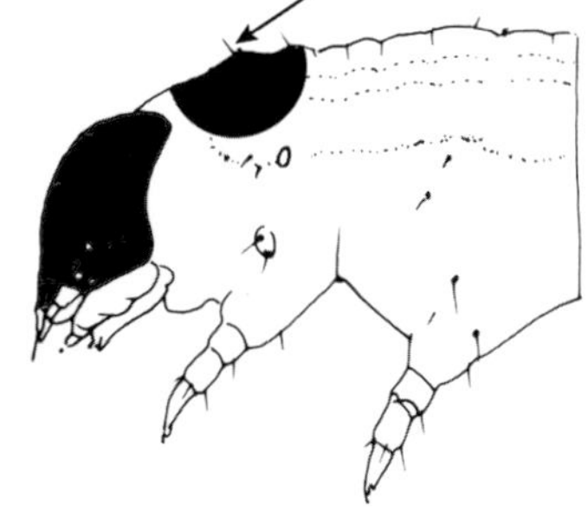

Figure 36. Key to caterpillars (Lepidoptera larvae) in California citrus. Adapted from Ebeling 1959, Peterson 1962, except anal comb and cabbage looper by L. L. Deitz from Bosik 1997, black scavenger caterpillar from Quayle 1938, pink scavenger caterpillar by C. Feller from Gorham 1991.

Cutworms

The citrus cutworm (*Egira* =*Xylomyges curialis*, family Noctuidae) is the most common cutworm pest of citrus. While very common in the 1990s, it has lessened in severity in recent years with reductions in the use of organophosphate and carbamate insecticides. Other cutworm species, mostly variegated cutworm (*Peridroma saucia*), are occasionally found on citrus but rarely cause economic damage. Consult the key (Figure 36) to distinguish citrus cutworm from other caterpillars in citrus.

Description and Biology

Females lay 40 to 225 eggs in a single or double layer, mainly on the upper side of young leaves. The milky white to dark eggs hatch in 5 to 10 days.

Larvae (the "cutworms") usually are found in citrus from mid-March to early May. Cutworm larvae are light green to pinkish or brown. All except the youngest larvae have a whitish stripe along each side. Their skin appears smooth to the naked eye; it lacks conspicuous hairs or tubercles. When disturbed, older larvae curl up and drop to the ground.

Citrus cutworm larvae mature in 3 to 6 weeks, then drop to the ground and form an egg-shaped pupal cell composed of soil particles. Pupal cells occur in debris or in soil, usually 1½ to 2 inches (4–5 cm) deep, mainly under the periphery (dripline) of the tree. The pupae remain dormant until the following spring.

The grayish citrus cutworm moths emerge from early January to the end of April and, soon after emerging, mate and lay eggs. Like most moths, adults rest during the day and fly and lay eggs at night. Citrus cutworm has only one generation per year, and it completes its entire life cycle on citrus.

Larvae of the beet armyworm and certain forms of cabbage looper (both pictured later) resemble the citrus cutworm. However, beet armyworm usually has a black spot on the side of the larva's body above the second pair of true legs. Cabbage looper has only 2 pairs of prolegs near the middle of its body, not the 4 pairs of midbody prolegs as on cutworms and beet armyworm. Beet armyworm and cabbage looper are rarely pests in citrus.

Damage

Citrus cutworm damage can be substantial because larvae eat leaves and blossoms and chew holes in young fruit. A few citrus cutworms can cause more damage than other caterpillar species because they move throughout the tree, taking a few bites from numerous small fruit. After petal fall, the cutworm larvae prefer to attack young fruit, resulting in

For many species of caterpillars, their adult moths can be monitored with pheromone-baited sticky traps. Trapping for adults can identify which species is abundant in the orchard and helps time monitoring and pesticide applications against larvae.

Because it is readily dislodged from foliage, citrus cutworm can be monitored with a sweep net. Insert new growth flushes into the net and shake vigorously. Larvae of certain other caterpillar species are monitored by timed-count searches of foliage.

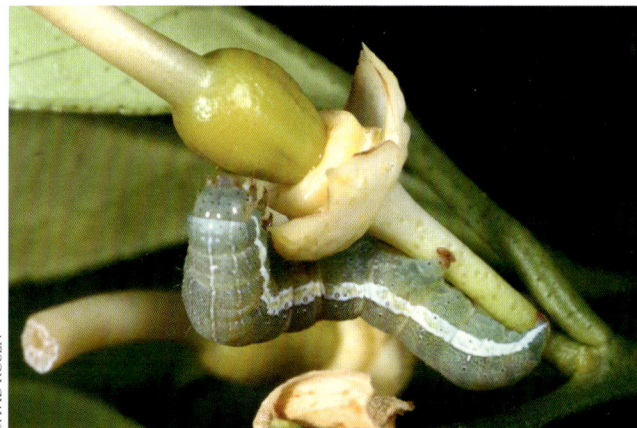
Citrus cutworm larvae feed on young citrus fruit. Just a few citrus cutworms can cause serious damage because cutworms often move around and take a few bites from many fruit.

Citrus cutworms feed on blossoms as well as young fruit. Larvae are usually light green in the first three instars and pinkish or brown in the fourth and fifth instars. All except the youngest larvae have a distinct whitish stripe along each side.

Young citrus cutworms feed mostly on the edges of tender leaves, rasping the surface.

feeding scars. Older fruit are rarely attacked.

Variegated cutworm usually becomes a pest only when cover crops or nearby weeds dry up or are destroyed. This causes large numbers of larvae to move from these alternate hosts where variegated cutworm prefers to feed.

Natural Enemies

Two larval parasites are highly effective in reducing the next year's cutworm population. An ichneumonid wasp (*Ophion* sp.) lays its eggs in nearly mature larvae. The parasite larva develops and consumes its host after the cutworm constructs its pupal chamber. This biological control is usually overlooked because *Ophion* larvae and pupae occur hidden in the cutworm's soil cell.

Another ichneumonid (*Banchus* sp.) lays its eggs in larvae and kills the cutworms before they can mature into pupae. The *Banchus* larva exits dead cutworms and forms a slender, black pupa.

Trichogramma spp. egg parasites and various predators discussed earlier in "General Predators" also kill some cutworms. In some orchards, a fungal pathogen sometimes kills up to about 25% of the cutworm pupae.

The citrus cutworm moth lays its eggs in clusters of 40 to 225, mostly on young leaves. The eggs are milky white at first, darkening later as the larvae develop inside.

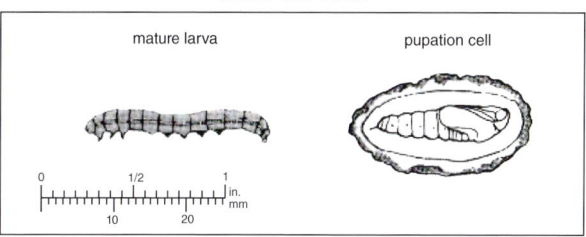

Monitoring

Use pheromone trapping of male moths combined with temperature monitoring to predict when to sample larvae and (if needed) when to treat cutworms. Place one trap per 10 acres in the orchard on January 15 and check it each week for moths. Using a biofix (degree-day accumulation start point) of the second week of consecutive moth flight in January, and a lower development threshold of 46°F, begin sampling for larvae at 250°D. You will see larvae emerging at about 350 to 400°D after the biofix. At about 400 to 500°D after the biofix, the cutworm population consists primarily of first and second instars. These young larvae are the stages most susceptible to insecticides.

To determine whether larvae are abundant enough to warrant control, use methods such as timed-count inspections of shoots (the number of larvae per hour of search) or the shaking of foliage into a sweep net (larvae per 25 net shakes). Monitor orchards weekly from early to mid-April through the month after petal fall. Before petal fall, the threshold for treatment is higher because the larvae are attacking foliage. Slow-acting insecticides can be safely used at this time. After petal fall, the larvae prefer to attack the small fruit, so the threshold for treatment is lower and faster-acting insecticides are needed. For current monitoring and management recommendations, consult the online *UC IPM Pest Management Guidelines: Citrus*.

Orange Tortrix

Argyrotaenia franciscana (=A. citrana)

The orange tortrix (family Tortricidae) is primarily a pest on Valencias and navels in southern California. It occurs on many plants, including grapes, goldenrod, strawberries, and willow.

Description and Biology

The female moth lays 50 to 150 eggs, overlapping them like fish scales on smooth surfaces such as stems, fruit, and the upper surface of leaves. Eggs are oval, flattened, and cream colored, as pictured later for omnivorous looper. The larvae feed inside webbed foliage and, like other tortricids, wriggle and drop suspended on a silken thread when disturbed. Orange tortrix larvae grow up to $3/5$ inch (15 mm) long and have a greenish to straw-colored body and a straw-colored or golden brown head and brown prothoracic shield (plate on top of the first segment behind the head). They most closely resemble omnivorous leafroller larvae. However, as illustrated in the key (Figure 36), the small tubercles (mounds) at the base of the bristles on the side and back of the omnivorous leafroller are chalky white. The tubercles on orange tortrix are brown, greenish, or yellow, not white.

Larvae pupate in dense silken cocoons within webbed foliage. Adults emerge in 8 days to 3 weeks, depending on temperature. Adults are variably colored moths, which may be brown, gray, orangish, or silverish with darker markings.

The citrus cutworm overwinters as a pupa (shown here) inside a cell built from soil particles. If cutting into this pupal cell reveals a maggotlike larva or an oblong, light brown pupa with a darker band around its midsection, the cutworm has been parasitized and killed by an *Ophion* sp. wasp.

The most important cutworm parasite is an *Ophion* sp. wasp. This adult *Ophion* lays its eggs in older cutworm larvae. The immature parasite does not fully develop until the cutworm drops and makes its pupal cell in soil.

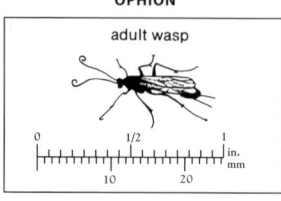

The variable body color of the orange tortrix larva is greenish to bright yellow or pale straw colored. Orange tortrix has a straw-colored to golden brown head and brown prothoracic shield. At the base of the bristles on tortrix larvae, the tubercles (mounds) are brown, greenish, or yellow.

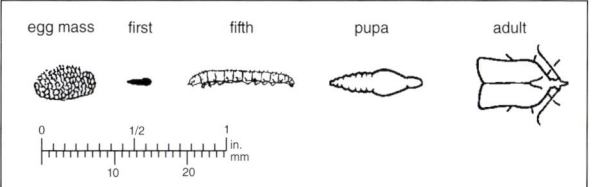

Young orange tortrix larvae often feed under the sepals, around the fruit button, scarring rinds. Later generations of orange tortrix feed when fruit are ripening and may chew holes in the rind.

At rest, adults are about ½ inch (13 mm) long and bell shaped. In coastal areas, orange tortrix can have three or more generations per year. Away from the coast, orange tortrix produces two or three generations.

Damage

First-generation orange tortrix larvae roll and chew young leaves, sometimes webbing groups of leaves together and feeding inside these nests. Second-generation larvae appear when leaves are hardening, and these larvae prefer to chew young fruit under the sepals and around the button. This feeding causes only superficial scars. Later generations of the tortrix feed among clusters of ripening fruit, eating holes into the rind that may become decayed. This damaged fruit usually drops within 1 to 2 weeks.

Natural Enemies

A braconid wasp (*Apanteles aristoteliae*), two ichneumonid wasps (*Exochus nigripalpis* and *Exochus* sp.), and two tachinid flies (*Actia interrupta* and *Nemorilla pyste*) are the most common parasites. Female wasps lay their eggs in tortrix larvae, while tachinid flies lay eggs externally, usually on or near the caterpillar's head. The parasite larvae feed inside and kill their host. *Apanteles aristoteliae* pupates in a white cocoon outside the killed caterpillar. *Exochus* pupates inside the dead caterpillar and leaves a round exit hole when the adult wasp emerges. Tachinids exit their dead host during its late larval or pupal stage, and each tachinid forms a dark reddish, oblong pupal case nearby.

Monitoring

Inspect trees for orange tortrix larvae throughout spring and summer, especially from May through July. At 7- to 10-day intervals, look for orange tortrix larvae and evidence of parasitism mainly on the south and east quadrants of trees. Treatment decisions are based on the number of larvae found per hour of search, with the threshold varying by cultivar and level of parasitism. For current monitoring, threshold, and treatment recommendations, consult the online UC IPM *Pest Management Guidelines: Citrus*.

Fruittree Leafroller

Archips argyrospila

The fruittree leafroller is a minor pest in citrus. Natural enemies help reduce its populations, and treatments are rarely needed.

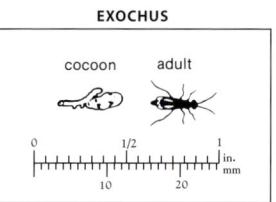

Exochus spp. are common parasites of orange tortrix. The adult wasp leaves a round exit hole when it emerges from a caterpillar it killed while feeding inside as a larva. *Exochus* adults are active wasps, about 5/16 inch (7 mm) long. They have a black head and body and brown legs.

Description and Biology

The fruittree leafroller (family Tortricidae) overwinters as eggs, mainly on twigs in the upper canopy. Eggs occur in flat masses, which are brown to gray before hatching, typically from the middle of March through late April. Old egg masses from which the fruittree leafrollers emerged are gray and brittle, and the eggs have ragged exit holes.

Fruittree leafroller larvae occur during spring. They are greenish and have a shiny brown or black head and a dark prothoracic shield (Figure 36). Larvae feed mostly on new growth flushes and tie or roll leaves or blossoms together with silken threads, often curling leaf terminals as they feed inside. Larger larvae construct a new leaf "nest" frequently, often daily. Most larvae pupate before petal fall, either inside their leaf nests or in thin silk cocoons on branch or trunk bark.

Moths emerge about 8 to 12 days after pupating. Adults are about ½ inch (13 mm) long and reddish brown with white and gold markings. After mating, females lay the overwintering eggs. Fruittree leafroller has one generation per year.

Damage

Fruittree leafrollers occasionally chew a hole in the rind of Valencia or other varieties of late-harvested citrus in spring. Larvae use silk webbing to tie leaves to fruit and bore inside. Damage tends to occur to mature fruit. Fruit chewing occurs when there is little available new leaf flush (which larvae prefer). This could occur because the flush has been consumed or in weak or drought-stressed trees that produced little flush. Chewed fruit is unmarketable and often decays or drops within 1 to 2 weeks.

Natural Enemies

Various general predators prey on small larvae. Certain wasp and tachinid fly species parasitize larvae, and *Trichogramma* spp. wasps parasitize the eggs.

Monitoring

To determine whether fruittree leafroller may be a problem, establish one or two permanent observation trees per site at five locations per block. Before mid-March, look on twigs and small branches in the upper one-third of the canopy for gray to brown, flat egg masses.

Monitor for leafroller larvae when the spring feather-leaf flush appears. Whether using timed counts or a counting frame, be sure to open leaf nests and count only nests that contain a live larva.

With timed counts, search the outer canopy of the south and east side of four trees at each of five sites per block (20 sample trees per block). Spend about 2 to 5 minutes per tree and count all the live leafroller larvae. Record the number of leafrollers per unit time and calculate the average number of larvae per hour of search.

Alternatively, monitor fruittree leafroller larvae with an L-shaped counting frame of two pieces of PVC pipe 20 inches (0.5 m) long connected at a right (90°) angle. As pictured, hold the frame against the outer flush and count the number of infested versus noninfested terminals within the frame. Take one frame sample from the northeast corner of 20 randomly selected trees in a diagonal through the block. Calculate the percentage of infested terminals.

If many new flush terminals are infested with a live larva, watch carefully for leaves being attached to mature fruit. Thresholds depend on how close the larvae are to pupating (shortly before which they stop feeding) and whether larvae

Fruittree leafroller larvae roll and tie leaves together with silk and feed inside. Their bodies are green, contrasting with their shiny black or brown head and prothoracic shield.

Fruittree leafroller egg masses are laid on twigs and small branches, ¼ to 1½ inch (6–37 mm) in diameter, in the upper one-third of the tree. The lower egg mass has exit holes chewed by the emerged larvae.

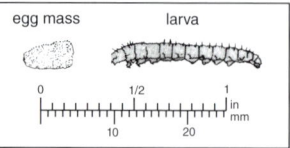

Fruittree leafroller can be sampled by calculating the percentage of shoots infested with live larvae within an L-shaped counting frame. The counting frame consists of two 20-inch (50 cm) pieces of PVC plastic pipe connected at a right (90°) angle. You imagine that the frame includes two additional sides to form a square, and count the leafrollers inside the square area.

The eggs of the omnivorous leafroller are disc shaped and laid in overlapping clusters. Amorbia and orange tortrix lay similar egg masses.

are expected to feed on fruit versus leaves. Harvesting the fruit before the susceptible period eliminates the problem of fruittree leafroller attacking mature fruit. For current monitoring, threshold, and treatment guidelines, consult the online UC IPM *Pest Management Guidelines: Citrus*.

Omnivorous Leafroller

Platynota stultana

The omnivorous leafroller is a rare pest of citrus in the San Joaquin Valley and southern interior of California. It feeds on many other fruit trees, grapes, row crops, ornamentals, and weeds. Natural enemies and insecticide applications for other pests often control omnivorous leafroller.

Description and Biology

Females lay disc-shaped eggs overlapping like fish scales on the upper surface of leaves and on fruit, resembling the egg masses of amorbia and orange tortrix. Omnivorous leafroller larvae also resemble those of other tortricids, especially the orange tortrix. But in comparison with other caterpillars in citrus (Figure 36), omnivorous leafroller has distinctly white tubercles at the base of the bristles. The main blood vessel along the back of larvae is often visible as a faint dark stripe. The larval body color ranges from cream to brown or greenish brown.

Larvae usually develop through five instars, but sometimes there are six molts. First instars are cream colored with a light brown head and prothoracic shield (plate on top of the first segment behind the head). Second and third instars have brownish black or black heads and shields.

On the omnivorous leafroller, the tubercles (mounds from which bristles arise along its back and sides) are characteristically chalky white. The main blood vessel along the back is often visible as a faint dark stripe. The color of omnivorous leafrollers' head and prothoracic shield varies from black to light or dark brown, depending on its age. The fifth instar (shown here) and sixth instar have a brown head and shield.

During the summer and fall, omnivorous leafroller larvae feed on new growth flushes and the peel of maturing fruit. Scarring of the peel, especially if caused by young larvae, is usually superficial; damaged leaves may have ragged edges.

COTESIA — adult wasp

This cocoon of *Cotesia* (=*Apanteles*) *medicaginis* can often be seen near parasite-killed caterpillars. The adult wasp lays its eggs in larvae of many lepidopterous species, including the omnivorous leafroller.

lay one or several eggs on or near the head of host larvae. The emerging maggots bore into the host and feed and pupate inside. *Elachertus proteoteratis* larvae are pale, maggotlike, and feed in a group attached on the outside of the host larva. They pupate on or near the caterpillar they killed. Other parasitic wasps include *Cotesia* (=*Apanteles*) *medicaginis*, which feeds inside caterpillars then emerges to pupate nearby in a white cocoon. *Trichogramma* spp. feed inside leafroller eggs, causing parasitized eggs to darken or leaving tiny round exit holes in egg shells. Parasites are most abundant and effective during midsummer.

Monitoring

Look for omnivorous leafroller larvae in the south and east quadrants of trees from spring though fall when monitoring for other pests. Look for small caterpillars under sepals of young fruit when you monitor for citrus thrips. During the summer, check to see if a significant proportion of any caterpillars present are parasitized, which indicates effective biological control. For current monitoring and treatment recommendations, consult the online *UC IPM Pest Management Guidelines: Citrus*.

Fourth instars have a brown shield, but the head can be either brown or black. The fifth and sixth instars have a brown head and prothoracic shield. Larval length ranges from less than 1/16 inch (1.5 mm) when newly hatched up to about 1/2 inch (13 mm) during the last instar.

The larvae roll and tie leaves together or tie leaves to fruit with silk threads and feed inside their webbed "nests." Mature larvae pupate inside silken cocoons within the rolled leaves. The adults are mottled brown, gray, and tan moths that appear bell shaped when at rest. Omnivorous leafroller can be present throughout the year, but it usually does not build to high populations. It has five to six generations per year.

Damage

In the spring, small omnivorous leafroller larvae spin webs and feed on new foliage. Later in the season, they tie leaves to fruit and feed under the sepals (calyx), leaving ring scars similar to those of citrus thrips. In the summer and fall, they tie leaves to the ripening fruit and chew the rind.

Natural Enemies

The most common parasites are tachinid flies and a tiny eulophid wasp (*Elachertus proteoteratis*). Tachinid females

A distinguishing feature of the amorbia larva is a black stripe on each side of the head and on the thorax, above the first pair of legs.

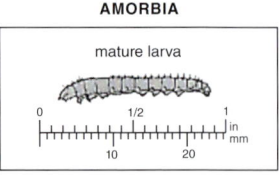

AMORBIA — mature larva

Adult amorbia are about 1 inch (25 mm) long, with variably colored forewings that are typically orangish to tan with dark markings. Amorbia and adults of other moths in the family Tortricidae (such as fruittree leafroller and omnivorous leafroller) are bell shaped when their wings are folded at rest.

Amorbia

Amorbia cuneana

Amorbia (family Tortricidae), also called the western avocado leafroller, is primarily a pest of avocado. It occasionally damages citrus that is grown next to avocados, mostly in southern California and the San Joaquin Valley.

The female moth lays pale green eggs on leaves in a flat, overlapping mass of 5 to 50 eggs. Larvae are yellowish green when young and develop through five instars. They turn darker green as they mature, and grow up to 1 inch (25 mm) long. Amorbia larvae can be distinguished from other caterpillars in citrus by the dark horizontal line on each side of amorbia's head and above the first pair of legs (Figure 36). As with larvae of orange tortrix and other tortricids, amorbia larvae wiggle violently and drop when disturbed.

Pupae are ½ to ¾ inch (12–19 mm) long and usually occur in rolled leaf shelters. Pupae initially are pale green, gradually turn tan, and are brown when mature.

Amorbia adults are bell shaped when their wings are folded at rest. Their variably colored forewings are typically orangish to tan with dark markings. Adult amorbia are about 1 inch (25 mm) long, about twice the size of the similar-looking orange tortrix adult. Amorbia has two to three generations per year, and adult flights typically occur in the early spring, midsummer, and autumn.

Damage

Larvae feed on new growth flushes, often rolling leaves or tying leaves to fruit. Amorbia larvae sometimes feed on young fruit at petal fall and later on maturing fruit. Feeding on young fruit causes circular calyx-end fruit scars similar to citrus thrips damage, except that amorbia scarring is deeper. Older larvae especially may also eat holes through the rind into the pulp.

Natural Enemies

Various natural enemies often keep amorbia populations below economically damaging levels. These include several species of tachinid flies and parasitic wasps, as discussed for orange tortrix and omnivorous looper. One of the most effective amorbia parasites is *Trichogramma platneri*, which causes parasitized eggs to turn black. Augmentative releases

This circular brown scar around the button end of this immature orange was caused by the larva of amorbia (western avocado leafroller) feeding under the calyx when the fruit was young.

Amorbia larvae tie leaves to fruit and feed on young or maturing fruit. Damaged fruit often develop decay at the feeding sites.

of the tiny *T. platneri* wasps are used to control amorbia in avocado, with releases timed to coincide with egg laying as determined by using the commercially available pheromone-baited trap for amorbia adults.

Monitoring

Look for amorbia around petal fall, especially in citrus near avocado groves. Search for small larvae under the fruit sepals when looking there for citrus thrips. Amorbia larvae produce webbing to protect themselves. Later in spring, look for webbing and leaf rolls in young foliage and feeding damage on young and mature fruit on the outside canopy. For current monitoring and management recommendations, consult the online *UC IPM Pest Management Guidelines: Citrus*.

Light Brown Apple Moth

Epiphyas postvittana

Light brown apple moth (family Tortricidae) feeds on citrus foliage and fruit in other parts of the world. It was first found in North America in 2007 on ornamental plants in coastal counties of California. In both appearance and behavior, the light brown apple moth is similar to other leafroller (tortricid) species.

The eggs are white, light green, or yellowish and are laid slightly overlapping each other in a mass. Each mass may contain from 2 to 170 eggs but typically has 20 to 50 eggs. Larvae develop through 5 or 6 instars and mature at a length of up to ¾ inch (19 mm) long. Larvae initially are pale yellow-green, then become darker green or light greenish brown as they mature. They have a darker green central stripe that may continue to the prothoracic shield. The head is light yellow-brown to dark brown, and the prothoracic shield (plate on top of the first segment behind the head) is greenish brown to indistinct with no dark markings. Larvae tie leaves together, or tie leaves to fruit, to create shelters within which they feed.

A tachinid fly has attached its eggs to the head of this amorbia larva. The emerging maggots will bore into the larva and feed inside.

Light brown apple moth larvae initially are pale yellow-green, then become darker green or light greenish brown as they mature. They have a light yellowish brown or pale greenish brown head and prothoracic shield, or the prothoracic shield may be indistinct. Larvae of the light brown apple moth can be confused with larvae of other tortricids, including amorbia, fruittree leafroller, and orange tortrix

Adult tachinids are stout, black flies with many distinct hairs. This adult tachinid and its oblong pupal case are next to the larger amorbia pupal case within which the parasite developed as a larva.

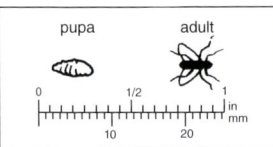

TACHINID PARASITE

pupa adult

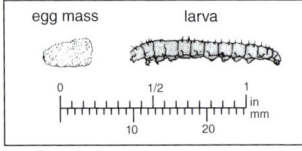

LIGHT BROWN APPLE MOTH

egg mass larva

There is a considerable variation in the coloration of the wings of light brown apple moths, especially on the males. This male is mostly reddish brown, while other males have a distinct two-tone wing coloration with a light brown or tan front half and a dark reddish brown rear end. Wings of either sex of light brown apple moths can also contain gray, pale yellowish, silverish, and blackish to purplish areas or mottling.

Western Tussock Moth

Orgyia vetusta

Western tussock moth is an occasional pest on citrus in southern California and in foothill orchards in the San Joaquin Valley. It occurs on many other orchard trees, including apple, avocado, cherry, prune, and walnut, as well as on live oaks, perennial lupine, and certain ornamentals.

Tussock moths overwinter as eggs that hatch about the time the spring growth flush is expanding. Eggs occur in a mass that the female covers with light brown to whitish gray hairs. Young larvae are black with long bristles. Late-instar larvae have numerous red and yellow spots and clumps of black, grayish, and white hairs. Western tussock moth is generally the only species in citrus that has larvae with prominent tufts of hairs on their prothorax and abdomen.

Mature larvae pupate in cocoons mainly on scaffold branches and trunks but also near trees on objects such as orchard heaters and fence posts. Cocoons are covered with light brown to whitish gray hairs and occur singly or side by side in a group.

Pupae occur in thin-walled silken cocoons between leaves webbed together. Pupae are about ½ inch (13 mm) long and change from green to brown to dark reddish brown as they mature.

The adults are moths about ¼ to ½ inch (6–13 mm) long and vary greatly in color, especially the males. The wings commonly are light brown, with variable patterns of darker brown. Wings can also contain gray, pale yellowish, and silvery areas and blackish to purplish mottling. Unlike other leafroller moths, males have an expanded outer edge of their wing (costal flap) that at rest folds up over the top of the wing.

If damage like that caused by amorbia or other leafrollers is unusually abundant, investigate whether light brown apple moth or another new pest may be the cause. To detect leafroller larvae, inspect leaves for the characteristic webbing at the midrib vein on the underside of leaves and between leaves. At flowering, check blossom clusters in the lower canopy for webbing and larvae. If fruit are present, look where fruit touch leaves or other fruit.

Adult male light brown apple moths can be detected using delta sticky traps baited with a commercial pheromone specific to light brown apple moth. To determine whether this pest is present, one trap per 5 acres is typically used, with at least one trap per growing location.

Because of quarantines, special management regulations may apply where light brown apple moth is found. Report to agricultural officials any suspected light brown apple moths found outside of areas where it is known to occur. For updated information, visit the UC IPM Web site (www.ipm.ucdavis.edu) or contact your county agricultural commissioner.

The oldest instar larva of the tussock moth has numerous red and yellow spots from which radiate gray to white hairs. They also have four dense tufts of white hairs on their back and two tufts of black hairs on their head.

Females are wingless, oblong, and covered with short whitish gray hairs. Males are typical-looking moths, mostly grayish or brown with blackish and white markings. Adults emerge and mate from late April through July. The wingless female lays 125 to 300 eggs by late summer in a single, hair-covered mass, often inside her empty pupal case. The average time from egg hatch in spring to the adult stage is about 75 days. The western tussock moth has one generation per year.

Young tussock moth larvae have just emerged from this egg case. Young larvae are black with long bristles.

Tussock moth females are hairy and wingless.

Telenomus californicus, an egg parasite of tussock moths, is often seen moving about the egg masses.

Damage

A high population of western tussock moth larvae can consume all the new spring growth flush, and the larvae may also chew newly set or young fruit. Their damage resembles that of citrus cutworm and katydids.

Natural Enemies

A small black wasp (*Telenomus californicus*, family Scelionidae) sometimes parasitizes many of the tussock moth eggs in a mass. Parasites of tussock moth larvae include the ichneumonid wasps *Hyposoter exiguae*, *H. fugitivus*, and *Iseropus stercorator orgyiae*.

A tiny beetle (*Trogoderma sternale*, Dermestidae) preys on eggs in southern California. Adult dermestids are common from March through September, and the bristly dermestid larvae have been observed throughout the year. If the surface of an egg mass scrapes off easily, masses have been colonized by egg predators, and predators or empty egg shells will be apparent.

Monitoring

Treatment is rarely needed. Where tussock moth may be a problem, look for egg masses or larvae during spring. Before mid-March, carefully inspect scaffolds and the trunk on all sides for egg masses, which often are laid on old pupal cases. New egg cases are buff colored; old ones are gray and weathered. Also check the egg masses for parasitism and predation as described above to assess the extent of biological control. To monitor caterpillars, use timed-count inspections of foliage during spring (larvae per hour of search). For current recommendations, consult the online *UC IPM Pest Management Guidelines: Citrus*.

Scavenger Caterpillars

Scavenger caterpillars are usually not pests as they feed mostly on dead plant material and sooty mold. Treatment is rarely needed.

The pink scavenger caterpillar (*Pyroderces rileyi*, family Cosmopterigidae) occurs sporadically in southern California coastal areas. The egg is oval, flattened, pale colored, and about 1/50 inch (0.5 mm) long. Eggs usually are laid singly and hatch within a few days. Larvae are mostly dark pinkish and have a light brown head, black mouthparts, and a dark brown prothoracic shield. When fully grown, larvae are only up to 1/3 inch (8 mm) long, much smaller at maturity than other species of Lepidoptera in citrus, except for citrus leafminer and peelminer. The yellowish brown pupa lies within a whitish cocoon that is about 1/5 inch (5 mm) long. The adult is a mottled brown and whitish moth with black markings and is about 1/3 inch (8 mm) long.

On orange and lemon trees, the caterpillar is mainly a scavenger, feeding on dry or decaying fruit, dead floral parts, and sooty mold. Larvae occur mainly among fruit clusters and under sepals. During the winter and spring, most larvae occur in mummified fruit on the ground or in the tree. During the summer, the larvae may nibble on the rind of ripe Valencias, often near the stem end or on the sides of fruit in a cluster. The feeding is usually superficial and does not cause appreciable damage. A heavy infestation can result in fruit drop in the orchard or fruit decay during storage.

The black scavenger caterpillar (*Holocera iceryaeella*, family Blastobasidae) also occurs occasionally on coastal citrus. It is a scavenger and generally does not cause damage. The oval eggs are about 1/50 inch (0.5 mm) long. The larvae are up to 2/5 inch (10 mm) long and are mostly dark reddish brown and grayish with a black or brown head and prothoracic shield. Pupae are dark brown and about 1/4 inch (6 mm) long. The adults are ashy gray moths about 1/3 inch (8 mm) long.

California Orangedog

Papilio zelicaon

The California orangedog (family Papilionidae), also called black anise swallowtail, is a native butterfly that feeds on citrus. Its native food is perennial anise or sweet fennel (*Foeniculum vulgare*) and related weeds in the family Apiaceae (Umbelliferae). Treatment usually is not needed.

The spherical eggs are commonly yellow and laid singly on plants. Immature stages vary greatly in color. Young larvae are mottled brown or black with white on top. Later instars are whitish green to bright green with alternating bands of black and yellow or orangish spots on each segment. Mature larvae are about 1½ inches (37 mm) long. When disturbed or threatened, swallowtail larvae can protrude a bright orange-red, fleshy forked scent gland (osmetarium) from the top of their thorax immediately behind their head.

CALIFORNIA ORANGEDOG

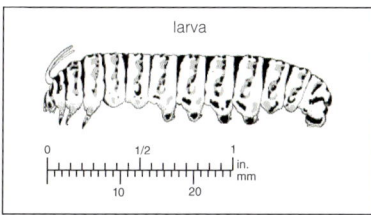
larva

The California orangedog adult (also called black anise swallowtail) is mostly black and yellow.

The appearance of orangedog (black anise swallowtail) larvae varies greatly with age. Young larvae are mostly mottled brown or black with white on top.

Late-instar California orangedogs are commonly whitish green to bright green with alternating bands of black and orangish or yellow spots on each segment. When disturbed, larvae sometimes stick out their orange-colored scent gland (as shown here) and give off a strong odor.

This cocoon of a *Hyposoter* sp. wasp resembles a small bird dropping. The shriveled skin of the host larva usually remains attached to the *Hyposoter* cocoon.

PINK SCAVENGER CATERPILLAR

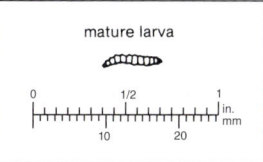
mature larva

HYPOSOTER

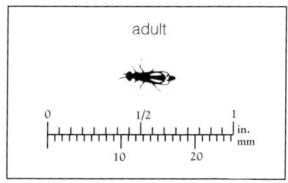
adult

Hyposoter spp. wasps are important parasites of many caterpillar species in citrus, including cabbage looper, orangedogs, and western tussock moth. This adult *Hyposoter exiguae* is laying her egg in a young beet armyworm larva.

At maturity, larvae form pale brown to gray or greenish papery pupal cases. Adults are mostly black and yellow day-flying butterflies.

Orangedogs feed on tender citrus leaves, occasionally defoliating young trees. They rarely cause economic damage in mature orchards. Parasites, especially a *Hyposoter* sp. wasp, are often highly effective at reducing populations. Because the anise swallowtail prefers to lay eggs on sweet fennel, fennel is sometimes interplanted in strips with citrus to act as an egg-laying trap crop, which is then mowed after the egg-laying peak in each generation. A sweet fennel trap crop also provides nectar as food for the adults of many natural enemies.

Giant Orangedog

Papilio cresphontes

The giant orangedog, also called giant swallowtail, was introduced from the southeastern United States and is now common in residential citrus in coastal southern California. It is a minor pest in citrus in certain other states, but it has not been reported in commercial citrus in California. No treatment is needed.

Females lay orangish, spherical eggs individually on tender citrus leaves. Larvae are blotchy gray, white, and brownish and grow to about 2 inches (5 cm) long. In comparison with the late instars of the California orangedog, mature giant swallowtail larvae are longer, have a saddle-shaped patch on top near their middle, and are not green. Like other swallowtails, when disturbed or threatened, giant swallowtail larvae can protrude their bright orange-red, fleshy forked scent gland (osmetarium) from on top behind their head.

The papery pupal cases are about 1⅝ inch (40 mm) long. Adults are mostly brown with white and yellow. In comparison with the adult California orangedog, the giant orangedog is larger and has more white than yellow on its wings. Adults can be very large, with a wingspan of about 3¼ to 5½ inches (85 to 140 mm).

Giant swallowtail (giant orangedog) adults are mostly brown with white and yellow and have a wingspan of 3¼ to 5½ inches (85–140 mm). In comparison with the California orangedog, the giant orangedog adult is larger and has more white than yellow.

Giant orangedog caterpillars are blotchy gray, white, and brownish and at maturity about 2 inches (5 cm) long. Superficially, a giant orangedog larva resembles a bird dropping.

Citrus Looper

Anacamptodes fragilaria

The citrus looper is a native looper or measuring worm (family Geometridae) that occurs at low levels in most citrus-growing areas. Looper larvae are easily recognized as they move in a characteristic looping or inchworm fashion.

The female moth lays about 100 pale green, spherical eggs singly on leaves. In about 1 week, eggs hatch into grayish

Citrus looper larvae move in a characteristic looping or inching fashion, pulling their rear forward as they arch up their back. Mature larvae are up to 1½ inches (37 mm) long with a grayish body.

larvae, which at maturity are up to 1½ inch (37 mm) long when their body is fully extended. Citrus looper larvae have one pair of prolegs on their sixth abdominal segment, as illustrated in Figure 36. Unlike most species of caterpillars in citrus, citrus loopers do not have any prolegs around what appears to be the middle of their abdomen.

Citrus looper pupal cases are not enclosed in silk. Pupae and adult moths at rest are about ⅗ inch (15 mm) long. Adults are mostly gray with white and black mottling. There are several generations per year.

Larvae consume new growth flushes. The very young larvae typically feed on the lower leaf surface along the leaf margin. Older larvae eat holes in leaves or consume leaves entirely and also feed on blossoms, young fruit, and (rarely) on mature fruit.

Citrus looper natural enemies include the braconid wasp *Apanteles praesens*. Treatment is rarely required when natural enemies are conserved.

Cabbage Looper

Trichoplusia ni

Cabbage looper (family Noctuidae, subfamily Plusiinae) is a rare pest in citrus. It feeds on many cultivated plants, including beans, peppers, and tomatoes, and it especially prefers vegetables and weeds in the crucifer (Brassicaceae) family.

Females lay round, greenish white eggs singly on the upper surface of leaves. Larvae are green, usually with pale yellow to white stripes on top of their back and one along each side. When mature, larvae are about 1½ inches (37 mm) long. Cabbage looper larvae differ from citrus looper by having two pairs of prolegs near the middle of their abdomen, on abdominal segments 5 and 6, as illustrated in Figure 36. The similar-looking citrus cutworm and beet armyworm have 4 pairs of prolegs near the middle of their body.

Cabbage looper larvae are greenish with pale stripes. Although larvae often move in a looping manner, they differ from citrus looper by having two pairs of prolegs that are more forward on their body, on abdominal segments 5 and 6.

Mature cabbage looper larvae typically curl a leaf and pupate inside. Pupae are about ¾ inch (19 mm) long, brown to greenish, and covered with thin silk webbing. Adults are mottled brown, gray, and silvery nocturnal moths about ¾ inch (19 mm) long when at rest.

A nuclear polyhedrosis virus (NPV) and about two dozen parasite species kill cabbage loopers. These include *Trichogramma* spp. egg parasites and larval or pupal parasites, such as the braconid wasp *Hyposoter exiguae* and several species of tachinid flies. The encyrtid *Copidosoma truncatellum* oviposits in cabbage looper eggs but does not kill the loopers until their mature larval or prepupal stage. Numerous tiny *Copidosoma* wasps then emerge from a single dead host. Treatment is rarely required when natural enemies are conserved.

Beet Armyworm

Spodoptera exigua

Beet armyworm (family Noctuidae) occasionally feeds on citrus foliage but rarely causes economic damage. Treatment is rarely needed.

Females lay spherical, pale greenish, pinkish, or white eggs in masses that are covered with whitish, cottony scales from the female's body. Larvae are usually dull green caterpillars, but their overall color can vary from pale to dark

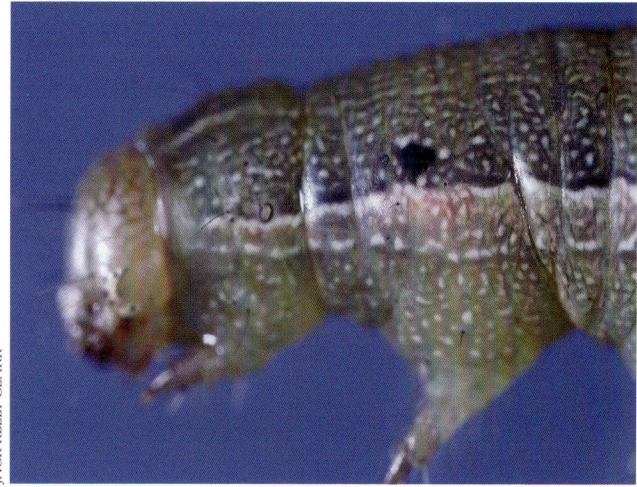

Beet armyworm resembles cabbage looper and citrus cutworm, but beet armyworms usually have a black spot on the side of the body above the second pair of true legs. The beet armyworm's body is usually dull green but can vary from pale to dark green with wavy, light-colored stripes running down the back and a broader pale stripe along each side.

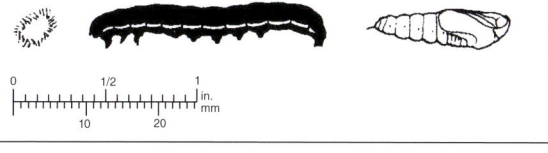

BEET ARMYWORM

green. They typically have many fine, wavy, light-colored stripes down the back and a broader pale stripe along each side. Larvae develop through five instars and grow up to 1½ inches (37 mm) long.

Beet armyworm larvae usually, but not always, have a dark spot on the side of the thorax above the second true leg (Figure 36). Similar-looking citrus cutworm larvae lack this dark thoracic spot. Similar-looking larvae of cabbage looper have two pairs of prolegs near the middle of their body, not the four pairs of midbody prolegs as on beet armyworm and citrus cutworm.

Beet armyworm pupae occur in a silk cocoon in litter or near the soil surface. The adult moth has a wingspan of about 1 inch (25 mm) and is mostly gray with black, brown, and white mottling.

Natural enemies include various predators and parasites, such as *Hyposoter exiguae*, an ichneumonid wasp that parasitizes larvae.

MINERS

Citrus leafminer and citrus peelminer are Lepidoptera species with larvae that feed just under the surface of plant tissue, where they create winding tunnels by the scissoring action of their feeding.

Citrus Peelminer

Marmara gulosa

Citrus peelminer (family Gracillariidae) is a pest in the Coachella and San Joaquin valleys. It tunnels in the fruit rinds of all types of citrus, but grapefruit, pummelo, and certain smooth-skinned navel varieties, including Atwood, Fukumoto, Barnfield, and Thompson Improved (Ti), are especially susceptible to damage. Citrus peelminer also feeds in leaves, stems, or fruit of many crops, ornamentals, and weeds, including beans, cotton, grapes, nuts, oleander, stone fruits, willows, and various vegetables.

Description and Biology

Adults are 1/12-inch-long (2 mm) dark moths with white bands or stripes on the wings and legs. Females deposit eggs singly on citrus stems and fruit. The pale larvae hatch, tunnel down into tissue under their egg shell, and mine just beneath the surface in the top layer of cells of stems or fruits. When populations are high, peelminers will also tunnel in citrus leaves, but this is rare.

Larvae develop through 5 or 6 instars over about 2 to 3 weeks. Their winding tunnels grow wider as they molt to

The citrus peelminer adult is a tiny, mostly dark moth. It has white bands or stripes on the wings and legs.

Citrus peelminer larvae are pale colored or whitish, except during the last instar, which is pink or reddish.

Mature citrus peelminer larvae form a flat silk sheet on a twig, leaf, or fruit. The larva places minute silk balls on the silk cover and then pupates.

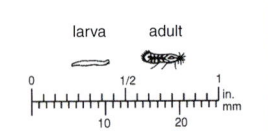

Citrus peelminer larvae feed in pale, winding tunnels just beneath the surface of rinds and green stems (sucker shoots). The mines often cross over each other. Most damage is to fruit or suckers in sheltered places, such as the side of fruit facing in toward the trunk.

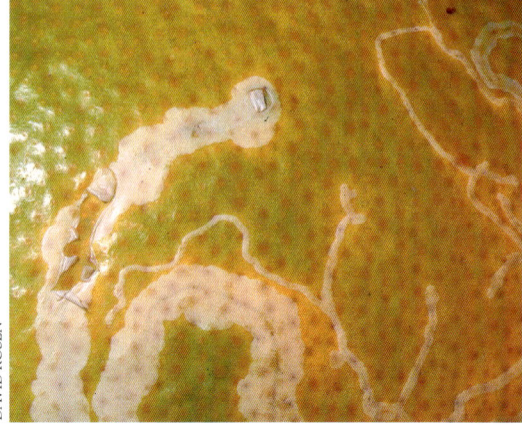
Unlike leafminer, citrus peelminer larvae leave no dark- or light-colored trail of excrement in their tunnels.

larger sizes. At maturity, larvae are less than 1/5 inch (5 mm) long and change from whitish to pink or reddish. The pink-stage, silk-spinning larva exits the mine and spins a flat silk covering on a fruit, leaf, or twig. It covers the silk case with minute silk balls and then pupates.

Citrus peelminer develops throughout the year on different host plants. During the spring, it attacks the stems of willows, oleander, and various tree crops such as walnuts and plums. During summer, it begins to attack the large grapefruit and pummelo fruit, cotton stems and bolls, grape vines, and bean stems. Later in the season, it moves to navel oranges. The time for one generation ranges from several months during winter to several weeks in summer, depending on temperature. Peelminer has six to eight generations per year, mostly from May to November.

Be sure to distinguish peelminer from the citrus leafminer. Citrus leafminer tunnels in new flush leaves, rarely in stems or fruit, while peelminer prefers fruit and stems. Citrus leafminer larvae leave a trail of dark- or light-colored frass (excrement) in their mine, but citrus peelminer does not leave an excrement trail. Citrus leafminer mines do not cross over each other, while citrus peelminer mines frequently criss-cross.

Damage

In the most susceptible varieties, up to 80% of fruit rinds may be mined. Damage to fruit is on the surface, but one mine can render the fruit unacceptable for fresh market. Presence of live peelminer larvae can be an export quarantine issue. Susceptible citrus varieties are at higher risk for damage if they are located next to crops where peelminer populations build up, such as beans and cotton. When those crops reach maturity and begin to dry, the moths emerge and move into neighboring citrus. Moths may migrate in substantial numbers because citrus peelminer is usually not an economic pest in other crops, and it is rarely controlled, except in citrus.

Natural Enemies

Eulophid parasitoids including *Baryscapus*, *Closterocerus*, *Hemiptarsenus*, *Nechrysocharoides*, and *Pnigalio* species have been found in citrus peelminer larvae. These native parasitoids move from other leafminer pests to citrus peelminer but generally produce low levels of parasitism. A native eulophid (*Cirrospilus coachellae*) has been shown to suppress peelminer populations in the Coachella Valley. The wasp deposits multiple eggs inside the mine of each peelminer larva. The hatching wasp larvae feed in groups attached to the peelminer in its mine. Parasites pupating in the mine can be seen through the leaf epidermis as small, black, oblong shapes surrounded by black dots (parasite meconial pellets or fecal material).

This tiny wasp (*Cirrospilus coachellae*) is an important parasite of citrus peelminer in the Coachella Valley. The female lays several eggs on each peelminer larva, and the maggotlike wasp larvae feed within the mine attached to the outside of their host.

When peelminer has been killed by *Cirrospilus coachellae*, about two to seven naked, black parasite pupae occur in each tunnel with the dead peelminer. Several of these oblong wasp pupae are visible here through the rind epidermis surrounded by their fecal material (the black dots). Evidence of past parasite activity includes exit holes near the remains of peelminer larvae.

Monitoring

Management of citrus peelminer with insecticides is difficult because female moths prefer to deposit eggs on fruit in the lower inside canopy and it is difficult to reach this area with foliar spray applications. Also, the fruit are rapidly growing and reducing the effect of the spray, and larvae are protected inside mines. If fruit sampling reveals mining, it is too late for insecticides to protect the fruit from damage. Monitor peelminer flights and development time using degree-days to time treatments that kill adults and eggs, preventing development of the fruit-damaging larvae.

Male moths can be monitored at the beginning of the season using sticky traps baited with artificial sex pheromone lures. The first-generation peelminer adult flight occurs in noncitrus host crops, usually during late March to early April. Using that first flight as a biofix and a lower developmental threshold of 55°F, peelminer completes a generation every 580°D. There are 7 potential flights in the San Joaquin Valley and 9 potential flights in the Coachella Valley. The first two flights of citrus peelminer following the biofix attack the stems of noncitrus hosts such as oleander, walnuts, willows, and various weeds. The third flight of moths lays eggs on pummelo and grapefruit varieties. The fourth or fifth flights begin to lay eggs on susceptible navel orange varieties. Citrus peelminer requires a minimum fruit diameter of about 3 inches (7.5 cm) for grapefruit and pummelos and about 2½ inches (6.3 cm) for navels for successful development of larvae.

If the grove is a less-susceptible variety of citrus fruit and has not experienced more than 5% of fruit damaged, avoid spraying with insecticides, because insecticides are only partially effective in reducing mining and may disrupt natural enemies. For susceptible varieties of citrus, time the insecticide treatments at the initiation of the third and fourth flights for pummelos and grapefruit and the fourth and fifth flights for navels. Citrus next to infested crops such as beans and cotton are at high risk for major attacks and will need treatment during these susceptible periods.

For current monitoring and management recommendations, consult the online *UC IPM Pest Management Guidelines: Citrus*. For more photographs and information on biology and identification, consult *Citrus Leafminer and Citrus Peelminer* (Grafton-Cardwell et al. 2008).

Citrus Leafminer

Phyllocnistis citrella

Citrus leafminer (family Gracillariidae) tunnels in leaves and can severely distort young shoots. Since its introduction into California in 2000, it has moved northward into southern and central California. In addition to all varieties of citrus, it mines closely related plants such as calamondin and kumquat.

The adults of citrus leafminer (shown here) and peelminer are each about 1/12 inch (2 mm) long. The adult leafminer is light tan, silver, or whitish. These tiny moths are shown here in comparison with the 1-inch (25 mm) square grid marked on a trap.

Leafminer larvae are a translucent yellowish green, except for the last instar, which is paler. Leafminer body segments have rounded edges, unlike the triangular-shaped segment edges of peelminer larvae.

The adult citrus leafminer has a black dot at the tip of its mostly light-colored wings.

Leafminer fourth-instar larvae (prepupae) roll and tie leaf edges with silk. Inside the rolled leaf edge, the pale prepupa (not shown) darkens into this brown pupal case, inside which develops an adult moth.

Description and Biology

Adults are tiny, light-colored moths. They have silvery and white iridescent forewings with brown and white markings and a distinct black spot on each wing tip. The moths are most active from dusk to early morning and spend the day resting on the underside of leaves. Soon after emerging from the pupal case, females emit a sex pheromone that attracts males. Females lay eggs singly and only on succulent new leaf growth. Eggs hatch in about 4 to 10 days.

Larvae develop through four larval instars in about 1 to 3 weeks. They create shallow, meandering mines in leaves as they scissor (feed) in the top layer of cells. As a larva grows, its mine becomes more visible and its excrement forms a thin, light or dark trail within the mine. Larvae molt and become larger, causing the mine to enlarge as they complete their three feeding instars. Mature larvae (fourth instars, also called prepupae) emerge, crawl to the edge of the leaf, roll the leaf edge, tie it with silk threads, and pupate inside.

Egg to adult development takes 2 to 7 weeks, depending on temperature. Leafminer can develop throughout the year, but it is most abundant during periods of abundant new leaf flush combined with warm temperatures. Leaf damage occurs in California primarily from summer through fall.

Another species, the citrus peelminer, does not leave a dark- or light-colored frass trail in larval mines. Citrus peelminer creates criss-crossing tunnels in stems and fruit rather than serpentine, nonoverlapping tunnels in new flush leaves. Also, the peelminer pupa is covered with pale silk balls, which leafminer lacks. For more photographs and information on biology and identification, consult *Citrus Leafminer and Citrus Peelminer* (Grafton-Cardwell et al. 2008).

Damage

Citrus leafminer larvae create shallow tunnels (mines) by scissoring the cells just under the lower or upper surface of young leaves, causing the foliage to curl and distort. Except on coastal lemons that produce multiple crops, citrus trees

Citrus leafminer leaves frass (excrement) in its mines, producing a dark- or light-colored trail. There is no obvious discolored frass trail left behind in peelminer tunnels (pictured earlier).

Citrus leafminer larvae tunnel mostly in succulent, flushing new leaves, causing terminal leaves to curl and distort.

This adult *Pnigalio* sp. parasitizes citrus leafminer. The wasp larva is a pale maggot that feeds attached to the outside of a leafminer larva, then pupates within the mine near the dried, shriveled skin of its dead host. This adult and its pupa and mature larva are about $1/12$ to $1/8$ inch (2–3 mm) long.

more than 4 years old generally tolerate leaf damage without any effect on tree growth or fruit yield. Citrus leafminer is likely to cause economic damage in nurseries and new plantings because the growth of young trees is retarded by leafminer infestations. However, even when citrus leafminers are abundant on young trees, trees are unlikely to die.

In Florida, citrus leafminer creates openings that allow entry of the pathogen that causes citrus bacterial canker. This disease causes leaf spots and rind scars, drop of leaves and fruit, and twig dieback. Help keep citrus bacterial canker out of California, as discussed in the chapter "Diseases."

Natural Enemies

Citrus peelminer and leafminer share many of the same eulophid wasp parasites, including *Chrysocharis*, *Cirrospilus*, and *Pnigalio* species wasps. Elsewhere in the world, as citrus leafminer becomes established for a longer period, its populations become fairly well controlled on mature trees by parasitic wasps.

Whether leafminer biological control will be effective in California cannot be determined at this time (2010). After an invasive pest's introduction, many years may be required before natural enemies adapt and disperse. For biological control to be effective, new natural enemy species may need to be introduced and crop management practices may need to be modified.

Monitoring

Male moths can be detected using sticky traps baited with commercial sex pheromone. Using a triangular trap with a small opening reduces the number of other insects caught in the stickem, thereby reducing the effort needed to examine trap catches. Citrus leafminer moths are easily identified in the traps as small light brown moths with a black spot on the tip of each of their wings. The traps can be used for identifying major flights and for timing insecticide applications if they are needed.

Treatments may need to be applied to young trees and nursery plantings. Mature trees (4 years or older) should not be treated with insecticides for this pest unless it is an unusual situation such as multiple-cropping coastal lemons. For more current monitoring and management recommendations, consult the online *UC IPM Pest Management Guidelines: Citrus*.

HEMIPTERA (HOMOPTERA)

Hemipteran insects have piercing or sucking mouthparts and incomplete metamorphosis: eggs hatch into immatures (nymphs) that develop gradually into adults without a pupal stage. Nymphs differ from adults primarily in size, lack of wings, and color. The order Hemiptera includes true bugs (suborder Heteroptera) and the Homoptera, an informal group including aphids, leafhoppers, mealybugs, and whiteflies (discussed below) along with scale insects (discussed earlier).

Controlling ants and dust and avoiding broad-spectrum pesticides toxic to natural enemies are important methods for managing most Hemiptera, especially aphids, mealybugs, scales, and whiteflies. Which crops or other plants are grown near citrus and how they are managed greatly influence the abundance of Hemiptera that have multiple host plants and readily migrate among hosts, such as glassy-winged sharpshooter and potato leafhopper.

Specific monitoring methods used for certain Hemiptera include pheromone-baited sticky traps and degree-day accumulation discussed earlier for California red scale, and branch shaking (beating), timed counts, or yellow sticky traps for glassy-winged sharpshooters as described below. Monitor most Homoptera species by inspecting plants for the insects themselves and any associated shiny honeydew, blackish sooty mold, and honeydew-feeding ants. For current species-specific monitoring and management recommendations, consult the online *UC IPM Pest Management Guidelines: Citrus*. These pests' identification, biology, and natural enemies are described below.

Mealybugs

Several species of mealybugs feed on citrus in California. These same species occur on a wide variety of fruit trees, field crops, and ornamentals. In most situations, mealybugs are well controlled by natural enemies, and insecticide application is rarely warranted.

The citrus mealybug (*Planococcus citri*) is the most common species, especially in coastal citrus. The Comstock mealybug (*Pseudococcus comstocki*) primarily occurs on lemons in the San Joaquin Valley. The citrophilus mealybug (*Pseudococcus fragilis*) and longtailed mealybug (*Pseudococcus longispinus*) are especially rare because of their effective parasites. The introduced pink hibiscus mealybug (*Maconellicoccus hirsutus*) is reported in California only in the Imperial Valley, where it is under good biological control.

Description and Biology

Mealybug nymphs and females are soft, oval, and covered with white, powdery wax. Their body is distinctly segmented and somewhat flattened, although this may be obscured by their wax coating. In most species, female adults and older nymphs have wax filaments around the body margin. Depending on the species, filaments can be especially prominent at their posterior end. Males are tiny, delicate, two-winged insects with two long tail filaments. Adult mealybug males are rarely seen.

For females and older nymphs, the pattern of wax covering their body and the thickness and length of marginal

These adult females and nymphs of citrus mealybug have a yellow orange body mostly covered by powdery wax. The filaments around the body margins are not much longer at the rear end than on the sides. Citrus mealybug often has a dark lengthwise stripe on its back where the wax is thinner.

The Comstock mealybug has a thicker wax cover than the citrus mealybug. Two filaments at its rear end are about one-quarter the length of the body, or somewhat longer.

In most species, female mealybugs lay orangish eggs in a wax-covered mass. Vine mealybug (shown here with its eggs) closely resembles citrus mealybug, but vine mealybug has not been reported in citrus in California. Because new species are periodically introduced, report any suspected new pests or unusual mealybug problems to the local agricultural agency or Cooperative Extension office.

Pink hibiscus mealybug females have a purplish to dark brown body but are commonly covered with white powdery wax. Unlike most other mealybugs, pink hibiscus mealybug lacks wax strands or filaments around the margin of its body.

Longtailed mealybug has well-developed waxy filaments around its margin. Its tail filaments are more than three-quarters of the body length.

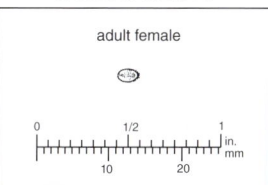

Identifying the mealybug is important when relying on species-specific natural enemies to provide biological control or when suspecting the mealybug of being an introduced exotic species. When positive identification is warranted, submit a sample to the local county agricultural commissioner or university Cooperative Extension office. Positive identification of the species usually requires an expert and microscopic examination of the types and arrangements of gland pores and setae. Examine at least several mature females in good condition when using this key. Populations may consist of mixed species, and the wax secretions can break off some specimens, leading to their misidentification. Immature mealybugs have different wax patterns than adults. Use this key for mature females only.

Features of a mealybug helpful in identifying the species.

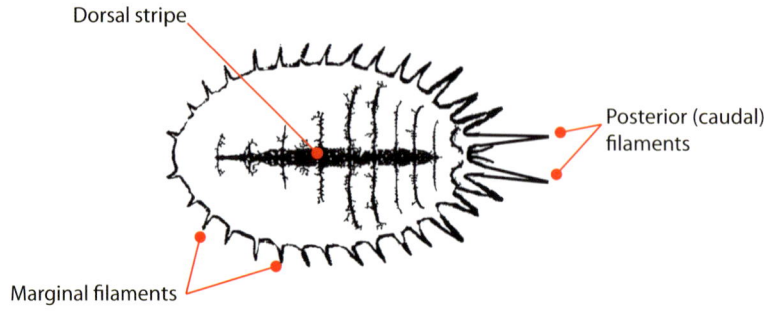

1. Wax filaments around body margin well developed?

Yes. Comstock, citrophilus, citrus, longtailed, or vine mealybug. **Go to 2.**

No, stop here. Pink hibiscus mealybug. Submit a sample for expert identification if this species is not known to be established in that area.

2. Caudal (tail) filaments greater than 3/4 of the body length?

Yes, stop here. Longtailed mealybug, which may also have a well-defined dorsal stripe.

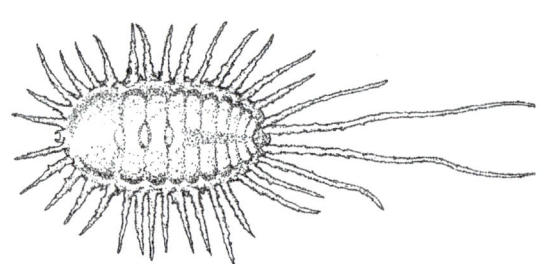

No. Go to 3.

3. Caudal filaments between 1/4 and 3/4 of body length?

Yes, stop here. Comstock or citrophilus mealybugs. These species cannot be reliably distinguished by appearance in the field. Both have a scarcity of wax on 4 spots on each segment, allowing the body color to show and giving the appearance of 4 dark longitudinal stripes along the back. Comstock mealybug does not establish in navels or tangerines, but it can occur in lemons. Citrophilus mealybug can be in any citrus. Both species are under excellent biological control in California citrus unless natural enemies are disrupted.

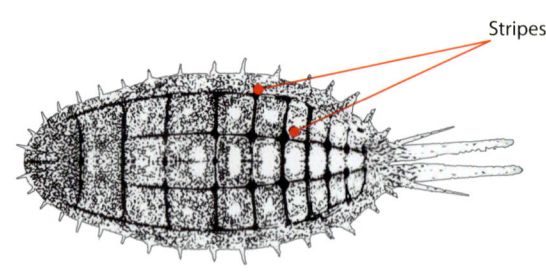

No. Go to 4.

4. Caudal filaments less than 1/4 of body length?

Yes. Citrus or vine mealybugs. Citrus and vine mealybugs cannot be reliably distinguished by appearance in the field. Vine mealybug has not been reported infesting citrus in California, but it is a serious pest of grape. Especially in interior growing areas or where citrus is grown near grape, if mealybugs resembling these are a problem, submit a sample for expert identification.

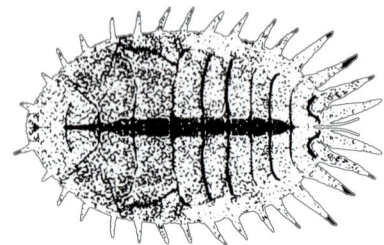

Figure 37. Identification key to the species of mealybugs in California citrus. Adapted from Gill 1982, Godfrey et al. 2002, longtailed mealybug by P. J. Hollyoak in Gorham 1991.

filaments can be used in the field to distinguish among species (Figure 37). For example, the filaments around the margins of citrus mealybug are not much longer at the posterior end. Citrus mealybug also often has a dark stripe down the center of its back where the body color is visible through the powdery wax. In comparison with citrus mealybug, the Comstock mealybug has thicker wax covering its body. At the posterior end, it has two longer wax filaments that are about one-quarter the length of the body to somewhat longer.

Over a 10- to 20-day period, each female can lay several hundred eggs in cottony egg sacs attached to leaves, fruit, or twigs. Longtailed mealybug is an exception in that it produces no external egg sac. Longtailed mealybug eggs hatch internally and nymphs emerge from the female.

In most species, newly hatched nymphs are light yellow to orange and free of wax. Once the mobile first instars (crawlers) settle to feed, they soon excrete a waxy cover. Mealybugs produce two to four overlapping generations per year, and all stages can occur in citrus throughout the year.

Damage

Mealybugs suck phloem sap, reducing tree vigor and excreting sticky honeydew on which blackish sooty mold grows. If a cluster of mealybugs feeds along a fruit stem, fruit may drop. Damage is most severe in spring and fall.

Citrus and citrophilus mealybug can occur in any citrus. Pink hibiscus mealybug also feeds on all citrus varieties, but in California this species has been reported only in Imperial County. Comstock mealybug does not establish in navels or tangerines, but it can occur in lemons.

Because natural enemies usually keep mealybug populations below economically damaging levels, promptly submit samples to agricultural officials for identification if mealybugs are unusually abundant, causing atypical damage, or appear to be an unfamiliar species. New mealybug species are periodically introduced, sometimes causing severe problems. For example, the introduced vine mealybug (*Planococcus ficus*), a serious pest on grapes, closely resembles citrus mealybug. Vine mealybug has not been reported on citrus in California, but it infests citrus and other crops elsewhere in the world.

Natural Enemies

When undisturbed by ants, dust, or insecticide treatment, parasites provide excellent control of the Comstock, citrophilus, and longtailed mealybugs and are very important in controlling citrus mealybug, as listed earlier in Table 7. Pink hibiscus mealybug, discovered in Imperial County in 1999, is also now well controlled by an introduced parasitic wasp (*Anagyrus kamali*, family Encyrtidae). Brown and green lacewings, pirate bugs, and syrphid flies discussed earlier in the "General Predators" section also feed on mealybugs.

Mealybugs, such as the ones shown here, prefer to feed in protected locations. When monitoring for mealybugs, separate and inspect fruit that touch leaves or other fruit.

The *Cryptolaemus* adult (left) and its mealybug-mimicking larva (right) feed voraciously on all mealybugs. The mealybug destroyer overwinters poorly in California. Purchased lady beetles are introduced in some orchards during spring to reproduce and provide biological control.

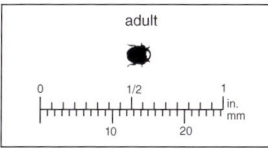

Mealybug Destroyer. The most important predator is the mealybug destroyer (*Cryptolaemus montrouzieri*). Larvae of this lady beetle closely resemble a mealybug, but at maturity *Cryptolaemus* larvae are about twice as large as an adult female citrus mealybug. The *Cryptolaemus* larva also has longer lateral wax filaments and moves faster. It can be recognized as a lady beetle larva if its wax is gently brushed away. Both the adult and larva feed voraciously on all stages of mealybugs but prefer mealybug eggs and young nymphs.

Mealybug destroyer adults reproduce well only when feeding in spots where mealybug eggs and wax are abundant. *Cryptolaemus* does not survive the winter well away from the coast but can be purchased and released on mealybug "hot spots" early in the spring. If a heavy population of mealybugs must be reduced quickly by pesticide application, apply an effective material with the shortest residual toxicity, then reestablish biological control by collecting or purchasing and releasing mealybug destroyers after any residue is no longer toxic to them. For specific recommendations, consult the online *UC IPM Pest Management Guidelines: Citrus*.

These oblong brown to orangish pupae belong to the citrus mealybug parasite (*Leptomastix dactylopii*). After emerging, the yellowish brown adult parasite leaves a round hole or hinged cap on the end of each pupal case.

Leptomastix. *Leptomastix dactylopii* (family Encyrtidae) is a common parasite of the citrus mealybug. The adult *Leptomastix* is a yellowish brown wasp less than ⅛ inch (3 mm) in length. The female lays its eggs in late-instar mealybug nymphs and young adults. One wasp larva feeds inside each host, developing through four larval stages and causing the mealybug to become a brownish to orange, barrel-shaped mummy. After pupating, the emerging adult wasp chews a round hole through the mummy and exits the mealybug it killed.

The woolly whitefly is named for the numerous fine, curly, waxy filaments that cover the third instar and pupa (fourth instar). This species is under complete biological control in southern interior and coastal growing areas of California. It is an occasional pest in the San Joaquin Valley and desert growing areas, where hot weather disrupts its natural enemies.

Female whiteflies lay eggs as shown here in circles, partial circles, or scattered haphazardly, usually on young succulent leaf flush. Named for their flylike shape and pale wax covering, adults of most species are similar in appearance to this woolly whitefly.

In the field, whiteflies are identified to species primarily by the appearance of their fourth instars (pupae). This pupa of the bayberry whitefly has a clear wax fringe around the body margin.

Giant whitefly pupae are distinctly elevated in profile, with edges perpendicular to the leaf surface. Some of these pupae are dark because they have been parasitized by *Entedononecremnus krauteri*. A round hole chewed by an emerging adult parasitic wasp is visible at the upper right.

Empty pupal skins from which a whitefly adult emerged are clear with a ragged or T-shaped opening. One of these bayberry whitefly skins (lower right) has a round exit hole from the emergence of an *Encarsia* parasite. The dark cast skin and meconia from the parasite are also visible inside.

The pupa of the citrus whitefly has no waxy fringe or filaments, but it does have a distinct Y-shaped marking.

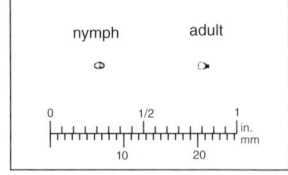

Whiteflies

Several whitefly species occur in citrus, but they are usually under effective biological control. The bayberry whitefly (*Parabemisia myricae*), citrus whitefly (*Dialeurodes citri*), and woolly whitefly (*Aleurothrixus floccosus*) are most likely to be seen. At least five other whitefly species can occur in citrus in California, but they are rarely or never abundant enough to warrant insecticide application.

Description and Biology

Whiteflies are named for the pale, powdery wax covering their body and wings. Female whiteflies lay oval, pale eggs, mostly on the undersides of young leaves. The first-instar nymphs are initially mobile (crawlers). After settling to feed, nymphs lose their legs and remain at the same spot until they mature into adults. Nymphs of most whitefly species are flattened and oval, and in many species they resemble nymphs of certain soft scales.

Although wing markings help distinguish certain species,

Adults

Ash whitefly, *Siphoninus phillyreae*	Bayberry whitefly, *Parabemisia myricae*	Mulberry whitefly, *Tetraleurodes mori*	Woolly whitefly, *Aleurothrixus floccosus*	Nesting whitefly, *Paraleyrodes minei*	Greenhouse whitefly, *Trialeurodes vaporariorum*	Citrus whitefly, *Dialeurodes citri*	Giant whitefly, *Aleurodicus dugesii*
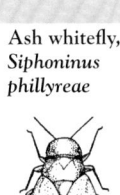	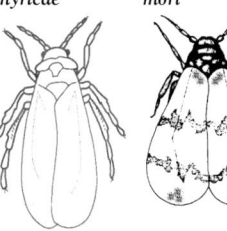	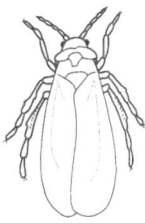	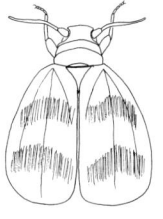	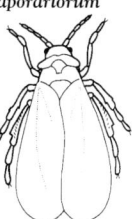			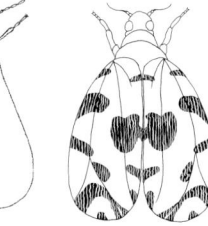
No marks on white wings.	No marks on pearl-colored wings.	Reddish to gray wing markings, including a blotch at the base and tip of each wing.	No marks on white wings.	Dusky blotches on forewings.	No markings on white wings. Wings usually held flat—parallel to top of its yellowish body. Often powdery wax around eggs.	No markings on wings, which are often held relatively flat. Often leaves light powdery wax where eggs are laid.	Grayish blotches or mottling on wings. Larger that other species, about 3/16 inch (4 mm) long. Often feeds in groups. Makes spiraling wax patterns during egg laying on leaves.

Pupae

	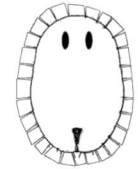	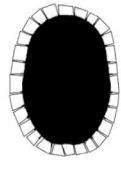	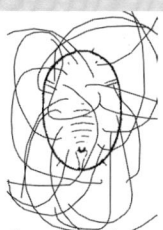	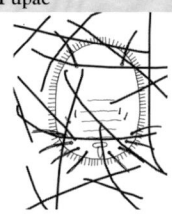	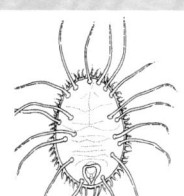	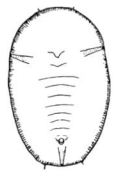	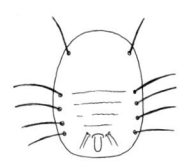
One broad band of wax, or two closely adjacent bands, down their back. The body is ringed with tiny tubes, which often have a liquid droplet at their end.	A pale yellowish to translucent body fringed with broad, flat ribbons of clear wax, which may easily be overlooked.	A black body fringed with broad, flat ribbons of white wax that do not curl over the body.	Covered with long, thin, curly wax filaments, which may entirely obscure the pupa's body.	Pale green to orangish with a fringe of many short filaments around the margin. Produces wax strands thicker than those of woolly whitefly, so that insects appear to be loosely covered with tiny spaghetti noodles.	Long submarginal waxy filaments and a marginal fringe of many short filaments. Elevated in profile, with edges parallel to the leaf surface.	Yellow to transparent, oval, and relatively flat in profile. No waxy fringe or filaments. Distinct Y-shaped pattern from three spiracular (breathing tube) furrows, one on each side and at the rear.	Several filaments around the margin and body covered with powdery wax. Distinctly elevated in profile with edges parallel to the leaf surface. Most prominent are long wax strands, ≥1 inch in length (not shown), hanging from infested leaves, giving foliage a white, bearded appearance.

Figure 38. Whitefly species that occur on citrus in California can be distinguished by pupal shape, filaments, waxiness, and sometimes by adult wing markings. *Sources*: Dreistadt 2002, Flint 1995, giant and nesting whiteflies by David H. Headrick.

The undersides of these citrus leaves are infested with wax-covered woolly whitefly nymphs. When whiteflies are abundant, it is usually because their biological control has been disrupted by ants, dust, pesticides, or weather.

adults of most whitefly species are similar in appearance. The fourth-instar or pupal stage is distinctive, so pupae are used in the field to distinguish whiteflies to species. Pupae (fourth instars) of the giant whitefly, nesting whitefly, and woolly whitefly are covered with long waxy filaments or strands, and this waxiness differs between these species (Figure 38). Citrus whitefly pupae lack waxy filaments and marginal fringes but have a distinctive Y-shape on the back.

After the distinctive fourth nymphal (pupal) stage, a winged adult emerges through a ragged or T-shaped split in the pupal skin. If the whitefly was parasitized, the empty skin has an oval or round hole chewed by the emerging adult wasp and the empty pupal case is often dark or discolored.

Damage

Whiteflies suck phloem sap, causing leaves to wilt and drop prematurely when whiteflies are numerous. Nymphs excrete sticky honeydew that collects dust and supports the growth of blackish sooty mold. Large infestations can almost blacken entire trees, including fruit, because of extensive sooty mold growth. Honeydew attracts ants, which disrupt the biological control of the whiteflies and other pests.

Natural Enemies

Natural enemies provide partial to complete biological control of all species of whiteflies in most locations when undisturbed by ants, dust, insecticide treatment, or weather. Tiny parasitic wasps are especially important, such as *Encarsia* and *Eretmocerus* species that attack the whitefly nymphs, as listed earlier in Table 7. Parasitism is enhanced by using alternate row pruning to stagger the subsequent new flush and provide refuge for parasites that emerge from late-instar nymphs and whitefly pupae that occur on older foliage. Green lacewings (*Chrysopa* and *Chrysoperla* spp.), pirate bugs (*Orius* spp.), and certain spiders discussed earlier under "General Predators" also help control whiteflies. Certain tiny black lady beetles (*Delphastus* spp.) are specialized predators that feed on all whitefly life stages.

As the *Amitus spiniferus* larva matures within a woolly whitefly, the parasite's body darkens to brown then black and becomes visible through the body surface of the host.

Various species of tiny wasps are the most important natural enemies of most whitefly species. This *Amitus spiniferus* (family Platygastridae) is laying an egg into a woolly whitefly nymph.

Cales noacki (family Aphelinidae) are tiny orangish wasps that parasitize early instars of the woolly whitefly.

Delphastus spp. are small, mostly black lady beetles that feed on many different species of whitefly.

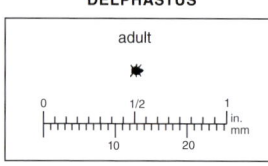

DELPHASTUS

The mulberry whitefly is occasionally seen on citrus. Its pupa is black with a white fringe. The other whitefly species in citrus have lighter colored to translucent bodies, except when they are parasitized by a certain dark species of wasps.

INTEGRATED PEST MANAGEMENT FOR CITRUS 161

Aphid feeding on new growth flushes commonly distorts leaves, but citrus can tolerate extensive leaf curling without reduced yield. Usually no control is needed. Aphid honeydew provides food for adult natural enemies when other food is not available. Be sure to identify the cause of distorted leaves, as other pests, such as the exotic Asian citrus psyllid, also curl and distort shoots.

Monitoring

Throughout summer, inspect plants for whiteflies where they may be a problem, such as when biological control is disrupted. Examine the underside of leaves directly above areas with shiny honeydew or blackish sooty mold. For more current recommendations, consult the online UC IPM Pest Management Guidelines: Citrus.

Aphids

Aphids cause curling of new flush leaves and produce honeydew. However, direct damage by aphids is rare for mature citrus because aphids are easily reduced by natural enemies, and citrus can tolerate significant foliar damage without effects on yield or vigor of the trees. Pesticide application may be warranted for young, vigorously flushing trees if aphid infestations are prolonged.

The most common species of aphids infesting citrus in California include the spirea aphid (*Aphis spiraecola*), the cotton (or melon) aphid (*Aphis gossypii*), and the black citrus aphid (*Toxoptera aurantii*). Spirea aphid is most common on the coast, and cotton aphid is most common in the San Joaquin Valley. All three species can transmit the *Citrus tristeza virus* and so indirectly affect citrus production. While their vectoring ability is fairly inefficient, they can be significant vectors of certain strains of the virus. In recent years, there has been an increase in the appearance of severe strains of *Citrus tristeza virus* in California, thus their role as vectors may increase with time. In other areas of the world, the brown citrus aphid (*Toxoptera citricida*) is a serious pest because it is a very efficient vector of *Citrus tristeza virus*, and the virus tends to increase in severity when transmitted by this aphid species. See the "Tristeza" section in the chapter "Diseases" for more information.

Description and Biology

Adult aphids are pear shaped with long legs and antennae. A pair of tubelike projections (cornicles) near their hind

Spirea aphid is the most common aphid in southern California citrus. Healthy spirea aphids are always bright green, but fungal diseases and parasitic wasps can cause their color to change.

The cotton, or melon, aphid usually occurs in colonies composed of several color forms. Individuals range in color from yellow, to green, to dull black. The puffy tan aphid (a mummy) is parasitized by a *Lysiphlebus* sp. wasp.

Both the black citrus aphid and the exotic brown citrus aphid have dark-colored bodies. Shown here are black winged and wingless adults and dark brownish nymphs of the black citrus aphid.

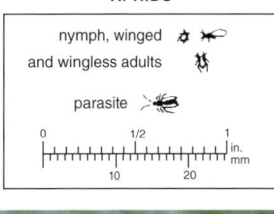

Cotton, or melon, aphid (shown here) and spirea aphid do not have a dark cell in their wings. Their mostly clear wings distinguish these species from the black and brown citrus aphids.

Winged black citrus aphids have a distinct black streak (darkened pterostigma cell) along the bottom of their wings. Be alert for the exotic brown citrus aphid, which also has a dark wing cell. These two species can be distinguished by their wing veins and antennal banding, as illustrated in the aphid key.

Parts of an aphid useful in distinguishing the species.

Read both descriptions and compare the specimen with the drawings provided before proceeding. Take several specimens through the key individually to verify your identification.

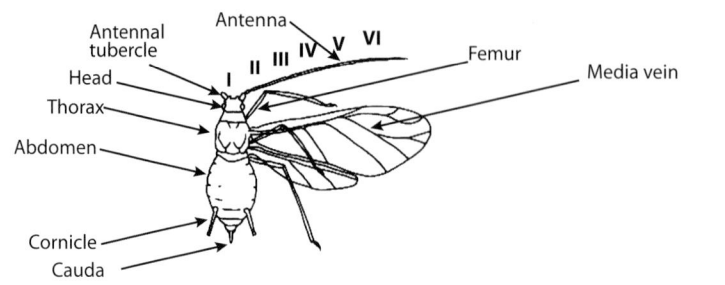

Key to Winged Aphids Potentially Infesting California Citrus

1A. Frontal antennal tubercles prominent, cornicles and cauda long. Go to 2.	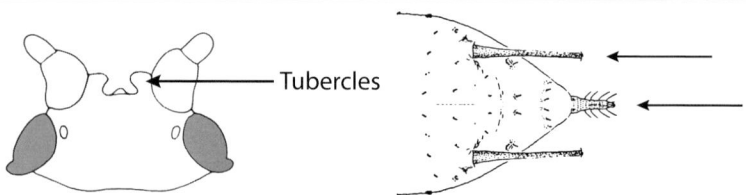	

1B. Frontal antennal tubercles not prominent, cornicles and cauda short. Go to 3.	

2A. Frontal antennal tubercles strongly converging (pointing inward). *Myzus persicae* (green peach aphid)	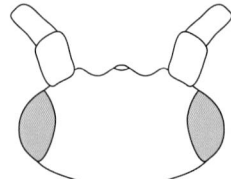 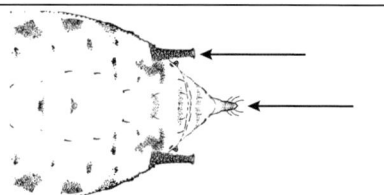

2B. Frontal antennal tubercles diverging (sloping outward). *Macrosiphum euphorbiae* (potato aphid)	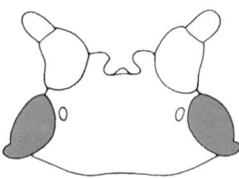

3A. Media vein branched once. *Toxoptera aurantii* (black citrus aphid)	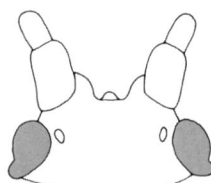 

3B. Media vein branched twice. Go to 4.	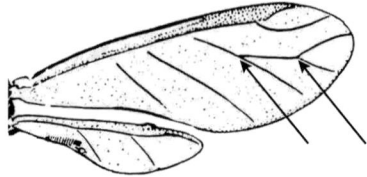

4A. Antennal segment III dark, segments IV-V banded. *Toxoptera citricida* (brown citrus aphid) (Not found in California as of 2010)	

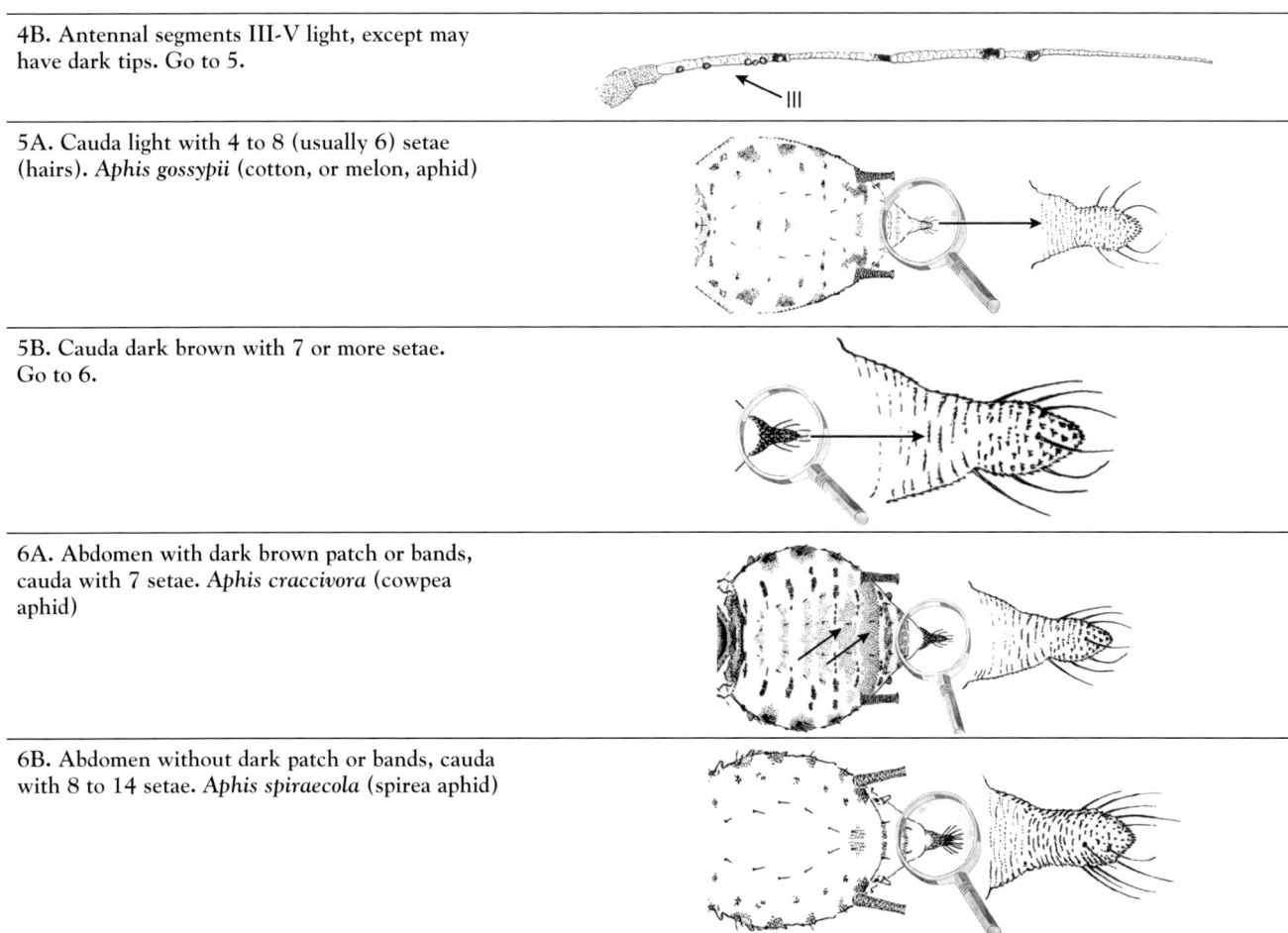

Figure 39a. Key to winged aphids in California citrus, including the brown citrus aphid not yet found in California.

end distinguishes most aphids from other insects. Aphid species resemble each other, but their color helps distinguish the species in citrus. Spirea aphids are always bright green. Cotton aphids vary from yellow to grayish green to dull black, and a colony of cotton aphids is usually composed of several color forms. Black citrus aphids have dark bodies.

Learn how to identify the aphid species in your orchard using the illustrated keys (Figures 39a, 39b). It is especially important that you learn to distinguish the black citrus aphid (already established in California) from the brown citrus aphid (an exotic pest found in Florida and Mexico). Promptly report any suspected brown citrus aphids or other new pests to agricultural officials.

Aphids have multiple overlapping generations, and the adults (either winged or wingless) give birth to live young without mating throughout most of the year. When weather is warm and they are feeding on succulent foliage, aphids may complete a generation in less than 2 weeks.

Damage

Aphids suck phloem sap from buds and the underside of leaves (mainly feather growth), producing honeydew on which blackish sooty mold grows. Aphid feeding causes leaves to curl, but citrus tolerates extensive leaf curling without reduced yield. Thus, treatments are not generally needed.

Black citrus aphid, cotton (or melon) aphid, and spirea aphid can acquire *Citrus tristeza virus* during feeding on an infected tree, then transmit the virus when they move and feed on a new tree. Because the aphid can transmit the virus faster than insecticides can kill it and because the transmission rate is fairly low, insecticide treatments are rarely applied for aphids. However, as severe strains of *Citrus tristeza virus* increase in California control of aphid populations may become part of the program to reduce disease spread. Consult the online *UC IPM Pest Management Guidelines: Citrus* for current recommendations.

Natural Enemies

Beneficial fungal diseases kill many aphids when conditions are warm and humid. Predators and parasitic wasps usually keep aphid populations below damaging levels when ants and dust are controlled. Aphid-feeding general predators include green lacewings and minute pirate bugs. Larvae of

Key to Nonwinged Aphids Potentially Infesting California Citrus

1A. Frontal antennal tubercles prominent, cornicles and cauda long. Go to 2.

1B. Frontal antennal tubercles not prominent, cornicles and cauda short. Go to 3.

2A. Frontal antennal tubercles strongly converging (pointing inward). *Myzus persicae* (green peach aphid)

2B. Frontal antennal tubercles diverging (sloping outward). *Macrosiphum euphorbiae* (potato aphid)

3A. Dorsal (top of) abdomen with dark patch. *Aphis craccivora* (cowpea aphid)

3B. Dorsum without a dark patch. Go to 4.

4A. Cauda with > 8 hairs (setae), body dark brown or black. Go to 5.

4B. Cauda with 4 to 11 hairs, body yellow to dark green. Go to 6.

5A. Antennal segments III and IV banded, segment VI light, cauda with 9 to 19 hairs. *Toxoptera aurantii* (black citrus aphid)

INTEGRATED PEST MANAGEMENT FOR CITRUS   165

5B. Antennal segments III–IV light, segment VI dark, cauda with ≥ 25 hairs. *Toxoptera citricida* (brown citrus aphid)

(Not found in California as of 2010)

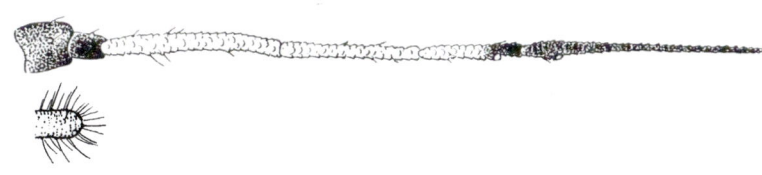

6A. Cauda light, with 4 to 8 (usually 6) hairs, femoral hairs short. *Aphis gossypii* (cotton, or melon, aphid)

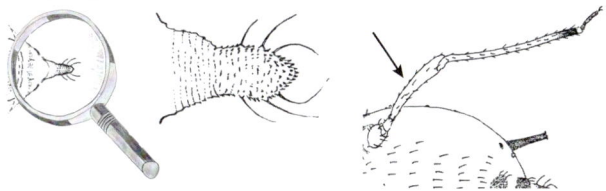

6B. Cauda dark, with 8 to 11 hairs, femoral hairs long and fine. *Aphis spiraecola* (spirea aphid)

Key adapted from E. E. Grafton-Cardwell undated, *Aphis* spp. and *Macrosiphum* adapted from Kono and Papp 1977, *Toxoptera* spp. from Ebeling 1959, Essig 1949.

Figure 39b. Key to nonwinged aphids in California citrus, including the brown citrus aphid not yet found in California. See Figure 39a for instructions on using this key.

Adult parasites chew a round hole when emerging from aphids they have killed. Sometimes the cut flap stays attached to the mummy as if hinged (photo top). This *Lysiphlebus* sp. and other wasps in the family Aphidiidae usually leave golden or tan mummies. Other wasp species (family Aphelinidae) typically form blackish mummies.

These dark spirea aphids have been killed by a naturally occurring fungal disease. Aphids are very susceptible to disease when they are abundant and conditions are humid.

Syrphid fly larvae feed mostly on aphids, such as these spirea aphids. Syrphid adults, also called hover flies or flower flies, resemble honey bees. An adult syrphid is pictured in "General Predators" earlier in this chapter.

This pale, oblong syrphid egg was laid near aphids on which the larva will feed. Unlike the similar eggs of brown lacewings, syrphid eggs do not have a knob on one end.

many syrphid flies and certain species of lady beetles are specialized aphid feeders. A moderate aphid population (about 40% of growth flushes infested) can be beneficial on mature trees because aphids and their honeydew provide a good food source for many natural enemies of other pests early in the season when other hosts are not available.

Asian Citrus Psyllid

Diaphorina citri

Asian citrus psyllid feeds on many plants in the family Rutaceae, preferring *Citrus* spp. and close relatives such as ornamental orange jasmine (*Murraya paniculata*). Originally from Asia, this exotic, aphidlike insect now occurs throughout Central and South America, the southeastern United States, and along the Mexico-California border. Learn to recognize Asian citrus psyllid and its damage, including symptoms of the citrus greening disease it vectors, and promptly report any suspected findings to agricultural officials.

Description and Biology

Adult Asian citrus psyllids are brownish, with a bright yellow-orange abdomen on egg-laying females. Adults are ⅛ to ⅙ inch (3–4 mm) long and readily fly a short distance when disturbed. Females lay bright yellow-orange, almond-shaped eggs on succulent shoots and "feather flush" leaves. The nymphs are flattened, yellowish orange, and excrete abundant white wax.

Damage

Asian citrus psyllid nymphs and adults suck phloem sap and excrete copious amounts of honeydew on which blackish sooty mold grows. During feeding, the psyllids inject a toxic saliva that permanently distorts leaves and shoots. New shoot growth that is heavily infested by psyllids does not expand or develop normally and is more susceptible to breaking off. Most important, this psyllid vectors *Candidatus* Liberibacter spp. of bacteria that cause Huanglongbing (citrus greening disease).

Psyllids acquire bacteria when feeding on infected hosts, then transmit the bacteria when they move and feed on other plants. These phloem-inhabiting pathogens cause yellowing of citrus shoots and an asymmetrical mottling of leaves. Affected fruit are often lopsided, fail to turn a mature color, and develop a highly acidic or bitter flavor that makes them unmarketable. Trees die within 3 to 5 years of infection. For more discussion see "Huanglongbing (Citrus Greening)" in the chapter "Diseases" and *Asian Citrus Psyllid* (Grafton-Cardwell et al. 2006).

Natural Enemies

Biological control is not relied on in California, where Asian citrus psyllids are quarantined and targeted for eradication with insecticides. In other areas of the United States, important natural enemies include parasitic wasps and generalist predators such as lacewings, minute pirate bugs, syrphids (hover flies), and various spiders. The ashy gray lady beetle (*Olla v-nigrum*) and multicolored Asian lady beetle (*Harmonia axyridis*) are the most important predators of Asian citrus psyllid nymphs in Florida. These lady beetles occur throughout California and are pictured earlier in the "General Predators" section. Both adults and larvae of these lady beetles feed on various soft-bodied pests, including psyllids.

A parasitic wasp (*Tamarixia radiata*) imported from Asia into Florida kills psyllid nymphs. The female wasp lays its eggs underneath the nymph, and the wasp larva feeds attached to the outside of older nymphs. The parasite

Asian citrus psyllid adults are brownish. They readily fly when disturbed. When resting, they tilt their rear end up at a 45° angle. Report any suspected Asian citrus psyllids to agricultural officials. This exotic pest vectors Huanglongbing, a bacterial disease lethal to citrus.

Asian citrus psyllid eggs and recently hatched nymphs are yellowish orange. These usually occur on the tips of growing shoots or in the crevices of unfolded "feather flush" leaves, where females prefer to lay their almond-shaped eggs.

The flattened, yellowish orange nymphs of Asian citrus psyllid excrete abundant white wax tubules. These nymphs are found on the new flush growth, where their feeding curls leaves and distorts shoots as pictured at the beginning of the chapter "Managing Pests in Citrus."

emerges as an adult from underneath the dead psyllid or through a hole that it chews in the body of the mummified psyllid.

Monitoring

Adult psyllids can be detected by visually inspecting plants and using yellow sticky cards to trap the winged adults. Examine new leaf shoots (feather flush) to detect the yellowish orange psyllid eggs and nymphs. Also watch for Huanglongbing symptoms as described above and in the chapter "Diseases." Report to agricultural officials any psyllids or suspected Huanglongbing symptoms found outside areas where they are known to occur.

Glassy-Winged Sharpshooter

Homalodisca vitripennis (=*H. coagulata*)

Sharpshooters and other leafhoppers (family Cicadellidae) are active insects that walk rapidly sideways or readily jump when disturbed. Glassy-winged sharpshooter feeds, reproduces, and is often abundant on numerous agricultural crops and ornamental plants. Although populations are rarely high enough to damage citrus, glassy-winged sharpshooter may need to be managed because of quarantines to protect grapes against the *Xylella* bacterium that causes Pierce's disease.

If the strain of *Xylella* that causes citrus variegated chlorosis were to arrive in California, glassy-winged sharpshooters would need to be managed to reduce the transmission of that disease in commercial citrus. Citrus variegated chlorosis causes severe yellowing between leaf veins, resembling certain nutritional deficiencies. Affected leaves frequently develop brown lesions. Infected trees may exhibit reduced vigor and growth and abnormal flowering and fruit set. Affected fruits are often small and hard with high acid content; they are not acceptable for either fresh market or juice. Report to agricultural officials any suspected findings of this disease or other exotic pests. See "Citrus Variegated Chlorosis" in the chapter "Diseases" for more discussion.

Description and Biology

The adult glassy-winged sharpshooter is about ½ inch (13 mm) long, larger than most other leafhopper species. Adults are mostly dark brown to blackish with a whitish or yellow abdomen. Their mostly dark head has numerous, tiny, ivory to yellowish spots on top. These head spots distinguish glassy-winged sharpshooters from smoke-tree sharpshooters, which have pale, wavy lines on their head.

Female glassy-winged sharpshooters lay eggs in masses of about 5 to 15 into foliage, just under the lower surface of young, fully developed leaves. When first laid, an egg mass

The adult glassy-winged sharpshooter is mostly dark brown with white and yellowish patches and spots.

Glassy-winged sharpshooter nymphs (photo right) are uniformly brown to olive-gray and wingless. Nymphs have prominent, bulging red eyes. Nymphs of the similar smoke-tree sharpshooter have blue eyes.

Glassy-winged sharpshooter causes a greenish blister in leaves where it has inserted an egg mass. The left leaf contains two groups of unhatched eggs. After eggs hatch, a permanent brown to gray scar remains in tissue, as in the right leaf.

This cluster of unhatched glassy-winged sharpshooter eggs inserted into foliage causes a slight bulge or blister in the leaf surface. The female covers the leaf blister with a white, chalky secretion that often is more visible than the leaf blister.

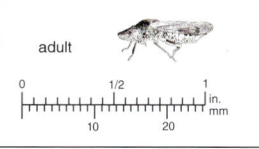

GLASSY-WINGED SHARPSHOOTER

Chalky, white excrement (sharpshooter rain) forms on plants infested with glassy-winged sharpshooters.

These glassy-winged sharpshooter eggs were killed by parasites, as evidenced by a tiny, round hole in each egg chewed by the *Gonatocerus* adult parasite that emerged. When eggs are not parasitized, glassy-winged sharpshooter nymphs leave a hard-to-see slit when they emerge.

This *Gonatocerus ashmeadi* is one of several tiny egg-parasitic wasps that help control glassy-winged sharpshooter. *Gonatocerus ashmeadi* is common wherever glassy-winged sharpshooter occurs in California.

resembles a greenish blister on the leaf. The female covers the leaf blister with a white, chalky secretion that often is more visible than the leaf blister. Eggs turn brown as they mature, and the nymphs hatch about 10 to 14 days after eggs are laid. A permanent brown to gray scar remains in leaf tissue after the nymphs emerge.

The glassy-winged sharpshooter has two generations per year in California. Eggs are laid in citrus during late March through April and again in late June through August. Eggs laid in spring produce nymphs that mature into first-generation adults beginning in mid-June. The number of adults, and the abundance of their eggs, continues to increase through July and August. Second-generation adults appear from late summer through fall. In addition to citrus, glassy-winged sharpshooter overwinters in various ornamentals, trees, and weeds.

Damage

Glassy-winged sharpshooters feed by sucking from the nutrient-poor xylem (water-conducting system) of the plant and must consume copious amounts of fluid to gain enough nutrition. During feeding, adults and nymphs excrete large amounts of liquid that dries and leaves a pale residue that gives fruit and foliage a whitewashed appearance.

Extremely high populations of glassy-winged sharpshooter can reduce fruit quality and yield, at least in coastal lemons and Valencias in southern California. Such very high populations are rare because of parasitism and treatments to reduce sharpshooter populations that might contaminate harvested citrus or move to infest grapes.

The glassy-winged sharpshooter is a serious pest of grapes because it vectors the Pierce's disease strain of *Xylella fastidiosa*. Different strains of this bacterium cause disease in other plants, such as almond leaf scorch and oleander leaf scorch. The bacteria multiply and block the water-conducting xylem, causing water stress and eventual plant death. There is no known cure for *Xylella* diseases. Because glassy-winged sharpshooters often overwinter in abundance in citrus, citrus acts as a source of sharpshooters for neighboring vineyards. Areawide treatment programs in citrus reduce the impact of glassy-winged sharpshooters on neighboring grapes. Glassy-winged sharpshooter reportedly vectors the *Xylella fastidiosa* strain that occurs on citrus elsewhere in the world and causes citrus variegated chlorosis.

Natural Enemies

Several tiny wasps (*Gonatocerus* spp., family Mymaridae) parasitize glassy-winged sharpshooter eggs in California. Where possible, avoid insecticides that can disrupt biological control because these parasites introduced in urban areas have significantly reduced sharpshooter populations, at least on ornamentals and unmanaged vegetation.

The remains of parasitized eggs are easily recognized by the tiny, round hole at one end of the glassy-winged sharpshooter egg through which the adult parasite emerged. The egg-parasitic wasp *Gonatocerus ashmeadi* is common wherever glassy-winged sharpshooter occurs in California. In the southern interior and coastal areas of California, *Gonatocerus walkerjonesi* can be a very effective parasite in the late summer when the second-generation eggs are deposited. Neither *G. ashmeadi* nor *G. walkerjonesi*, however, are normally present at high levels during the first generation of glassy-winged sharpshooter egg laying.

Monitoring

Glassy-winged sharpshooter monitoring methods vary by location, time of year, and the purpose for monitoring. Yellow sticky traps for adults are used to detect whether glassy-winged sharpshooters have been introduced into new areas and to assess whether they are abundant or migrating around orchards.

Use branch beating or timed counts to assess the need for and effectiveness of suppression treatments to reduce sharpshooter spread from citrus to *Xylella*-susceptible neighboring crops such as grapes. During cool weather (winter, early spring), beat or shake one group of branches on each of 20 citrus trees per 10-acre (4 ha) block. Count the number of glassy-winged sharpshooter adults and nymphs that you dislodge onto a collecting sheet placed beneath branches. During warmer weather, especially when egg masses are present (during April and June–August), conduct a timed search. During a 3- to 5-minute examination of each of 20 trees per 10-acre block, count the number of adults, nymphs, and live egg masses observed.

Use sweep net shakes to determine whether trees need to be treated (disinfested) immediately before harvest so that fruit can be shipped from a generally infested region (such as southern California) to an uninfested area for packing. To detect mobile stages of glassy-winged sharpshooter, stuff citrus foliage into a sweep net and shake vigorously. Inspect the contents of the net for any live glassy-winged sharpshooter adults or nymphs.

Monitoring recommendations may change, for example, if glassy-winged sharpshooter becomes more widely established or new pathogenic strains of *Xylella* are introduced into California. For current recommendations, consult the online *UC IPM Pest Management Guidelines: Citrus*.

Potato Leafhopper

Empoasca fabae

The potato leafhopper (family Cicadellidae) is most likely to be a pest of citrus in San Joaquin Valley orchards near the leafhopper's alternate hosts, including cotton, tomatoes, and pastures. Potato leafhopper breeds in large numbers on wild plants and field crops, and during late summer and fall the leafhoppers sometimes migrate to overwinter in citrus trees.

Potato leafhoppers are greenish, slender, wedge-shaped insects with bristlelike antennae and rows of spines along the hind legs. Adults are about 1/8 inch (3 mm) long, and both adults and nymphs move rapidly forward, backward, and from side to side.

Beet leafhopper (*Circulifer tenellus*) is a similar-looking species that is pale green to gray or light brown and feeds on many of the same plants as potato leafhopper. Beet leafhopper is the primary vector of the phytoplasma that causes stubborn disease in citrus. However, managing leafhoppers has not been found to reduce the incidence of stubborn disease.

Leafhoppers feed by puncturing plant cells and sucking their liquid. Potato leafhopper feeding on citrus rinds causes yellowish to light brown, roundish scars on fruit. The scars are particularly apparent on green fruit and resemble thrips oviposition scars, except leafhopper scars are more clustered and do not have darkened centers.

Biological control is apparently not important in managing potato leafhopper. Where leafhoppers are a problem, see the online *UC IPM Pest Management Guidelines: Citrus* for recommendations.

Leafhoppers have one or more rows of spines along their hind legs. This potato leafhopper is a slender green insect that can be a pest of citrus in the San Joaquin Valley, usually in orchards near cotton, tomatoes, and pastures or near the foothills.

Potato leafhopper feeding causes roundish discoloring on fruit, and the injured blotches are typically clustered in groups. Scars usually develop on older fruit during late summer or fall, when leafhoppers migrate into citrus.

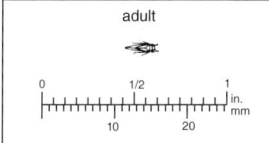

LEAFHOPPER

This nymph with banded antennae is the damaging stage of the forktailed bush katydid. During spring, it feeds on young fruit.

Eggs of the forktailed katydid are usually inserted into the edges of old, tough citrus leaves. Katydid eggs are relatively flat, with an oblong to oval shape that becomes visible when an infested leaf is held against bright light.

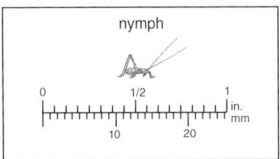

Forktailed katydid nymph feeding on young fruit causes scars that are deeper than damage from citrus thrips or mechanical injury. Where leaves have small amounts of fresh chewing damage during spring, inspect nearby fruit for forktailed katydid feeding.

Forktailed katydid damage is typically a single circular scar, usually in the midsection of fruit. Damage often occurs on the sheltered side of fruit facing in toward the trunk. This injury occurred when a nymph chewed the small, immature fruit as in above photo.

CHEWING INSECTS

Ants and brown garden snail (discussed later in this chapter), caterpillars (see above), and several types of insects discussed below chew the bark, fruit, leaves, or roots of citrus. These pests' identification, biology, natural enemies, and monitoring are discussed below. For current monitoring and management recommendations, consult the online UC IPM *Pest Management Guidelines: Citrus*.

Forktailed Katydid

Scudderia furcata

The forktailed katydid (order Orthoptera), also called forktailed bush katydid, chews young fruit and leaves. The angularwinged katydid (*Microcentrum retinerve*) also feeds on citrus but is not an economic pest because it chews only leaves and is less common.

Description and Biology

Forktailed katydid females usually lay their oblong or oval eggs into the edges of older citrus leaves. Nymphs hatch in April or May, develop through 6 instars, and take 2 to 3 months to mature. Katydid nymphs are wingless and have black-and-white-banded antennae. The early instars are black and white and become green when they molt to the later instars. Katydids have only one generation per year and overwinter as eggs laid during summer or fall.

Angularwinged katydids are larger than forktailed katydids, and the adults have broader wings. Both nymphs and adults have a more humpbacked appearance than forktailed katydid. Angularwinged katydid nymphs do not have the distinct light and dark antennal bands that are present on forktailed katydid. Angularwinged katydid females lay gray, oval eggs in a group of two overlapping rows on the surface of twigs and leaves.

Damage

The forktailed katydid causes economic damage because nymphs chew young fruit just after petal fall. As the damaged fruit expand, a circular, scabby scar and rind tissue distortion develop around the feeding site. Forktailed katydids usually take a single bite from a fruit and then move to feed on another nearby fruit. In this way, a few katydids can damage a large quantity of fruit in a short time. Nymphs and adults also eat holes in new flush and maturing fruit, creating injury that resembles damage by citrus cutworm.

Natural Enemies

Parasitic wasps attack katydid eggs, and various predators eat nymphs and adults. However, natural enemies are not always sufficiently effective to prevent economic damage because after petal fall, relatively few katydids can scar many fruit in a short amount of time.

Monitoring

Before petal fall, katydid nymphs will be feeding on the newly expanding leaf flush. Look for katydid nymphs starting in April before young fruit are present. Approach the tree slowly, because katydids have excellent eyesight and will hide behind leaves or jump away if they see you moving quickly. Look for chewed leaves and search in that area for katydids. If you find forktailed katydids, consult the online *UC IPM Pest Management Guidelines: Citrus* for recommendations. The tolerance for katydid is higher before petal fall, when they are feeding on leaves, than after petal fall, when they are feeding on fruit. Slower-acting insecticides can be used before petal fall, when the risk for damage is low.

Grasshoppers

The devastating grasshopper (*Melanoplus devastator*), valley grasshopper (*Oedaleonotus enigma*), and other species of grasshoppers (order Orthoptera) become pests when they migrate in large numbers from nearby vegetation and feed on young trees.

Description and Biology

Grasshoppers are robust, elongate insects. Commonly they are brown, gray, green, or yellowish with greatly enlarged hind-leg femurs adapted for jumping. Grasshoppers have relatively short antennae, which distinguishes them from katydids and most other Orthoptera, which have long antennae.

Most species of grasshopper have only one generation per year and overwinter as eggs. Eggs hatch when soil warms in spring, usually about April. The nymphs feed on almost any species of nearby green plant, molting five or six times before becoming adults. Adults live and feed for 2 to 3 months, during which time females typically deposit elongate pods of about 20 to 100 eggs in the topsoil of undisturbed areas.

Nymphs and adults readily move, and each individual typically feeds on several different plants during its lifetime. As vegetation is consumed or dries when the rainy season ends, grasshoppers migrate to succulent plants. Adults, sometimes in a large swarm, can fly several miles a day. Nymphs readily jump and walk to other plants or are carried by wind.

Damage

About April, nymphs begin feeding on grasses and other vegetation in unmanaged areas. Grasshoppers become economic pests when large numbers migrate into orchards and extensively chew young tree foliage. Mature trees are not harmed by grasshopper feeding.

Grasshopper populations vary from year to year. Grasshoppers become more numerous after warm, moist

Grasshoppers have greatly enlarged hind-leg femurs adapted for jumping, and the winged adults are good fliers. Large numbers of grasshoppers, such as this gray bird, or vagrant, grasshopper (*Schistocerca nitens*), occasionally migrate into citrus and seriously damage young trees.

Leaves on this young citrus tree were extensively chewed by grasshoppers. This tree is on the edge of an orchard adjacent to unmanaged vegetation, from which grasshoppers migrated when the weeds dried.

springs produce abundant vegetation in uncultivated areas, favoring grasshopper survival.

Natural Enemies

Grasshoppers are eaten by arboreal predators such as birds and robber flies (family Asilidae) and soil-dwelling egg predators such as blister beetles (Meloidae). In unmanaged vegetation, a combination of predators, parasites, and pathogens (bacteria, fungi, and protozoa) can cause grasshopper populations to crash. However, once large numbers of grasshoppers have migrated in to feed on young trees, biological control appears to be of little benefit in preventing economic damage to citrus.

Monitoring

Do not take control action based solely on damage. Caterpillars, earwigs, Fuller rose beetle, and snails also chew leaves. Where grasshoppers are abundant enough to be problems on young trees, they can be observed during the day feeding openly and flying or jumping among plants.

Monitor for grasshoppers in uncultivated areas near young trees if migrating grasshoppers have been a problem in the past or you believe grasshoppers may become a problem, for example, because spring weather has been rainy and warm. Monitoring of adjacent vegetation may be warranted to assess whether to manage grasshoppers there before they migrate. Grasshoppers can be difficult to manage once large numbers move onto young trees. For current recommendations, consult the online *UC IPM Pest Management Guidelines: Citrus*.

European Earwig

Forficula auricularia

Earwigs (family Forficulidae) feed on dead and living insects and other small organisms. They also feed on succulent plant parts and occasionally damage buds and leaves on young or newly grafted trees. Earwigs also chew young fruit, but in most orchards other pests are much more common causes of rind scarring.

The introduced European earwig is the most common of several earwig species that can occur in citrus. Adults are about ¾ inch (19 mm) long, reddish brown, and have a pair of prominent tail appendages that resemble forceps. Adults have wings under short, hard wing covers, but earwigs seldom fly.

Females lay masses of 30 or more eggs in the soil. Nymphs are whitish and remain in the soil until their first molt. Second instars and older nymphs are dark and resemble small, wingless adults. Earwigs generally have one or two generations per year and can be active year-round.

Earwigs feed mostly at night. Common hiding places during the day include bark crevices, mulch, topsoil, protected (touching) plant parts, and under trunk wraps. Earwigs can be especially problematic on young trees with trunk wrappers or cardboard guards. The cause of damage can be difficult to determine because it resembles that from other chewing pests that hide during day and feed at night, including brown garden snail and Fuller rose beetle.

Natural enemies are not known to be important in earwig management in citrus. If earwigs are damaging young trees, removing trunk wrappers where pests hide when wraps are no longer needed can eliminate an earwig problem. To determine whether earwigs are the cause of damage on young trees, lift and shake or sharply tap trunk wrappers. Look for earwigs dropping to the ground, where they quickly scurry for cover. On bearing trees with chewing damage, place a white sheet or other light-colored surface on the ground beneath lower, outer canopy foliage. Vigorously shake the foliage, then quickly check the collecting surface for dislodged earwigs and other pests.

European earwig chewing causes irregular scars on small fruit. Damage is most prevalent on young trees with trunk wraps, on lower canopy fruit of trees that are not skirt pruned, and where leaf litter or other debris on soil provides pest shelter.

Earwigs sometimes chew buds and leaves on young or newly grafted trees. However, earwigs are rarely abundant enough to warrant treatment. Removing trunk wraps usually eliminates the problem on young trees.

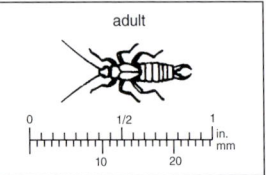

EARWIG

Diaprepes Root Weevil

Diaprepes abbreviatus

Diaprepes root weevil (family Curculionidae), also called citrus root weevil or Diaprepes, occurs in Florida, Texas, and the Caribbean region. Since 2005, introduced populations have been found in southern California.

Adult Diaprepes are strong fliers but tend to remain for life on a good host unless they are disturbed or inadvertently moved by people with infested material. If you find a weevil resembling *Diaprepes abbreviatus* in an area where it is not known to occur, place adults in a small jar filled with rubbing alcohol and take them to your local county agricultural commissioner for positive identification.

Description and Biology

Adult Diaprepes are colorful weevils about ⅜ to ¾ inch (10–19 mm) long. These snout beetles typically have black, lengthwise streaks on their back alternating with variable colors, commonly gray, orange, or yellow. Females oviposit on leaves in a gelatinous mass containing about 24 to 240 eggs. Each egg is oval and about 1/25 inch (1 mm) long. Newly laid eggs are white, but they turn grayish before they hatch about 7 to 10 days after oviposition.

The emerging grublike larvae drop to the soil, where they may wander on the surface before boring underground to feed on roots or the root crown. Larvae develop through 10 or 11 instars over about 4 to 16 months before they mature to a length of about 1 inch (25 mm). Mature larvae form a cell in soil and pupate inside. Adults emerge from the soil and move to leaves, where they feed, mate, and lay eggs. Adults can live for months. Development time from egg to adult ranges from about 5 to 18 months, depending on temperature.

Diaprepes root weevil and Fuller rose beetle have similar life cycles, and their immature stages strongly resemble each other. Adults of both species are nocturnal, hiding during the day, and their larvae live in soil, so these insects are easily overlooked. Many other invertebrates also chew leaves. Make sure that you correctly identify the cause of any damage before you take action. For more information,

Adult Diaprepes root weevils have dark streaks on their back but are otherwise variably colored, ranging from gray to yellow to orange and black.

Diaprepes lays clusters of oblong, pale eggs on foliage. Eggs are typically in a rolled leaf or between two leaves glued together. Report to agricultural officials any findings of the exotic Diaprepes root weevil outside of areas where it is known to occur.

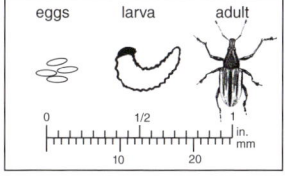

DIAPREPES ROOT WEEVIL

Diaprepes larvae feed underground, chewing the crown and large roots of citrus and many other plants. The immatures of Fuller rose beetle are very similar to those of Diaprepes, so positively identify any weevils in your orchard to determine the appropriate action.

Adult Diaprepes root weevils chew fruit and foliage, causing ragged-edge leaves and the blackish weevil feces (excrement) shown here.

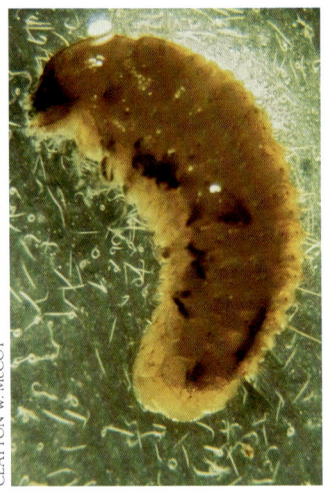

Many tiny entomopathogenic nematodes are emerging from this Diaprepes root weevil they killed. Where Diaprepes is established, entomopathogenic nematodes cause significant larval mortality in sandy, moist soils.

Diaprepes adults emerging from soil can be monitored with this Tedders ground trap. This trap consists of a pyramidal stand that guides the flightless weevils to climb upward through an inverted funnel, where they become captured in a collecting container at the top.

check online at www.ipm.ucdavis.edu and consult publications such as *Diaprepes Root Weevil* (Grafton-Cardwell et al. 2004).

Damage

Diaprepes feeds on more than 270 species of plants in 59 families and can damage or kill citrus and many ornamentals. Larvae chew roots, sometimes girdling the root crown. Before plants die, the foliage may fade and wilt on severely injured trees. Adults prefer to chew young, tender leaves from the leaf edges, causing irregular or semicircular notches. There also may be dark frass (weevil droppings) near foliage with chewing damage.

Natural Enemies

Where Diaprepes root weevil is established and eradication is not feasible, entomopathogenic nematodes provide some control when applied in sandy, moist soils. Introduced wasps that parasitize weevil eggs, and generalist predators that feed on eggs and recently emerged larvae (before larvae burrow into soil), may also be important where Diaprepes is established.

Monitoring

Inspect canopies for feeding damage characteristic of adult weevils, including chewed shoots and succulent leaves with notched edges or semicircular areas of missing tissue along the leaf margins. Look near chewed foliage for dark frass (weevil droppings) and leaves that are folded or stuck together. Open any glued leaves to inspect them for Diaprepes egg clusters. Shake chewed foliage over a light-colored collecting surface (such as a white sheet) to dislodge weevils, which will drop and behave as if they were dead.

Tedders traps and cone traps can be placed on the ground beneath infested trees to monitor the adults as they emerge from their underground pupal cells. If you know Diaprepes is present, you can also scrape away soil at the base of infested plants and look for chewed roots and root crowns and the soil-dwelling weevil larvae and pupae.

Nurseries may be required to follow special inspection and management procedures to ensure that all plants they ship are weevil-free. For more information, contact the local county agricultural commissioner or check online at www.ipm.ucdavis.edu.

Fuller Rose Beetle

Pantomorus cervinus

Fuller rose beetle (family Curculionidae) does not cause economic damage in mature citrus. The presence of its eggs on exported fruit was a quarantine concern in the past. At present, this beetle is rarely of concern except when high populations feed on replants or topworked tree buds.

Adults are grayish to brown, flightless weevils (snout beetles), about ⅜ inch (9 mm) long. In comparison with Diaprepes, Fuller rose beetles are usually smaller and adults lack distinct black streaks and contrasting, variable colors. The shape and orientation of the head distinguish Fuller rose beetle from two other grayish to brown snout beetles that occur in California citrus orchards (Figure 40). These similar-looking species, cribrate weevil and vegetable weevil, do not damage citrus.

All Fuller rose beetles are females that reproduce without mating, producing one generation per year. Yellowish eggs are laid in a mass of several dozen on fruit, especially underneath the sepals, or in cracks and crevices in the tree. When eggs hatch, the yellowish larvae drop to the ground and live in the soil, feeding on citrus roots for 6 to 10 months. They pupate in the soil, and the adults emerge 1½ to 2 months later. Adults are flightless and reach the canopy by climbing up the trunk or branches that touch the ground or vegetation.

Damage

Adults chew along the margins of citrus leaves, creating notches and ragged, sharp edges. Adult feeding is not a concern, except perhaps when common on topworked trees (where the beetles will feed on new buds) or on a young tree planted in a mature orchard (where populations that have built up on established trees move to feed

Adult Fuller rose beetles are brown to grayish snout beetles (weevils). Their hard body narrows toward the head, which has elongated mouthparts.

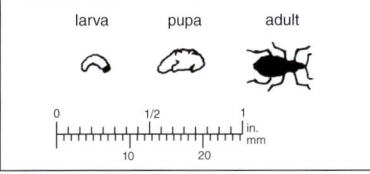

Fuller rose beetle often lays its oblong yellowish eggs on the underside of citrus fruit sepals or in sprinkler heads. Each egg is 1/25 inch (1 mm) long and hatches in about 3 to 13 weeks, depending on temperatures. When parasitized by *Fidiobia citri*, eggs become dark gold colored.

extensively on young trees). Larvae feeding on roots does not damage citrus.

Natural Enemies

An egg parasite (*Fidiobia citri*, family Platygastridae) is commonly present on trees with weevils. The *Fidiobia* adult is a mostly black wasp less than 1/25 inch (1 mm) long. The female wasp lays one egg inside each weevil egg. *Fidiobia* can parasitize up to 50% of the eggs in each Fuller rose beetle egg mass.

To detect parasitism on trees with weevil chewing, clip fruit stems 2 inches (5 cm) from fruit. Hold the stem and twist off the button. Look for masses of oblong weevil eggs on the underside of the calyx and on the fruit where it was covered by the calyx. In contrast to the pale color of unparasitized eggs, parasitized eggs are a dark gold color and may persist long after unparasitized eggs have hatched.

Monitoring

If Fuller rose beetle is a concern, inspect lower foliage for leaf-edge notching from July to November. If chewing is found, shake or beat those branches onto a sheet or tray placed beneath foliage. Fruit can be inspected as described above to look for beetle eggs under sepals and to assess the extent of egg parasitism.

ANTS

In some situations outside of citrus, ants are important natural enemies of insect pests and provide other benefits, such as aerating soil, recycling nutrients, and pollinating plants. In citrus, several ant species are pests because they disrupt biological control, damage plants and irrigation systems, and bite or sting people.

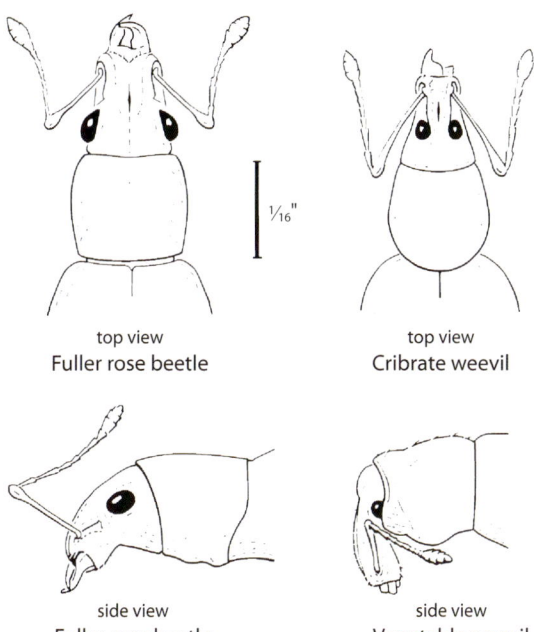

Figure 40. Fuller rose beetle adults can be distinguished from two other nondamaging species of grayish to brown snout beetles that occur in California citrus orchards. When viewed from the top, the Fuller rose beetle head is shaped differently (has more nearly parallel sides) and has bulging eyes in comparison with the cribrate weevil, which has a teardrop-shaped head and closely spaced eyes. Viewed from the side, the Fuller rose beetle's snout is less sharply pointed downward to the ground than that of the vegetable weevil. Adapted from Morse et al. 1987.

Honeydew-feeding ants, such as these Argentine ants tending a black scale, protect pest Homoptera from natural enemies. Ants thereby increase the populations of aphids, mealybugs, soft scales, whiteflies, and other pests.

Description and Biology

Ants are sometimes confused with termites, but ants have a constricted (narrow) waist and elbowed antennae (except males), and the winged reproductives (alates) have hind wings that are shorter than the forewings. Termites, on the other hand, have a broad waist, straight and beadlike antennae, and wings that are similar in size.

Adult ants are divided into three castes: winged males that die after mating, queens that remove their wings after mating, and workers that are sterile females. Workers excavate the nest, defend the colony, forage for food, and care for the brood (eggs, larvae, and pupae). Many ants nest in the soil or beneath rocks, trees, or other objects.

What species of ants are common can differ by location, although the southern fire ant can occur in all major growing areas. The Argentine ant occurs in citrus mostly in coastal and southern interior growing areas. It has been expanding its distribution and may have the potential to occur wherever citrus grows. The native gray ant is common in Central Valley citrus. Other *Formica* spp. can occur almost anywhere in California but are usually not observed where Argentine ants are present. Red imported fire ant occurs in parts of southern interior and coastal California and in a few scattered locations in the San Joaquin Valley. Report this invasive exotic pest to agricultural officials if it is found outside of locations where it is known to occur.

Since the biology and control of each species differs, correct identification is critical. Two key identification characteristics are the number of antennal segments and nodes (projections) on the petiole (the first portion of the abdomen). A 10× magnifier can be used to observe these characteristics and others, as illustrated in Figure 41. Alternatively, consult publications such as *A Key to the Most Common and/or Economically Important Ants of California* (Haney, Philips, and Wagner 1983) or *Key to Identifying Common Household Ants* (online at www.ipm.ucdavis.edu).

Damage

Argentine ants and fire ants can plug up irrigation sprinklers. Fire ants may directly damage plants by chewing twigs and tender bark of young trees. Fire ants may also bite or sting people working in the orchard, which may cause allergic reactions in a small percentage of the victims.

Some ants feed on honeydew excreted by phloem-sucking insects, such as aphids, citricola and other soft scales, mealybugs, and whiteflies. Argentine ants and native gray ants aggressively protect these sources of food by attacking those pests' natural enemies, thereby interfering with biological control. For example, ants can virtually eliminate the parasite *Metaphycus luteolus*, allowing brown soft scale populations to increase rapidly on trees. The most effective black scale parasite (*Metaphycus helvolus*) is usually scarce on trees heavily infested with ants. Populations of the citrus mealybug and woolly whitefly also increase when ants are present because ants kill or chase away their parasites. Ants also attack the predaceous larvae of dustywings, lacewings, syrphid flies, and some species of lady beetles.

Ants can also contribute to population increases of other pests that do not produce honeydew because ants will attack almost any parasite or predatory insect they encounter, regardless of its host. For example, ants interfere with *Aphytis* parasites and several lady beetles that help to control California red scale and citrus red mite.

Natural Enemies

Ants do not have effective natural enemies in citrus, but competition with other ant species can limit their populations. To improve the effectiveness of biological control against other pests, prevent ants from climbing trees by skirt pruning canopies and applying sticky polybutene-based materials on top of tree wrap around the trunk. Avoid broad-spectrum sprays for ants and other pests. Control ants without damaging the natural enemies of other pests by applying insecticide baits. Early spring or summer, when ant populations are just beginning to increase, are generally the best times to apply bait. To determine which bait to use, first identify the pest ant species, then consult the online *UC IPM Pest Management Guidelines: Citrus*.

Argentine ant workers are uniformly light to dark brown. They occur above ground in dense trails of many workers. The Argentine ant is the most important pest ant in coastal and southern interior California. It aggressively attacks parasites and predatory insects, disrupting biological control of many other pests, such as the brown soft scale shown here.

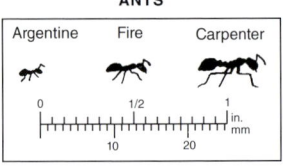

INTEGRATED PEST MANAGEMENT FOR CITRUS 177

This key includes 14 ant species that may occur in California citrus. Species considered to be major pests are Argentine ant (illustrated at right), native gray ants, southern fire ant, and red imported fire ant.

Read both descriptions and compare the specimen with the drawings provided before proceeding. Take several specimens through the key individually to verify your identification.

(Diagram of ant labeled: Antenna, Head, Thorax, Petiole, Abdomen)

1A. Ant with one node on petiole (Fig. 1). The single node in some ants is hidden by the abdomen (Fig. 7).

Go to step 2.

1B. Ant with two nodes on petiole (Fig. 2).

Go to step 10.

Fig. 1 — One node
Fig. 2 — Two nodes

2A. Thorax is smooth and evenly rounded when viewed from the side (Fig. 3).

Go to step 3.

2B. Thorax is uneven in shape when viewed from the side (Fig. 4).

Go to step 4.

Fig. 3 — Thorax smooth and rounded
Fig. 4 — Thorax uneven in shape

3A. Circle of hairs present on the tip of the abdomen (Figs. 5 and 9). Large, up to 0.5 inch (13 mm) long, black or reddish to dark brown.

Carpenter ant

3B. No circle of hairs at tip of abdomen (Fig. 6). About 0.13 to 0.25 inch (3–6 mm) long; brownish black head, red thorax, and velvety black abdomen.

Velvety tree ant

Fig. 5 Carpenter ants *Camponotus* spp. — Circle of hairs at tip of abdomen
Fig. 6 Velvety tree ant *Liometopum occidentale* — No circle of hairs at tip of abdomen

Carpenter ants and velvety tree ants nest in wood or soil.

4A. Node hidden by abdomen (Fig. 7). Dark brown to shiny black, 0.13 inch (3 mm) long. Gives off a strong odor when crushed.

Odorous house ant

4B. Node erect (Fig. 8).

Go to step 5.

Fig. 7 Odorous house ant, *Tapinoma sessile* — Node hidden by abdomen
Fig. 8 — Node erect

Figure 41. Identification key to ant species.

Figure 41. Identification key to ant species *(continued)*.

5A. Tip of abdomen with circular opening fringed with hairs (Fig. 9). **Go to step 6.** **5B.** Tip of abdomen with slit-shaped opening (Fig. 10). **Go to step 7.**	Fig. 9 — Circular opening Fig. 10 — Slit-shaped opening
6A. 3 distinct ocelli on top front of head, above eyes (Fig. 11a). Antennal scape short, barely extending beyond the head (Fig. 11b). Body about 0.1 to 0.2 inch (2.5–4.5 mm) long and variably colored, typically a mix of gray and reddish brown. **Native gray ants**	Fig. 11a Fig. 11b Native gray ants, *Formica* spp.
6B. Ocelli absent or indistinct. Thorax when viewed from above narrows distinctly in middle, hourglass shaped (Fig. 12a). Antennal scape long, extending beyond rear of the head by more than half its length (Fig. 12b). Body about 0.08 to 0.2 inch (2–4 mm) long, light to dark brown or black. **Small honey ant**	Fig. 12a Fig. 12b Small honey ant, *Prenolepis imparis*
7A. Top of thorax smooth without cone-shaped projection (Fig. 13). **Go to step 8.** **7B.** Top rear of thorax has cone-shaped or pyramid-like projection (Fig. 14). **Go to step 9.**	Fig. 13 — No projection on thorax Fig. 14 — Projection on top of thorax
8A. Top of thorax with several prominent, upright hairs (Fig. 15). About 0.07 to 0.1 inch (1.8–2.5 mm) long; color varies from yellow to light brown. **California rainbow ant** **8B.** Top of thorax without prominent upright hairs (Fig. 16). About 0.09 to 0.1 inch (2.2–2.6 mm) long, uniformly light to dark brown. **Argentine ant**	Fig. 15 California rainbow ant, *Forelius* (=*Iridomyrmex*) *pruinosus* Fig. 16 Argentine ant, *Linepithema humile* (=*Iridomyrmex humilis*)

9A. Uniformly brown ant about 0.06 to 0.08 inch (1.5–2 mm) long (Fig. 17). **Pyramid ant**	Fig. 17 Pyramid ant *Dorymyrmex (=Conomyrma) insanus*	Fig. 18 Bicolored pyramid ant *Dorymyrmex (=Conomyrma) bicolor*
9B. Orange or reddish brown head and thorax with darker brownish black abdomen. About 0.08 to 0.1 inch (2–3 mm) long, somewhat larger than pyramid ant (Fig. 18). **Bicolored pyramid ant**		
10A. One pair of spines on thorax (Fig. 19). Go to step 11.	Fig. 19 (One spine on thorax)	Fig. 20 (No spines on thorax)
10B. No spines on thorax (Fig. 20). Go to step 12.		
11A. Beardlike fringe of long hairs on the underside of head (Fig. 21). No grooves on top surface of head and thorax. A reddish, large ant about 0.22 to 0.4 inch (5.5–8.7 mm) long. **California harvester ant**	Fig. 21 California harvester ant *Pogonomyrmex californicus* (Beardlike fringe of hair)	Fig. 22 Pavement ant *Tetramorium caespitum* (Grooves)
11B. No beardlike fringe beneath head. Top surface of the head and thorax sculptured with many parallel grooves (Fig. 22). Brown to black ant, about 0.1 to 0.13 inch (2.5–3 mm) long. **Pavement ant**		
12A. Very small eyes (Fig. 23). Worker ants are all the same size, tiny, 0.08 inch (2 mm) or less; yellow to brown in color. **Thief ant**	Fig. 23 Thief ant *Solenopsis molesta* (Small eyes)	Fig. 24 Southern fire ant, *Solenopsis xyloni*, or Red imported fire ant, *S. invicta*. These species are difficult to distinguish from one another. Consult an expert for identification. (Large eyes)
12B. Large eyes (Fig. 24). Variable in size, with workers in the same trail ranging from about 0.07 to 0.2 inch (1.5–6 mm) long. Usually darker than thief ants, with some black markings on the abdomen. **Fire ants**		

Note: The apparent thorax of an ant actually consists of the thorax plus the first abdominal segment (or propodeum), which is fused to it at the posterior end. The apparent abdomen thus starts with abdominal segment 2.

Adapted from Reynolds et al. 2001, Ward 2005. Illustrations by A. D. Cushman and S. H. DeBord in Smith 1965 and Gorham 1991, and by R. R. Snelling in Ebeling 1978.

Monitoring

Check trees for ants in the spring when honeydew-producing insects appear. Ants often have swollen, almost translucent abdomens if they are collecting honeydew. Examine the ground for trails of ants and for the soil mounds and tunnel openings of underground ant nests. Periodically inspect for ants and bark damage under the trunk wraps of young trees.

Argentine Ant

Linepithema humile (=*Iridomyrmex humilis*)

The Argentine ant is the most common ant pest in coastal and southern interior citrus. Workers collect honeydew and create mass foraging trails on trees and on the ground. Workers are about 1/10 inch (2.5 mm) long and vary in color from light to dark brown. The Argentine ant has one node on the petiole, an uneven thoracic profile, and a 12-segmented antenna lacking a distinct terminal club (Figure 41).

Colony size varies, and the number of queens ranges from one to hundreds, with about 15 queens per 1,000 workers. From fall through spring, Argentine ants nest in sunny locations and often aggregate in large supercolonies. With the onset of hot weather, they disperse into smaller colonies and during the summer nest under trees and in other shady spots. After mating in late winter or the spring, the queens lay numerous eggs; thus ant numbers can increase enormously by midsummer and early fall.

Native Gray Ant

Formica aerata

Formica aerata is common in Central Valley citrus. Other gray ants (*Formica* spp.), also called field ants, occur throughout California but are relatively uncommon when the Argentine ant is abundant. Gray ants may return to orchards when the Argentine ant has been controlled. Native gray ants are a mix of gray and reddish brown. Workers are 1/10 to 1/5 inch (2.5–5 mm) long, considerably larger than the Argentine ant. Native gray ants move fast in irregular patterns and usually occur singly, and rarely in dense trails.

Southern fire ant workers are bicolored with a yellowish red head and thorax and a dark swollen part of their abdomen. Look for ants regularly under any wraps on young trees. This ant feeds on bark and can seriously damage young trees by girdling the root crown and trunk.

Most ants nest underground, where workers tend the immatures and queens. This exposed southern fire ant nest reveals the pale pupae and larvae, winged reproductives, as well as wingless worker ants.

Native gray ant workers are typically a mix of gray and reddish brown. They are common in Central Valley citrus. In comparison with Argentine ants, gray ants are larger and usually forage individually, not in dense trails.

Southern fire ants girdled the bark around the base of this young citrus tree. Unless trees are regularly inspected for pests beneath trunk wraps, growers may be unaware of this problem until trees die.

Red imported fire ant workers are mostly dark reddish brown. As with the southern fire ant, red imported fire ants are variable in size; large and small workers occur together in the same trail. When disturbed, these highly aggressive ants often attack and sting in large numbers. If red imported fire ant is not established in your area, report suspected infestations to agricultural officials.

Southern Fire Ant

Solenopsis xyloni

The southern fire ant can occur in all major citrus-growing areas. It may feed directly on tender twigs, bark, and leaves, sometimes girdling trees. If disturbed, southern fire ants can bite or sting, but they are not as aggressive as red imported fire ants.

Fire ant colonies make loose mounds or numerous scattered craters on the soil surface above their underground nests. When workers forage for food, large and small individuals occur together in the same trail. The workers are $\frac{1}{16}$ to $\frac{1}{4}$ inch (1.5–6 mm) long and bicolored with a yellowish red head and thorax and a dark abdomen. As with the red imported fire ant, workers have a ten-segmented antenna with a two-segmented terminal club and a two-segmented petiole. Due to extensive overlap in their appearance with that of imported fire ants, it is important to consult an expert for correct identification.

Regularly monitor bark under trunk wraps on young trees for these ants or their damage. Without regular trunk inspections, ant damage can be overlooked until hot weather increases trees' need for water. Leaves then turn yellow and drop, or the entire canopy collapses because trunks and root crowns have been girdled by fire ants. Ant-damaged bark often appears honeycombed, and the injured tissue may contain sap and dirt. To confirm the cause of bark damage during hot weather, monitor for ants in the morning or late evening when ants are active above ground.

Red Imported Fire Ant

Solenopsis invicta

Be especially alert for this highly aggressive ant, which has been found in scattered locations in southern and central California. Red imported fire ants readily run up any object that touches their mound, and they often attack and sting in large numbers.

Workers range in size from $\frac{1}{16}$ to $\frac{1}{5}$ inch (1.5–5 mm) long. They can be keyed to the genus *Solenopsis* using Figure 41. However, due to extensive overlap in their appearance with that of the native southern fire ant, consult an expert for correct identification.

Fire ant colonies are easily introduced in soil and plants arriving from infested areas. Inspect incoming soil and plants with soil to avoid introducing fire ants, especially if material is arriving from infested areas. If red imported fire ant is not established in your area, report suspected infestations to the county agricultural commissioner, or telephone 1-888-4FIREANT toll-free, or visit the www.fireant.ca.gov Web site.

MITES

Mites are tiny arthropods that, unlike insects, do not have antennae or wings. After hatching from the egg, mite development varies by type (taxonomic family) of mite. Spider mites (family Tetranychidae) and predatory mites (Phytoseiidae) hatch from eggs and develop through one 6-legged larval and two 8-legged nymphal stages before

Pale stippling and bleaching of the upper leaf surface is a characteristic symptom of spider mite feeding. Citrus red mite has been feeding on these navel orange leaves.

Table 10. Pest Mites in California Citrus, Their Importance, and When to Monitor for Mites and Their Damage.

Spider mites (family Tetranychidae)

Common name	Scientific name	Most susceptible varieties	Growing region and pest status			
			Coastal	Southern interior	San Joaquin Valley	Desert
citrus red mite	*Panonychus citri*	all	occasional pest late summer and fall		common pest in March through petal fall and occasional pest in the fall	occasional pest in spring; rare pest in late fall
hydrangea mite	*Tetranychus kanzawai*	late-harvested cultivars: grapefruit and Valencias	not reported as present	common pest during mid-summer to fall	not reported as present	occasional pest in summer and fall
Lewis spider mite	*Eotetranychus lewisi*	none	rare pest on oranges and lemons		not a pest	
sixspotted mite	*Eotetranychus sexmaculatus*	none	minor, in spring and early summer	rare	rare or not a pest	
Texas citrus mite	*Eutetranychus banksi*	all	rarely or not a pest	common pest fall to spring	occasional fall pest on early-harvested varieties; sometimes also seen in spring	common pest fall to spring
twospotted spider mite	*Tetranychus urticae*	varieties with fruit that grow in clusters or remain on the tree over winter	occasional pest summer and fall	rare pest summer and fall	occasional pest in summer through spring	rare pest
Yuma spider mite	*Eotetranychus yumensis*	fall-harvested mandarins in the San Joaquin Valley; grapefruit and lemons in the Desert	rarely or not a pest	rare pest	occasional pest during the summer on fall-harvested mandarins	occasional pest, mostly in late winter and spring on grapefruit and lemons

Other types of mites (families Eriophyidae, Tarsonemidae, and Tenuipalpidae)

Common name	Scientific name	Most susceptible varieties	Growing region and pest status			
			Coastal	Southern interior	San Joaquin Valley	Desert
broad mite	*Polyphagotarsonemus latus*	coastal lemons	occasional pest late July through early October on lemons		rarely or not a pest	
citrus bud mite[1]	*Eriophyes sheldoni*	lemons	occasional pest anytime of year, most often on fall blooms and fruit that develop from them[1]		rarely or not a pest	
citrus rust mite (silver mite)	*Phyllocoptruta oleivora*	oranges and lemons	occasional pest on oranges and lemons, mostly during fall to spring and rarely in summer	occasional pest on oranges and lemons, mostly in summer	rarely or not a pest	rarely a pest; in spring and fall when a problem
flat mite[2]	*Brevipalpus lewisi*	all	not a pest	minor pest summer and fall[2]		common pest in summer[2]

1. Economic importance uncertain; causes misshapen lemons less often in southern interior orchards in comparison with coastal lemons.
2. Accentuates other types of rind damage (e.g., scarring from citrus thrips feeding, leafhopper oviposition, and wind rubbing).

becoming 8-legged adults as illustrated in the "Citrus Red Mite" section. Broad mite, rust mite, bud mite, and flat mite differ from spider mites and predatory phytoseiid mites in that they are much smaller and (except for flat mite) have fewer immature stages.

Most tetranychid mites produce obvious silk webbing, hence the group's common name: spider mites. The heavy web spinners of this group include the hydrangea mite, twospotted spider mite, sixspotted mite, and Yuma spider mite. Citrus red mite produces no obvious webbing, and Texas citrus mite produces relatively little silk.

Spider mites suck plant juices from individual cells of leaves, reducing their photosynthetic capability and causing the foliage to discolor and in extreme cases to fall off of the tree prematurely. When they attack fruit, it is discolored and, if the trees are stressed by high mite populations, yield can be reduced. Most spider mites cause only cosmetic injury to rinds. The common citrus red mite can be tolerated in relatively high numbers on healthy trees, except when feeding precedes hot, dry (Santa Ana condition) winds. Natural enemies (including predatory mites) and weather often keep spider mites below damaging levels.

In addition to the effects of cultural practices and pesticide use as described below, the species of mites that may become pests differ by growing region and time of year, as summarized in Table 10. When fruit is harvested and crop variety can also be important. For example, broad mite is usually a pest only on coastal lemons. Mandarins and other fruit that grow in clusters provide shelter where fruit touch, thereby increasing problems with twospotted mite, which overwinters in protected places on fruit.

Dusty conditions can cause mite outbreaks and disrupt biological control of pest insects. Drive slowly and only when necessary through the orchard. Water roads during times of frequent traffic. Monitor dusty locations regularly for mites and other pest problems.

Spider Mites

Seven species of spider mites are at least occasional pests in California citrus. Citrus red mite and twospotted mite are the most common pest mites, but the damaging species vary greatly by cultivar, location, and orchard growing conditions and management practices.

Spider Mite Management

Spider mite feeding is more likely to damage citrus trees that are water stressed. Make sure trees are appropriately irrigated, particularly in the late summer and early fall when poor water penetration is common, especially in San Joaquin Valley orchards.

Conserve predators, especially predaceous mites, that help keep spider mite populations below economically damaging levels. Control ants, which disrupt biological control by attacking predatory insects including lady beetles and lacewings that feed on mites. Minimize dust, which interferes with natural enemy effectiveness and can cause mite outbreaks. Drive slowly and only when necessary through the orchard, water orchard roads, and plant windbreaks. Where a ground cover is maintained, dust is usually not a problem, but there is an increased risk of frost damage.

Monitor orchards and, if treatment is needed, use narrow-range oil or selective miticides (acaricides) whenever possible. The most effective miticides are often different from the insecticides used to control insects. Many broad-spectrum insecticides can cause outbreaks of spider mites even though some of these insecticides claim to control mites. In addition to their adverse effect on natural enemies, certain insecticides can increase mite populations by stimulating mite reproduction (hormoligosis) or by altering host plant physiology.

If applied in a way to achieve excellent coverage, narrow-range oil can be very effective on mites (except for mites that feed protected within distorted plant parts). Extensive research on the use of oil sprays against various mite and scale insects has resulted in the development of recommendations that use specific rates and timing of treatments on different varieties of citrus in different regions of California. Use these situation-specific rates to achieve expected pest control and limit the potential for leaf or fruit damage or drop due to phytotoxicity. For species-specific recommendations, consult the online *UC IPM Pest Management Guidelines: Citrus*.

Citrus Red Mite

Panonychus citri

In the past, citrus red mite was considered to be a major pest. However, research on San Joaquin Valley navels and coastal lemons demonstrated that citrus can tolerate high

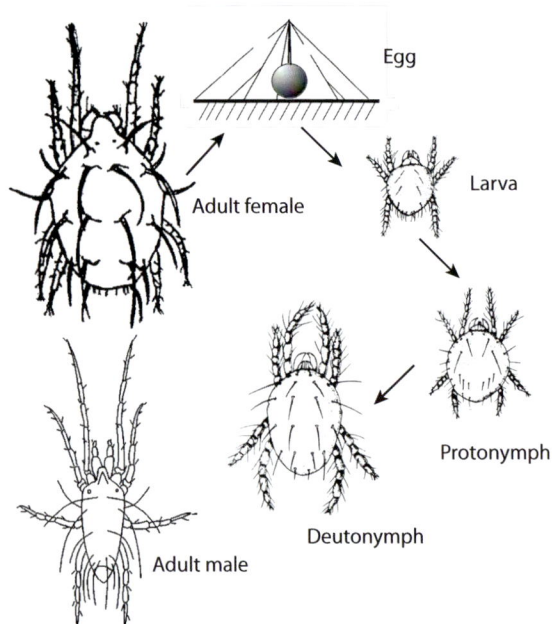

Figure 42. Citrus red mite life cycle and stages. Tiny fibrils resembling guy wires (tension cables attached to the top of utility towers, as shown above) radiate from the tip of the egg stalk. Texas citrus mite eggs are also reddish with a stalk but lack fibrils extending from the stalk. *Sources:* Adults from Childers and Fasulo 1995; immatures from Ebeling 1959, Newcomer and Yothers 1929.

The citrus red mite has large white bristles arising from prominent, red bumps on its back and sides. Viewed from above, this adult female has an oval-shaped body.

The citrus red mite male body tapers toward the rear, so its overall body shape is somewhat triangular. At maturity, the adult male is smaller than the adult female.

This recently emerged citrus red mite larva (left) is only slightly larger than a mite egg. Citrus red mite eggs are reddish and have a stipe (thin stalk) arising from the top center. With a good hand lens, you may see tiny fibrils resembling guy wires extending from the stipe to the leaf surface.

Citrus red mite nymphs are red to dark purplish and look like small adults.

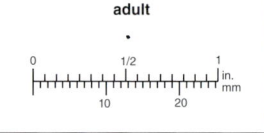

populations (8 adult female mites per leaf). One possible exception is during fall and winter Santa Ana conditions in southern California, when dry winds combined with moderate red mite densities can cause substantial leaf drop. Natural enemies (including a virus disease), temperatures above 90°F (32°C), and low humidity reduce citrus red mite populations. Growers using broad-spectrum insecticides see citrus red mite outbreaks more frequently than growers relying on natural enemies and soft pesticides for insect control.

Description and Biology

The adult female citrus red mite is tiny and oval shaped when viewed from above; the male is smaller and tapered at the rear end. Both adults and nymphs are red to dark purplish. Males are often found near molting female deutonymphs, waiting to mate with them when they emerge as adults. Each female lays from 20 to 50 eggs at a rate of 2 or 3 a day, depositing them on both sides of leaves. The egg hatches into a six-legged larva, which molts into a protonymph, then a deutonymph, then into an adult (Figure 42). Preceding each molt, mites enter a brief quiescent stage, in which they are unmoving. The life cycle from egg to egg-laying adult can be as short as 12 days.

Citrus red mite overwinters primarily as eggs, although all stages can be present year-round. Eggs can occur anywhere, but are often found along the midrib leaf vein. Populations increase in the spring, late summer, and early fall in response to new leaf growth; citrus red mites prefer to feed on young leaves but will also infest fully expanded mature leaves and fruit. In cooler districts on young orange and lemon trees, mites may be numerous throughout the year because the repeated growth flush provides preferred food almost continuously.

Damage

The citrus red mite extracts cell sap from leaves and fruit, causing bleaching or pale stippling of surfaces. In severe infestations, leaf stippling enlarges to become dry, necrotic areas. Severely damaged leaves may drop and twigs may die back. In coastal and southern California growing areas, during Santa Ana winds in the fall, low levels of citrus red mite in combination with the wind can result in blasting or burning and drop of leaves.

The mite stippling or silvering of green fruit usually disappears as fruit develop their mature color. When large populations feed on nearly mature fruit, the silvering may persist and reduce the grade of the fruit in the packinghouse. High mite populations can also contribute to fruit sunburn during hot weather.

Mite damage in southern California appears mostly in late summer or fall. In the San Joaquin Valley, mite damage is most prevalent in the spring, although low populations in the fall have been associated with leaf drop. In the desert valleys, citrus red mite is an occasional problem in spring and a rare problem in the fall. Citrus red mite is more of a problem when trees are water stressed, conditions are hot and dry, and after broad-spectrum pesticides have been applied.

Natural Enemies

Predaceous mites and insects and a virus play a significant role in regulating citrus red mite populations. *Euseius* species of predatory mites are the most common predators of citrus red mite in most growing areas. Predatory mites attack eggs and immature stages of the citrus red mite. The spider mite destroyer and sixspotted thrips are the most important mite predators in the desert valleys, but in other growing areas they are less commonly observed in comparison with *Euseius* species. These natural enemies are discussed below under "Mite Predators." Predaceous midges (*Feltiella* spp.) and (less commonly) a rove beetle (*Oligota oviformis*) discussed below prey on citrus red mite in coastal areas. Dustywings (primarily *Conwentzia barretti*) and green and brown lacewings pictured earlier under "General Predators" also prey on citrus red mites.

A virus disease specific to citrus red mite is widespread in citrus-growing areas. When hot, dry temperatures exceeding 90°F (32°C) occur for several consecutive days and high mite populations are present, virus epidemics occur and rapidly reduce citrus red mite populations. Symptoms of virus infection include mites that walk stiffly and curl their legs under their body. As they die, their bodies disintegrate and sometimes leave reddish brown to black watery spots on fruit or leaves. If diseased mites are mounted on a slide and examined under a polarizing microscope, virus crystals that shine in the polarized light are found in the cell nuclei.

Monitoring

In southern interior and coastal areas, depending on the local situation, consider monitoring from late summer through fall, especially in locations prone to fall Santa Ana winds. In the San Joaquin Valley, survey each orchard in February to determine whether mites are present. Scan several leaves per tree at various sites and use a hand lens to check a few leaves for eggs and immature stages. In March, or as soon as mites are detectable, begin regular monitoring and continue until red mite numbers decline below 1 per leaf and petal fall has occurred.

During the times recommended above, sample citrus red mite every 2 weeks:

- Collect a total of 100 fully expanded leaves from throughout the orchard.
- Select leaves from just inside the shady region of the tree.
- Determine the average number of pest mites per leaf by dividing the total number of mites found by 100.
- Count the number of active stages of predatory mites and calculate their average per leaf by dividing the total number of predatory mites by 100.
- Also note the presence of virus-infected citrus red mites.

Thresholds vary by location, time of year, and local conditions. For example, in San Joaquin Valley navel oranges, economic loss is not likely to occur if citrus red mite densities do not exceed eight mature females per leaf by 2 to 4 weeks after petal fall. Vigorous, well-irrigated trees are bet-

A virus often infects and kills citrus red mite. Infected mites move slowly, curl their legs under the body, and finally die.

Twospotted mites usually have two irregular, dark blotches on the sides of their body. The rest of their body typically ranges in color from yellowish to gray or green. Their eggs are translucent (clear and shiny) when laid, then turn opaque (whitish or milky) before hatching. The round eggs often occur among, or attached to, silk strands.

The twospotted mite female (top left) turns pink to orange when preparing for and just after its winter resting stage (diapause) and at other times of the year when populations are overcrowded or stressed. These nymphs are yellowish to green, colors typical for this species.

Description and Biology

Twospotted spider mite adults and immatures usually have two irregular dark blotches on the sides of their body, with the rest of their body ranging in color from yellowish to gray or green. Females turn pink to orange during the fall through early spring and enter a resting stage (diapause) during the winter. All stages can also turn pink to orange at other times of the year when they are overcrowded or stressed.

Twospotted mite eggs are spherical and translucent to opaque. Immature twospotted mites develop through three stages (larva, protonymph, and deutonymph) before becoming adults, as illustrated earlier for citrus red mite (Figure 42). If temperature and food supply are favorable, a generation can be completed in 7 days.

All stages can be present throughout the year, but twospotted mite overwinters primarily as females in protected places on the tree, such as in the navel of navel oranges, under the sepals, and where fruit touch, such as on the rinds of Murcott mandarins that grow in clusters. If the weather is mild, the mites will feed and reproduce throughout the winter. The number of twospotted spider mites increases in late spring and peaks during the summer. Mites usually first become obvious on the underside of leaves. As their populations build, mites also become abundant on the upper side of leaves and on fruit and cover infested parts with conspicuous, fine webbing.

Damage

Light infestations cause yellow or brown spots between leaf veins, usually appearing in mid- to late summer. Clusters of

ter able to tolerate mites than drought-stressed trees. Low to moderate populations are considered to be beneficial as they provide food for natural enemies such as the predatory mites needed for citrus thrips control. High temperatures and the virus disease reduce mite populations in June and July, so no treatment is generally required during summer. Fall populations in combination with water stress can sometimes lead to significant defoliation and fruit drop. For current monitoring methods and the threshold and treatment recommendations for your situation, consult the online UC IPM *Pest Management Guidelines: Citrus*.

Twospotted Spider Mite

Tetranychus urticae

The twospotted spider mite, also called twospotted mite, is an occasional pest on citrus, particularly in the San Joaquin Valley. Its damage potential varies from year to year and is related to water stress, heat, and disruption of its natural enemies by dust or broad-spectrum pesticides. Twospotted mite is usually a greater pest of deciduous fruit and nut trees, ornamentals, and vegetable crops than of citrus.

Feeding by twospotted mites can result in brown necrotic areas between major leaf veins. Damage becomes most apparent late in the season.

When twospotted mite populations are high, they produce abundant webbing that can cover leaves and fruit. The fruit rind may be bleached or silverish in appearance.

Adult female (top and left) and immature (right) Texas citrus mite are oval and tan to orange with dark blotches on the body. Texas citrus mites characteristically rest with their legs extended directly forward and backward, often along the midrib of the leaf.

Texas citrus mite eggs change from yellow when first laid to reddish brown before hatching. From the top (as shown) eggs appear round. Viewed from the side, eggs are somewhat flattened or disc shaped. Texas citrus mite eggs have a thin projecting stalk, but unlike citrus red mite, no fibrils extend from the stalk's tip to the leaf.

The adult male Texas citrus mite (left) has long legs and a triangular, brownish orange body with dark blotches. In order to mate with her, this male is guarding an immature female (quiescent deutonymph) that is about to molt into an adult.

In addition to the pale chlorosis typical of spider mite feeding, high populations of Texas citrus mite cause premature leaf drop, often leaving the leaf petioles remaining on stems.

dried, brown leaves and profuse webbing indicate a heavy infestation. If compounded by water stress, high populations of twospotted mite can cause leaf and fruit drop. When populations are heavy, fruit can look pale and stippled and not color up properly when ethylene treatments are used, which may reduce the grade in the packinghouse.

Fruit that grow in clusters or remain on the tree during winter or long after maturity are more susceptible to injury from twospotted mite. Where mites overwinter on fruit, such as between touching clusters of mandarins, damage can develop during late winter (beginning about late February in the San Joaquin Valley) when females feed on rinds as the weather warms. In comparison with navel oranges, Valencias often have higher twospotted mite populations. Valencias mature later and can be on the tree in late spring and early summer, well after the earlier harvest of navels.

Natural Enemies

Predators can provide substantial control of the twospotted mite, except when disrupted such as by broad-spectrum pesticides. These natural enemies include the *Stethorus* spider mite destroyer lady beetle and sixspotted thrips, which are commonly observed where mites occur at high density. Predaceous midges (*Feltiella* spp.) and (less commonly) a rove beetle (*Oligota oviformis*) prey on twospotted mite in coastal areas. These specialized natural enemies are discussed below under "Mite Predators." The *Euseius* predatory mites also feed to some extent on this webbing mite species, but they prefer the citrus red mite. Lacewings and minute pirate bugs discussed earlier under "General Predators" also feed on twospotted mite.

Monitoring

Check for twospotted spider mites when you monitor citrus red mite in late winter and early spring. Continue monitoring for twospotted mite occasionally during summer and more closely in late summer and fall. Look for yellow to brown feeding spots on foliage, particularly in the last growth flush. High populations in summer and fall may warrant treatment. If mites are a problem, consult the online *UC IPM Pest Management Guidelines: Citrus* for current control recommendations.

Texas Citrus Mite

Eutetranychus banksi

Texas citrus mite is a common pest in the southern interior and desert valleys from fall through spring. It is an occasional pest of citrus in the San Joaquin Valley during spring and fall. In all these growing areas, populations decrease in summer but increase from September through December. In the San Joaquin Valley, populations decrease again when weather becomes cold and wet, such as during the first overnight period of dense valley fog. Texas citrus mite is sometimes observed in spring in the San Joaquin Valley, especially following insecticides that disrupt biological control.

After hatching from eggs, Texas citrus mites develop through three stages (larva, protonymph, and deutonymph) before becoming adults, as illustrated earlier for citrus red mite (Figure 42). The entire life cycle takes approximately 3 weeks when temperatures are warm.

All life stages, including eggs, tend to occur along the midrib and lateral veins. Eggs are somewhat flattened or disc-like and range in color from yellow to reddish brown. Adults and nymphs are tan to orange or brownish green with irregular dark blotches on their body. Females and nymphs are round to oval and somewhat flatter than citrus red mite or Yuma spider mite. Males are more triangular and slender than females and have longer legs. Characteristically, Texas citrus mites extend their legs straight forward and straight backward parallel to the leaf surface, and colonies produce little webbing.

Damage

Texas citrus mite feeds primarily on leaves, where it can cause significant stippling and leaf drop. Leaf drop from Texas citrus mite is unique because the leaf blade falls to the ground while the petiole remains attached to the tree. Damage often begins in the tops of trees and progresses downward, primarily on new flush on the periphery of the trees. Leaf drop can result in sunburn, and if mite feeding causes extensive defoliation, fruit will drop from the trees.

In the San Joaquin Valley, damage is usually limited to early-harvested navels, where a combination of warm temperatures in fall and drought stress or deficit irrigation allow mites to thrive.

Natural Enemies

Predators of Texas citrus mite include sixspotted thrips, the *Stethorus* spider mite destroyer, and (except in the desert valleys) *Euseius* predatory mites discussed below under "Mite Predators." Lacewings and minute pirate bugs (*Orius* spp.) discussed earlier in the "General Predators" section also prey on Texas citrus mite.

Monitoring

In southern interior and desert valleys, look for Texas citrus mite from fall through spring. In the San Joaquin Valley, check for Texas citrus mite during spring if broad-spectrum insecticides have been used. Treatment may be warranted if significant amounts of leaf drop occur.

During fall in the San Joaquin Valley, look for Texas citrus mite from September through December on trees under drought stress or receiving deficit irrigation and on cultivars that bear early-harvested fruit, especially navels. Treatment may be warranted if leaves in the outer canopy at the tops of trees begin to defoliate and cold weather is not anticipated for a period of several weeks. Treatments are not needed if defoliation is limited to the leaves on the extremities of the fall flush that will naturally freeze or be pruned off during winter.

No official treatment thresholds exist at this time (2010). For current information, consult the online *UC IPM Pest Management Guidelines: Citrus*.

Yuma spider mite adults, nymphs, and eggs are peach or salmon colored and appear shiny. Many translucent empty egg shells and nymphal cast skins are also shown here. Yuma spider mite feeds mostly on the underside of leaves, where it produces substantial silk webbing.

Yuma Spider Mite

Eotetranychus yumensis

Yuma spider mite is an occasional pest of citrus in the inland valleys and desert areas of California. In the Coachella and Imperial valleys, it occurs on grapefruit and lemons, mostly in winter and late spring. In the southern San Joaquin Valley, it primarily occurs on mandarins during summer.

After hatching from an egg, Yuma spider mite develops through three immature stages (larva, protonymph, and deutonymph) before becoming an adult, as illustrated earlier for citrus red mite (Figure 42). Its body shape resembles that of citrus red mite, but Yuma spider mite is lighter in color (pale straw to dark pink) and shinier due to a relative lack of setae (hairs). Yuma spider mite produces heavy webbing

Yuma spider mite causes bleaching or pale stippling of citrus fruit and leaves, as shown on these Clementine fruit in the lower San Joaquin Valley in September.

on the underside of leaves and on fruit. The mites feed, and females lay peach-colored eggs, beneath this silk.

Yuma spider mite sucks fluids from leaf and fruit cells, causing discoloration and in rare cases defoliation. Feeding on green fruit causes stippling and bleaching of the surface that, in severe cases, can cause the fruit to be pale and discolored. Fruit discoloration in the southern San Joaquin Valley has most commonly been reported on early-harvested Clementine mandarins, where stippling results in pale fruit that do not color properly during postharvest ethylene treatments.

Natural enemies of the Yuma spider mite include sixspotted thrips, *Stethorus* the spider mite destroyer, and (except in the desert valleys) *Euseius* predatory mites discussed below under "Mite Predators." Lacewings and minute pirate bugs (*Orius* spp.) discussed earlier in the "General Predators" section also feed on Yuma spider mite.

Monitor Yuma spider mite by looking for stippling of leaves and fruit associated with large amounts of fine webbing. In the San Joaquin Valley, check for Yuma spider mite in mandarins during July and August. In the Coachella and Imperial valleys, monitor from fall through late winter.

Damage is usually not severe enough to warrant treatment. On bearing trees, treatment may be warranted if fruit stippling is sufficient to inhibit proper fruit coloring and beneficial organisms are not already reducing mite densities. On young trees treatment may be warranted if leaf drop appears imminent. If Yuma spider mite is a problem, consult the online *UC IPM Pest Management Guidelines: Citrus* for control recommendations.

Sixspotted Mite

Eotetranychus sexmaculatus

The sixspotted mite is a minor pest on citrus in some coastal growing areas. It also is occasionally a problem on avocado and tropical plants such as ornamental frangipani (*Plumeria* spp.). Sixspotted mites tend to be most abundant in spring and early summer, when a generation takes 3 to 4 weeks to complete. Populations decrease during dry, windy weather in the fall, but all life stages can be present and reproduction continues throughout the winter in coastal areas.

Sixspotted mite hatches from an egg and develops through three immature stages (larva, protonymph, and deutonymph) before becoming an adult, as illustrated earlier for citrus red mite (Figure 42). Adults and immatures have a lemon yellow to pale green or orangish body. They usually have three pairs of dark spots and are somewhat smaller than twospotted mites.

Sixspotted mites feed on the underside of leaves along the midrib or larger veins and form small colonies that cover themselves with silk webbing. A depression develops on the leaf underside where a colony has settled, and this feeding site becomes apparent as a slight bulge or blister on the upper leaf surface. The infested area may turn pale or yellow, and the leaf often becomes distorted. Leaf drop may occur with relatively few mites present.

Predaceous mites, sixspotted thrips, predaceous midges (*Feltiella* spp.), and (less commonly) a rove beetle (*Oligota oviformis*) and other predators discussed below under "Mite Predators" usually keep sixspotted mites under control. In areas protected from Santa Ana winds, this mite may occasionally require treatment.

Lewis Spider Mite

Eotetranychus lewisi

The Lewis spider mite is a rare pest of citrus in coastal areas of California. Sometimes it also damages ornamental poinsettia (*Euphorbia pulcherrima*).

Lewis spider mites hatch from eggs and develop through three immature stages (larva, protonymph, and deutonymph) before becoming adults, as illustrated earlier for citrus red mite (Figure 42). Lewis spider mite adults are smaller than twospotted mite adults. Adults and immatures vary in color from light yellow to reddish or white and have dark blotches

Sixspotted mite's body color varies from lemon yellow to orangish or pale green. It usually has six dark blotches on its body, only some of which are visible in this side view.

In interior southern California, hydrangea mite feeds on late-harvested fruit, such as this grapefruit rind it has bleached. Hydrangea mite produces copious cottony webbing on fruit where they touch other fruit or leaves. Webbing around the tops of fruit may form a cottony tent that spreads down and outward from the stem onto the tops of fruit.

along the lateral margins. The spherical eggs are pale orange to white and have a slender stalk projecting from their top. Lewis mites produce extensive webbing on fruit and leaves.

The Lewis spider mite feeds only on the fruit, mainly in depressions at the stylar (bottom) end of maturing fruit, causing a silvering of lemons or russeting (brownish discoloring) of oranges. Usually the damage is not severe enough to warrant treatment. Harvesting removes most of the infestation.

Hydrangea Mite

Tetranychus kanzawai (=*T. hydrangeae*)

Hydrangea mite, also called the Kanzawa spider mite, is increasingly common on citrus in the southern interior area and has been reported in the desert valleys. It damages late-harvested cultivars, commonly grapefruits and Valencias. In California, it was first noted on grapefruit in Hemet in 2002. It is an important pest in Asia and has been introduced in scattered locations throughout the world. Hydrangea mite feeds on dozens of crop and ornamental species.

Hydrangea mites hatch from eggs, and immatures develop through three stages (larva, protonymph, and deutonymph) before becoming adults, as illustrated earlier for citrus red mite (Figure 42). Eggs are light brown to orangish. Unlike the similarly colored eggs of citrus red mite, hydrangea mite eggs have no protruding stalk (stipe) and no radiating fibrils ("guy wires") on top. Nymphs are translucent to yellowish, resembling the overall color of some forms of twospotted mite. Hydrangea mite adults are reddish, like adults of citrus red mite or diapausing (overwintering) twospotted mites.

In citrus, hydrangea mite feeds on fruit, bleaching rinds. It produces copious fine webbing mostly on fruit, such as where fruit touch other fruit or leaves. When populations are high, webbing may extend down and outward from fruit stems so that the tops of fruit are covered with a cottony tent.

Look for hydrangea mite from summer through fall. Infestations often begin on the stylar end of fruit, so look there for bleached tissue and mite eggs, nymphs, and webbing.

Broad mites often feed and lay eggs in depressions on fruit. Eggs are mostly translucent but appear to be dimpled or covered in white speckles because of pale raised areas on the egg surface. Adults and nymphs are yellowish, and (as shown here) the female has a white stripe on her back. To see broad mites, you need a quality hand lens or a binocular dissecting microscope.

JACK KELLY CLARK

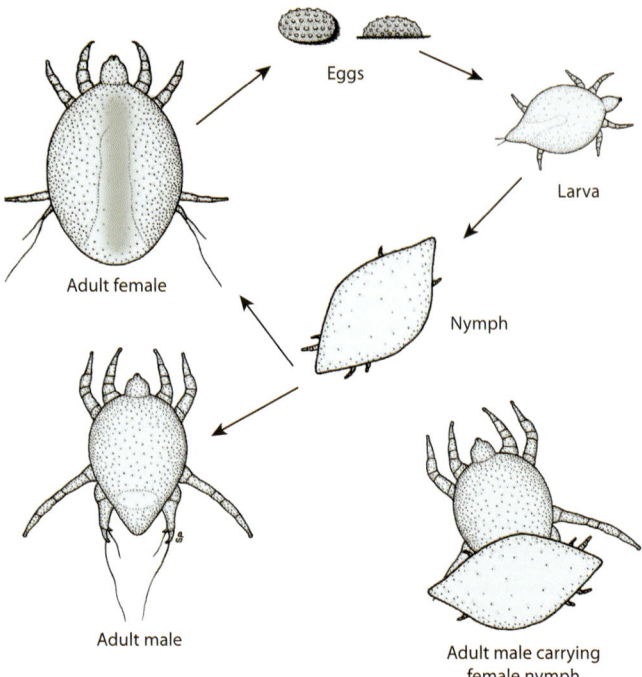

Figure 43. Citrus broad mite life cycle and stages. Unlike other types of mites, on broad mites (family Tarsonemidae) the ends of the rear legs are skinny or threadlike, especially on females. Males have more stout legs for clasping. Males often carry a female nymph on their back and mate immediately once she matures. See Figure 44 for a more detailed illustration of the male. Adapted from Smith, Beattie, and Broadley 1997.

Other Types of Mites

Broad mite, bud mite, rust mite, and flat mite are much smaller than spider mites and predatory phytoseiid mites. These species belong to mite families other than Tetranychidae and do not produce silk. These other mites have fewer development stages than spider mites, except citrus flat mite (family Tenuipalpidae), which develops though the same five stages as tetranychids.

Of these other mite species, certain ones often occur together on the same fruit or leaves, such as rust mite and broad mite and other mites in the Tarsonemidae family that resemble broad mite but are not citrus pests. Be sure to correctly identify which mite species are present and the cause(s) of any damage so you can select effective control actions.

Broad Mite

Polyphagotarsonemus latus

Broad mite (family Tarsonemidae) is an occasional pest of coastal lemons. Broad mite feeds on many plant species. But because it prefers warm and humid conditions, in California it is primarily a pest in greenhouses and ornamental nurseries. Damage in coastal lemons occurs from late July through early October, especially on trees with Argentine ants. Broad mite often occurs with citrus rust mite, and rust mite is usually the more abundant species.

Broad mite has four life stages: egg, larva, nymph, and adult (Figure 43). Adults and immatures are yellowish with a white stripe on the back of females. Adults are relatively wide but so small that you need at least a 20× hand lens to see them, or preferably a binocular dissecting microscope.

Adults have eight legs, and the rear pair is long and skinny at the end, especially on females. Because of their unusual rear legs (Figures 43, 44), tarsonemids are also called thread-footed or thread-legged mites. The larva and inactive nymph (also called a pseudopupa) have six legs. Eggs are mostly translucent but appear to have numerous white speckles or dimples because the egg surface is covered with pale, raised areas. Broad mites often occur in depressions on fruit where the females lay their eggs.

Broad mites suck fruit and leaf cell sap and prefer to feed on young fruit, up to about 1 inch (2.5 cm) in diameter. They feed mostly on the inward-facing side of young lemons inside the canopy, causing scabby scars or shallow cracks on rinds. When populations become dense, broad mites also feed on shady surfaces of fruit in the outside canopy. Although most damage occurs on the fruit, broad mites sometimes feed on young expanding leaves, causing them to curl and crinkle. This leaf cupping and distortion can resemble injury caused by glyphosate (Roundup) herbicide phytotoxicity.

If rinds develop scabby or shallow cracked scarring on sheltered surfaces as described above and tiny mite eggs that are mostly translucent with white speckles are present, this is usually sufficient to diagnose a broad mite problem.

Which predators feed on broad mite has apparently not been studied in California. No treatment thresholds have been developed for broad mite in citrus as of 2010. If high and increasing populations warrant treatment, consult the online *UC IPM Pest Management Guidelines: Citrus*.

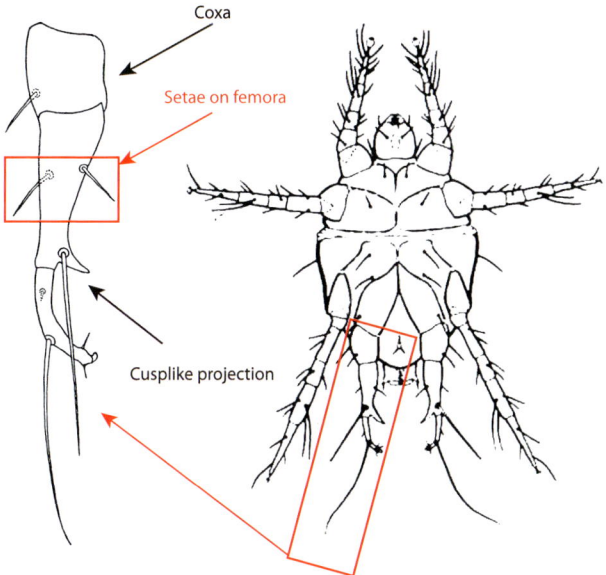

Broad mite, *Polyphagotarsonemus latus*

Figure 44. Citrus broad mite adult male underside (right). Broad mite can usually be identified by its habit of feeding openly on fruit, the whitish stripe down the back of females, and the whitish speckled appearance of its eggs. However, other nonpest species of tarsonemids also occur on citrus and can be reliably distinguished based on features such as an underside view of the male's rear leg (shown at left). The broad mite male has a cusplike projection at the inner tip of its femur (long leg segment), so the end of its rear leg appears clawlike. Two setae (hairs, circled) occur mid femora almost opposite each other, and the coxa (basal leg segment) has relatively parallel sides. Characteristic of mites in this family (Tarsonemidae), the rear legs are skinny or threadlike at their ends, especially on females (not shown). Observing these features requires careful preparation of specimens and a high-powered microscope (preferably a phase-contrasting binocular microscope with about 100× or greater magnification). Illustrations from Ewing 1939, Fasulo 2007.

Distinguishing Broad Mite from Other Tarsonemids

Other Tarsonemidae can be found in California citrus but are rarely, if ever, pests. These other tarsonemid species typically feed on honeydew, sooty mold, dead scales, and other decaying organic matter. They usually are observed only in hidden locations, such as under fruit sepals or beneath empty scale insect covers.

Because of their tiny size, reliably distinguishing the species of tarsonemid may require submitting samples to an expert, who will examine under magnification characters such as differences in the arrangement of hairs (setae) and leg shape, which is clawlike at the end of the rear legs of male broad mites, as illustrated in Figure 44. However, broad mite is usually the only tarsonemid that will be found openly on fruit surfaces. Broad mite's habit of feeding openly on fruit, the whitish stripe down the back of females, and the whitish speckled appearance of its otherwise nearly transparent eggs generally distinguish broad mite from other tarsonemid species in citrus.

Broad mite feeding on succulent shoots causes expanding young leaves to curl. This leaf cupping and distortion can resemble injury caused by glyphosate (Roundup) herbicide phytotoxicity.

Broad mites prefer to feed on the inward-facing side of young fruit that are growing sheltered within the inner canopy. As fruit grow, this earlier injury becomes visible on the rind as a patchwork of scabby scars or shallow cracks adjacent to areas of unblemished tissue.

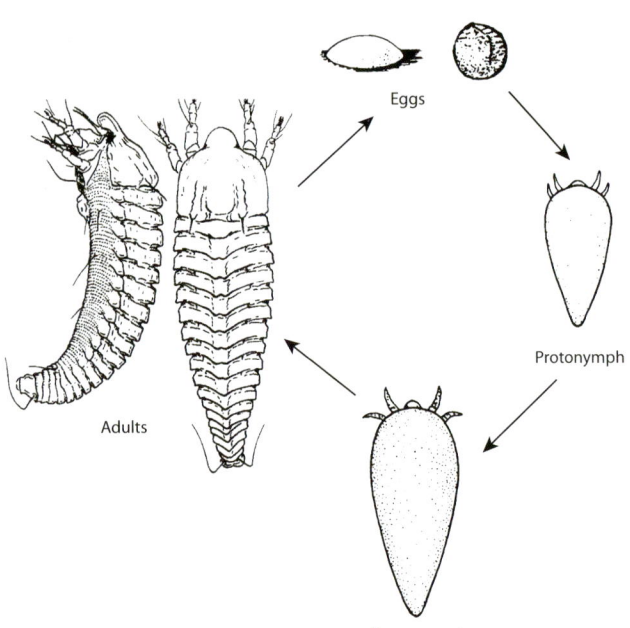

Figure 45. Eriophyid mite life cycle and stages. Citrus rust mite (shown here) and citrus bud mite are carrot- or wedge-shaped species that are smaller than spider mites. These mites (family Eriophyidae) have only two pairs of legs, which protrude near their head (the wider end of their body). Eriophyids develop through four life stages: the adult and egg (both shown in side and top views) and two nymphal stages. Adults from Keifer 1952, immatures from Smith, Beattie, and Broadley 1997.

Citrus Bud Mite

Eriophyes sheldoni

Citrus bud mite (family Eriophyidae) is primarily a pest of coastal lemons. It also occurs in the southern interior of California.

Eriophyids are tiny, elongate mites that are wedge or carrot shaped, tapering at their posterior end. They develop from the egg stage through two nymphal stages into adults, all of which have four legs around the wider end of the body near the mouth (Figure 45). The female lays about 50 eggs mostly in the bud scales of recent growth. These mites feed in the buds, killing them or causing rosette-like growth of the subsequent foliage and distortion of flowers and fruit. Fall blooms, and the fruit that develop from them, are more likely to suffer damage. Bud mite feeding may or may not reduce yield or fruit quality.

To detect bud mites, check the buds at leaf axils on green twigs throughout the orchard from midspring to autumn (a leaf axil is the narrow or inside angle where a leaf petiole and stem meet). Dissect these axillary buds under a dissecting binocular microscope or use a 20× hand lens to determine the percentage of buds infested with one or more live mites. Alternatively, collect one green fruit, about 1¼ to 2 inches (3–5 cm) in diameter, from each of 50 trees scattered throughout the orchard. Twist off the base of the stem and determine whether the underside of the sepals and button or where they covered the fruit are infested with live bud mites.

A fruit infestation of 15 to 20% indicates a bud infestation of about 45 to 50%. However, levels as high as 80% bud infestation have failed to cause consistent or predictable economic losses, and no bud mite threshold has been established. If a reduction in bud mite populations is desired, treat 2 to 3 months before the bloom that is to be protected. For current monitoring and treatment recommendations, consult the online *UC IPM Pest Management Guidelines: Citrus*.

Bud mites feed and lay their eggs inside developing buds. These mites are tiny and elongate.

Buds damaged by bud mite feeding develop into distorted flowers (shown here), leaves, or shoots. Feeding in buds during the fall and winter causes this damage to the spring bloom and the resulting fruit.

Bud mite feeding on young fruit caused these lemons to become misshapen as they grew. Bud mite is usually a pest only on coastal lemons.

These tiny rust mites are pale and wedge shaped.

Citrus rust mite feeding on lemon results in a relatively uniform silvering or bleaching of the rind, so on lemons this pest is called silver mite.

On mature oranges, citrus rust mite feeding causes brownish scarring (russeting) of the rind.

Citrus Rust Mite (Silver Mite)

Phyllocoptruta oleivora

This species is known as rust mite on oranges and silver mite on lemons. It is an occasional pest in coastal and southern interior California and a rare pest in the desert valleys.

Rust mite (family Eriophyidae) is about the same size and has the same wedge shape as bud mite, but rust mite is yellow and darker. After hatching from the egg, it develops through two nymphal stages into adults, all of which have four legs around their head (wider end) (Figure 45). A generation may be completed in 1 to 2 weeks in the summer. Development slows or stops during winter, depending on temperatures.

Citrus rust mite tends to occur together with broad mite, but rust mite is usually more abundant. Both species thrive in warm, humid conditions. Rust mite feeds on the outside, exposed surface of fruit that are ½ inch (13 mm) or larger. The damaged rind surface becomes silvery on lemon, rust brown on mature oranges, and black on unripe (green) oranges. Rind damage by rust mite resembles that from broad mite, except that rust mite damages somewhat larger fruit and the scarring is generally more uniform and contiguous than the patchy damage of broad mite. Rust mite damage occurs mostly from late spring to late summer, although damage is possible throughout the year. Predators such as *Stethorus* and green and brown lacewings reportedly feed on rust mite but apparently do not provide control.

When damage occurs differs among the growing regions, as summarized in Table 10. In general, monitor rust mite from early spring through summer. On orange trees, look for rust mites on young foliage in early spring and look for rust mites on fruit from late spring through summer. On lemon, rust mites are mostly on fruit throughout the season, so look there. To identify previous infestations, check fruit on the outside of the canopy for scarred rind tissue. To assess current season levels of rust mite, examine small green fruit on the inside canopy. Especially examine protected places on fruit, such as the stylar (bottom) end, under magnification. Once you find one or more infested fruit and if rust mites were a problem the previous year, watch the orchard closely. Threshold levels depend on last year's rust mite problems and current market conditions. If mite populations are increasing or scarring appears, treatment may be warranted. Consult the UC IPM Pest Management Guidelines: Citrus for treatment recommendations.

Flat Mite

Brevipalpus lewisi

The flat mite (family Tenuipalpidae) is a common pest of citrus in the desert regions and a minor pest in the San

The citrus flat mite adult is less than 1/50 inch (0.5 mm) long and often pinkish or salmon colored. Flat mites feed on rind tissue damaged by any cause, such as leafhoppers or thrips, causing that rind injury to develop into more obvious scars.

Joaquin Valley and southern interior, although there have been increasing observations in the San Joaquin Valley. It also damages pistachio, pomegranate, and (rarely) grapes.

Flat mite develops through five stages: egg, larva, protonymph, deutonymph (or nymph), and adult. Adults and immatures are flattened and vary in color but are often pink or reddish. The tiny spherical eggs are reddish. The flat mite is fairly heat tolerant, so populations persist during the hot summer. During warm weather, one generation is completed in about 2 weeks.

The flat mite is usually a secondary invader, feeding on rind tissue damage of any type and extending it (such as rind injury by leafhopper feeding or thrips feeding or egg laying). Flat mite feeding causes scabbing of the leafhopper injury, which would otherwise disappear as the fruit change color.

The predatory phytoseiid *Typhlodromus citri* preys on flat mite. However, the importance of flat mite biological control is largely unknown.

Pesticide treatments are generally not needed for flat mite, and no treatment thresholds have been established as of 2010. Treat when high mite levels appear and monitoring of fruit scarring indicates a need.

In Central and South America, related *Brevipalpus* spp. mites vector *Citrus leprosis virus*. Because it is a potential virus vector, flat mite may become a serious pest if *Citrus leprosis virus* is introduced into California. Leprosis results in discolored lesions on fruit, leaves, and twigs, which leads to peeling bark, leaf and fruit drop, twig dieback, and eventual tree death, as discussed in the "Citrus Leprosis" section in the Diseases chapter. If unusual or severe damage resembling leprosis develops, promptly report the problem to agricultural officials.

Mite Predators

The most important natural enemies of tetranychid mites are predators that specialize on them, including *Euseius* predatory mites, the spider mite destroyer (*Stethorus*), and sixspotted thrips. In most growing areas, predaceous mites are the most important predators of pest mites. However, in the desert valleys, sixspotted thrips and the spider mite destroyer are the important mite predators. In addition to these species, in coastal citrus orchards, predaceous midges (*Feltiella* spp., family Cecidomyiidae) are important natural enemies of spider mites, and a predatory rove beetle (*Oligota oviformis*) is sometimes found. Other natural enemies that feed on mites, including brown and green lacewings, dustywings, and minute pirate bugs, are discussed earlier under "General Predators." The whirligig mite (*Anystis agilis*), a relatively large, reddish species named for its fast circular motions, preys some on mites but prefers to feed on citrus thrips. Whirligig mite is described and pictured earlier in the "Thrips" section.

Euseius Predatory Mites

Euseius species predatory mites (family Phytoseiidae) are the most important natural enemies of citrus red mite, because they live year-round in the citrus trees and their populations increase when the citrus red mite populations increase. When feeding on citrus red mite, *Euseius* prey mainly on larvae and small nymphs. They also feed on citrus thrips, the nymphs of scales and whiteflies, pollen, leaf sap, and other tetranychids, but mite species that produce significant webbing are not preferred prey. Because they have a wide array of food, not just mites, *Euseius* do not always reduce mite

This shiny, pear-shaped *Euseius* predatory mite adult (left) is the most important predator of the citrus red mite. *Euseius* adults and nymphs become reddish after feeding on citrus red mites.

A predaceous mite egg (bottom) is oblong or oval. It is somewhat larger than the round spider mite egg next to it.

populations below the economic threshold, but they make a significant impact on their densities.

Several *Euseius* species occur in California citrus, and their appearance and biology are virtually the same. *Euseius tularensis* is important in California's Central Valley. *Euseius stipulatus* is prevalent in southern interior and coastal growing areas. *Euseius hibisci* occurs along the coast.

The *Euseius* female is about the size of the citrus red mite female. However, *Euseius* is pear or teardrop shaped, shiny, and translucent to reddish. *Euseius* nymphs look like the adults, except smaller. Adults and nymphs are white or yellow when feeding on pollen, leaf sap, or citrus thrips. *Euseius* turn reddish after feeding on citrus red mites. Depending on their diet, at 78° to 80°F (26° to 28°C) the development time from egg to adult is 6 to 10 days. Females live about 1 month, during which they lay from 12 to 36 eggs.

Euseius predators overwinter on succulent shoots inside the tree canopy, and their populations increase when new leaf flush develops in spring and fall. Sometimes this increase lags behind citrus red mite population development. *Euseius* occur primarily along the midvein on the underside of shaded leaves infested with spider mites and under the calyx of the developing fruit where pest thrips feed. During the heat of the day, they tend to move to shaded leaves.

The adult *Stethorus*, the spider mite destroyer, is shiny black and has a very finely pitted surface covered with pale, minute hairs. This small lady beetle feeds on all stages of pest mites.

The whitish, oval egg of *Stethorus* is much bigger than the round eggs of citrus red mite visible nearby.

Euseius predators move quickly when exposed to bright light. Collecting leaves containing *Euseius* and spider mites and examining the leaf undersides in sunlight allows this shiny, fast-moving predator to be distinguished from the sluggish pest mites. Predator mite eggs are clear, oval or oblong, and about twice the size of the round spider mite eggs.

Spider Mite Destroyer

Stethorus picipes

The *Stethorus* adult is a very small, shiny black lady beetle about $\frac{1}{16}$ inch (1.5 mm) long. Females lay tiny, pale, oblong eggs, usually scattered singly among spider mite colonies. Larvae develop through four progressively larger instars, are dark gray to brownish, and covered with numerous fine

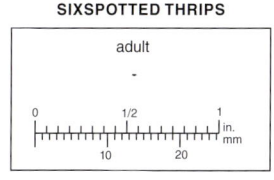

This sixspotted thrips feeds almost entirely on mites. Adults are yellowish with three dark blotches on each forewing.

The sixspotted thrips larva is translucent white to yellowish. In comparison with similar-looking pest thrips, this predator is common only where mites are abundant and can often be seen feeding on mites, as shown here.

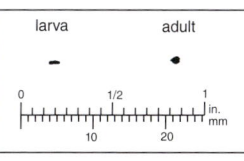

The spider mite destroyer larva is dark gray to brownish and covered with numerous fine hairs. This *Stethorus* is feeding on twospotted mites.

hairs. Pupae are oblong or oval and covered with short spines. Depending on maturity, pupae vary in color from black to brown, gray, or reddish.

Adults and larvae feed on all stages of pest mites, and each individual consumes about six mites per day. During warm temperatures, the spider mite destroyer can complete one generation (egg to adult) in about 3 weeks. Females typically lay about 100 to 200 eggs. *Stethorus* are especially significant in reducing mite numbers once mite populations have developed to moderate levels. *Stethorus* are specialized predators on mites. They are highly effective at finding where mites are abundant, and they voraciously feed in those patches of high mite populations.

Sixspotted Thrips

Scolothrips sexmaculatus

Adults and larvae of this predatory thrips are slender and mostly pale. Adults are about ⅛ inch (3 mm) long and have the long fringes on the margins of their wings that characterize thrips, as illustrated on page 195 and earlier in "Identifying Thrips Species." Adults can be distinguished from other mostly pale thrips by the three dark spots on each wing cover. Sixspotted thrips adults and larvae feed mostly on eggs and immature stages of mites. In citrus, sixspotted thrips adults feed mainly on the twospotted spider mite and other webbing mites, but they also feed on citrus red mites when they become abundant. Sixspotted thrips are specialized predators on mites. They are voracious feeders that can reduce mite populations very quickly.

Larvae of predaceous midges, such as this *Feltiella occidentalis*, feed on all stages of spider mites, mostly in coastal areas. Larval color varies from whitish or pale yellow to darker brown, orange, or pink. The head and mouthparts are at the more narrow (tapered) end of the body (lower left, somewhat hidden in this photo).

This flattened cocoon of a predaceous midge resembles that of dustywings pictured earlier in "General Predators," except this *Feltiella* cocoon is smaller. This cocoon occurs next to a leaf vein and among the silk webbing and empty body skins of spider mites on which the midge larva fed.

Predaceous Midges

Feltiella species

Midges (family Cecidomyiidae) are true flies, and some species have larvae that are predaceous on small arthropods such as mites. At least two mite-feeding species, *Feltiella acarisuga* and *F. occidentalis* (previously called *F. acarivora*), occur in California, mostly in coastal growing areas. *Feltiella* larvae feed on all stages of spider mites, including twospotted spider mite and sixspotted mite.

After hatching from eggs, predaceous midges develop through three larval instars and a pupal stage before emerging as adults (Figure 46). Adults are about 1/12 inch (2 mm) long, with a light brown or orangish body and long legs.

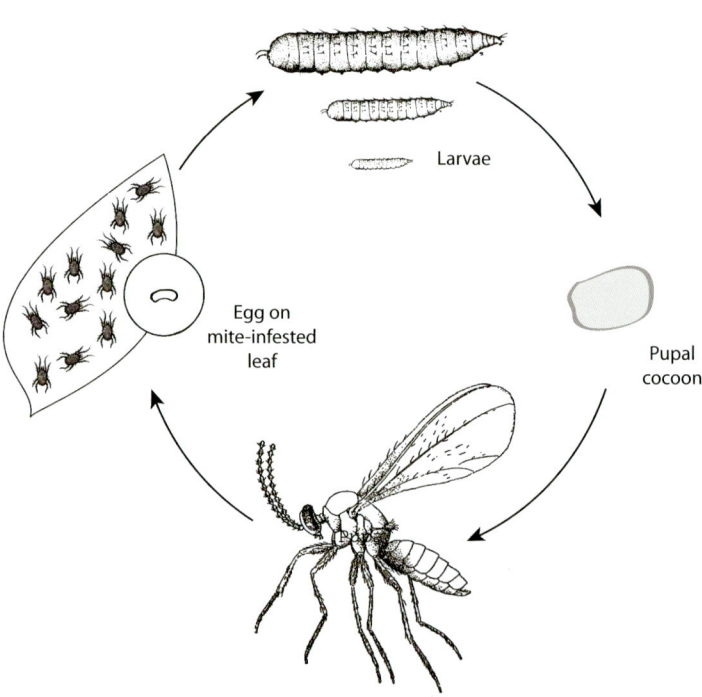

Figure 46. Predaceous midge (*Feltiella occidentalis*) life cycle and stages. *Feltiella* females lay eggs singly on leaves. Larvae develop through three progressively larger instars before pupating in a flat, silk cocoon. After emerging from the pupal case, adults mate and females seek leaves infested with spider mites, where they oviposit. Adult and larvae from Quayle 1912.

Females lay translucent, oblong to sickle-shaped (curved) eggs about 1/100 inch (0.25 mm) long. Larvae are maggotlike and at maturity are less than 1/8 inch (3 mm) long. Larval color varies from whitish or pale yellow to darker brown, orange, or pink, depending on larval age and prey. Pupation occurs beneath a flattened, white silk cover about 1/12 inch (2 mm) long. In coastal areas, all stages of *Feltiella* are usually present year-round, with one generation requiring about a month.

Oligota is a specialized predator of spider mites sometimes seen in coastal citrus. Adults of this tiny black rove beetle have a pointed abdomen that characteristically curves upward at the rear. Their head is bent downward, under the body, so the head is often not apparent, as in this photo.

The elongate, pale larva of the *Oligota* rove beetle has a black spot near the tip of the abdomen.

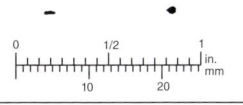

Predatory Rove Beetle

Oligota oviformis

Oligota oviformis, a predatory rove beetle (family Staphylinidae), occurs sporadically in coastal citrus in California. Both adults and larvae feed almost exclusively on spider mites of all stages, including citrus red mite and twospotted mite. After hatching from an egg, *Oligota* develops through four larval instars and a pupal stage before becoming an adult (Figure 47). Development from egg to an egg-laying adult occurs in about 6 weeks at warm temperatures, during which each *Oligota* can consume several hundred spider mites.

The black adult is about 1/25 inch (1 mm) long and, as with other rove beetle species, has shortened wing covers that expose the adult's abdominal segments when viewed from above. Characteristically, *Oligota* adults have a pointed abdomen that curves upward at the rear end and a head that is bent downward or under the body so the head is not apparent when viewed from above (Figure 47). Females lay eggs singly, usually on the underside of mite-infested leaves. The oval eggs are yellowish to pale orange and about 1/75 inch (0.3 mm) long. Larvae are yellowish and covered with fine spines. Mature fourth instars are about 1/10 inch (2.5 mm) long. At maturity, larvae reportedly drop from the tree and pupate in an oval, loosely woven silk cocoon in litter. The pupa itself is roughly oval and has visible abdominal segments and developing legs and wings. It is dark yellowish, orange, or reddish, then turns black immediately before the adult emerges.

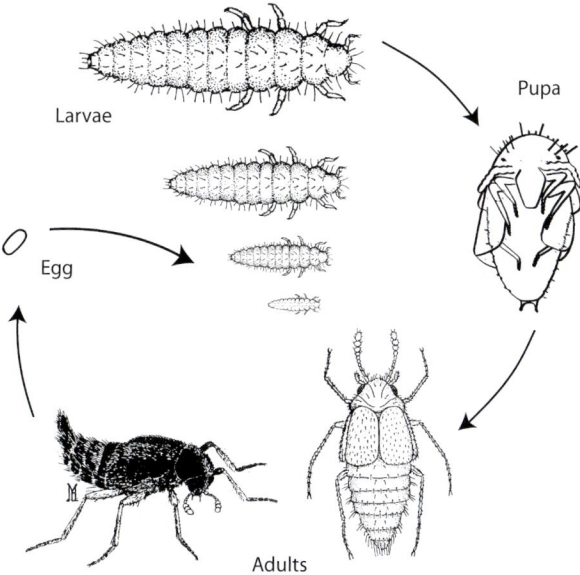

Figure 47. Predatory rove beetle life cycle and stages. The blackish adults have shortened wing covers, revealing their abdominal body segments (lower right, top view). The pointed rear of their abdomen is characteristically curved upward, and the head is bent downward (lower left, side view). After hatching from an egg, *Oligota* larvae develop through four instars. The pupa (shown here removed from its silk cocoon) has visible appendages and is mostly orangish or red but turns black immediately before the adult emerges. Adults and larvae from Quayle 1912; pupa from Moore, Legner, and Badgley 1975.

Other Invertebrate Pests

Exotic Fruit Flies

Fruit flies (family Tephritidae) are frequently introduced into California, and some species have repeatedly been eradicated. Inadvertent reintroduction from areas where these pests are established, such as Mexico and Hawaii, is a constant threat. Tephritid fruit flies attack most deciduous and subtropical fruit, including citrus. Exotic pest species include the Caribbean fruit fly (*Anastrepha suspensa*), Mediterranean fruit fly (*Ceratitis capitata*), melon fly (*Dacus =Bactrocera cucurbitae*), Mexican fruit fly (*Anastrepha ludens*), and oriental fruit fly (*Bactrocera dorsalis*).

Mediterranean fruit fly and other Tephritidae develop through four life stages. Displayed here (left to right) are the eggs, larva (a third instar), pupa, and adult.

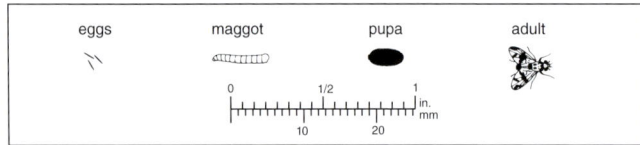

Tephritid adults are distinguished by the color patterns on their body and wings. The adult melon fly has a mostly orange-brown body. Its wings are mostly clear, except for brown spots along the veins.

An adult tephritid is about 1/4 inch (6 mm) long, considerably larger than the vinegar fly (common fruit fly) also shown here. The adult Mediterranean fruit fly (photo left) has a blackish thorax marked with silver. The abdomen is mostly tan with darker stripes. There are two light brown bands across each wing and another along the outer (distal) front edge, plus gray flecks scattered near the wing base.

The adult Oriental fruit fly's body color is variable, but typically it has bright yellow or white markings on a mostly dark thorax and black or dark markings on a mostly brown abdomen. The markings usually include a dark T shape on top of the abdomen.

The adult Caribbean fruit fly has a yellow-tan body with a black spot on the top rear of the thorax (at the scutellum). The clear wings have broad brownish bands.

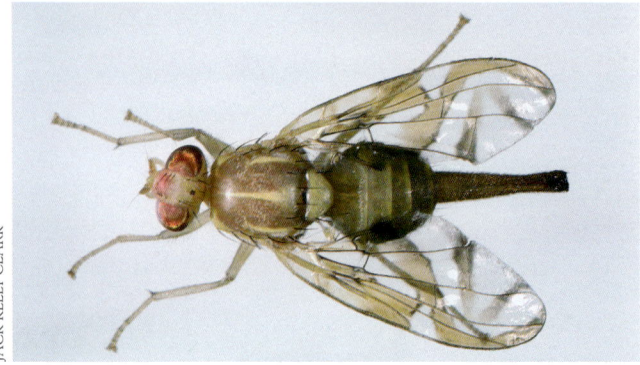

The Mexican fruit fly has a pale orange-yellow body with two or three whitish stripes along the thorax. *Anastrepha* spp. (including the Caribbean and Mexican fruit flies) have clear wings with brown markings that include an S pattern across the wing. The Mexican fruit fly female (shown here) has a long ovipositor that is one-half the body length or longer.

Adult tephritids are about ¼ inch (6 mm) long, considerably larger than the vinegar flies or common fruit flies (*Drosophila* spp., family Drosophilidae) that are attracted to fermenting fruit and are commonly used in laboratory studies. Tephritid adults are distinguished to species by the color patterns on their body and wings. For example, *Anastrepha* spp. (including the Caribbean and Mexican fruit flies) have clear wings with brown markings that include an S pattern across the wings, as pictured for Mexican fruit fly. The Oriental fruit fly adult has black or dark markings on a mostly brown abdomen, which usually includes a dark T"marking on top of their abdomen. Previously introduced pest species established in California include apple maggot (*Rhagoletis pomonella*), olive fruit fly (*Bactrocera oleae*), and walnut husk fly (*Rhagoletis completa*). See the online UC IPM *Pest Management Guidelines* on those crops for pictures of these species.

Female tephritids lay eggs in the fruit rind. The legless fly larvae feed in pulp, directly damaging the fruit and introducing fruit decay microorganisms. Mature larvae may pupate in the fruit but typically emerge and pupate in litter near the soil surface and overwinter there as pupae. Depending on seasonal temperatures, host plant, and the pest species, tephritids have one to many generations each year.

Parasitic wasps are important in reducing populations of some established fruit fly species. Fruit fly biological control is not important in California citrus because these are quarantined pests. Eradication treatments and restrictions on shipping fruit are imposed where citrus-feeding tephritid fruit flies are found.

Specially baited traps are used to detect adults, but tephritids are sometimes attracted and caught in sticky traps used for other pests. Unless you are confident the species found is already known to be established, such as olive or walnut fruit flies caught in citrus growing near those crops, take any tephritid flies you find to an expert for identification. Report to agricultural officials any fly maggots found in citrus fruit.

Brown Garden Snail

Cornu aspersum

The brown garden snail (class Gastropoda, family Helicidae, and formerly named *Cantareus aspersus* and *Helix aspersa*) is an introduced species that can extensively damage fruit and young trees. Snails are especially problematic in nurseries and in orchards where no-till weed control and sprinkler or low-volume irrigation create an ideal environment for snail development.

Description and Biology

The brown garden snail is about 1¼ inches (3 cm) in diameter at maturity. The mostly brown shell can include black, tan, and yellow and is patterned in bands, flecks, and swirls. Each snail has both male and female parts (is hermaphroditic) but mates with another snail before laying eggs. Depending on local climate and moisture, each mated adult lays a batch of eggs about once a month, each batch consisting of up to 80 eggs laid in a loose group in a shallow depression in the topsoil. Each egg is about ⅙ inch (4 mm) long and spherical or teardrop shaped with a protuberance at one end. Egg color ranges from pale milky or translucent to light or dark brown or orange.

Snails are active and feed mostly during the night and early morning when it is damp. In southern California, particularly along the coast, young snails are active throughout the year. In the San Joaquin Valley, brown garden snails are active primarily in late winter and spring. During hot, dry periods, the snails seal themselves off with a parchmentlike membrane and remain in the tree or soil. Mature snails hibernate in the topsoil during winter.

Damage

The brown garden snail chews fruit, succulent leaves, and (in nurseries) the bark on young trees. Fruit damage appears as circular chewed areas in the rind. Damaged leaves have large chewed areas along the margins. Moist slime, shiny trails from dried slime, or dark snail excrement is often visible around damage.

The brown garden snail chews ripe and ripening fruit, young leaves, and (in nurseries) the bark on young trees. Fruit damage appears as circular chewed areas in the rind.

Brown garden snail eggs are spherical or teardrop shaped and range in color from pale milky or translucent to light or dark brown or orange.

Natural Enemies

The predatory decollate snail (*Rumina decollata*, family Subulinidae) is widely distributed in southern California. It feeds on brown garden snail and other mollusks. The decollate's shell is elongated, tapered, and about 1 inch (25 mm) long when mature. While the shell is growing, the tip becomes brittle and breaks off. The decollate snail is self-fertile and lays about 500 eggs during its lifetime. The eggs are smaller than those of the brown garden snail and have a brittle shell. The decollate snail lives primarily in litter on the soil and emerges only to feed or to escape heavy rain or irrigation water. Besides attacking the brown garden snail, decollates feed on decomposing leaves, fallen or bruised fruit, and emerging seedlings. They have not been observed to feed on sound fruit or citrus leaves.

Decollate snails are commercially available for introduction into areas where they are not yet present. While not always consistently effective, introducing the decollate snail may reduce brown garden snail populations to insignificant levels in 4 to 10 years. Decollate introductions are legal only in certain counties, so check with the county agricultural commissioner or wildlife officials before introducing them. Do not release decollate snails near where poison baits are used because baits kill both the pest and predator snails. Decollate introductions are compatible with pruning tree skirts to make it more difficult for snails to attack low-hanging fruit and applying trunk barriers such as copper foil or an effective copper compound to trunks to repel snails that climb trunks. Decollate snails do not climb trees, thus they will not be affected by pruning or trunk barriers. For detailed recommendations on snail control and decollate introductions, consult the online *UC IPM Pest Management Guidelines: Citrus*.

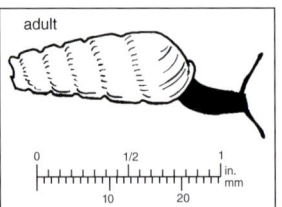

DECOLLATE SNAIL
adult

The decollate snail can be an effective predator of brown garden snails. The predatory decollate's shell is elongated, and the tip breaks off as the shell grows.

Nematodes

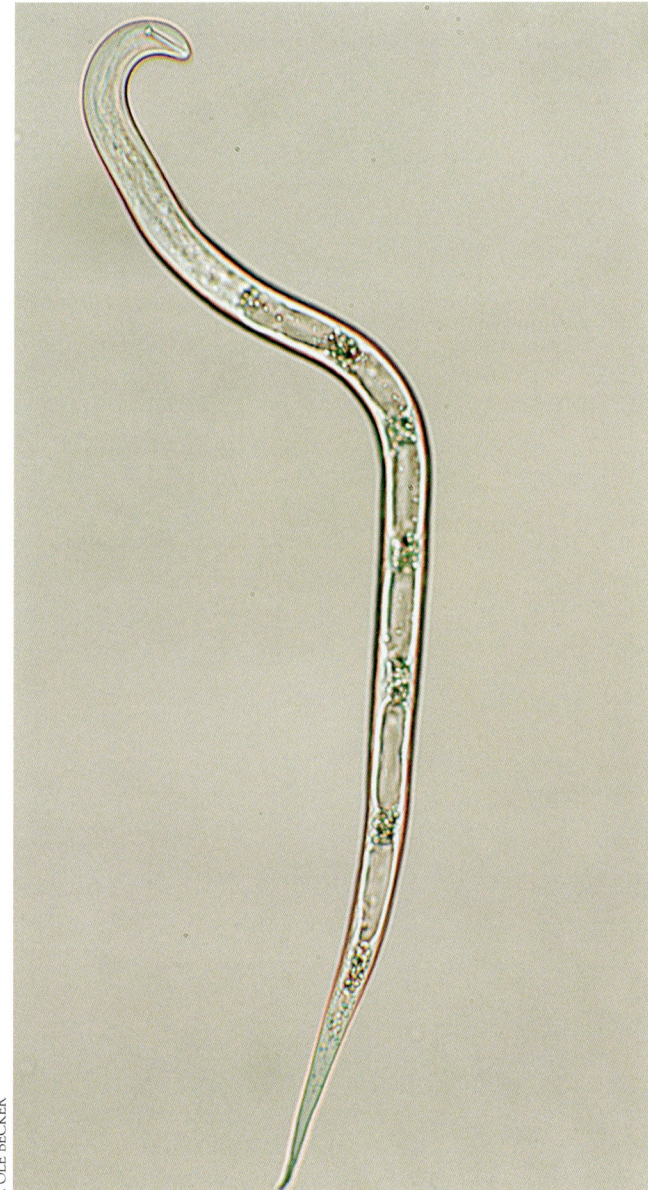

Nematodes are typically tiny (usually microscopic) unsegmented roundworms. Depending on the species, nematodes feed on bacteria, fungi, insects, plants, vertebrates, or other nematodes. Plant-parasitic nematodes live in soil and plant tissues but must feed on living cells. Most plant-parasitic nematodes feed on or in roots. Plant-parasitic nematodes can be distinguished from other species by their syringelike stylet (mouthpart) with which they penetrate plant cells. At the base of the stylet within the nematode's head is a muscular bulb that pumps contents out of plant cells through the stylet.

Citrus nematode (*Tylenchulus semipenetrans*) is the major nematode pest of citrus in California. It is present in most citrus orchards and in all soil types. It also parasitizes grape, lilac, olive, and persimmon. Citrus nematode feeds with the anterior portion (front end) inside the root. Its reproductive posterior portion (rear end) remains outside in the soil.

The sheath nematode (*Hemicycliophora arenaria*) is a relatively minor pathogen that occurs on citrus in the Coachella Valley and on some native desert plants. It has a broad host range and thrives well at high temperatures and at low moisture levels. Several other nematode species are important pests of citrus in Florida or elsewhere in the world.

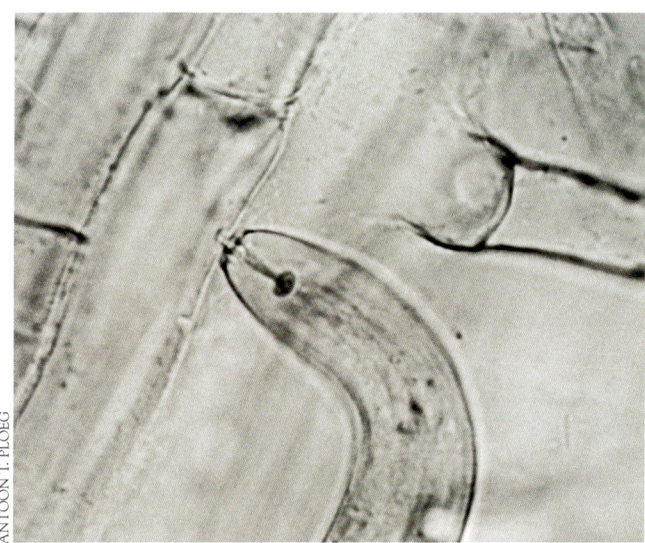

Plant-parasitic nematodes can be recognized by their syringelike stylet (mouthpart) with a muscular bulb at the base for pumping out plant cell contents. These feeding structures are visible in the head of the root lesion nematode (*Pratylenchus* sp.) shown here feeding attached to a root. Root lesion nematodes are not pests of citrus in California.

Description and Biology

The citrus nematode is 1/60 inch (0.4 mm) long or less at maturity. Adult sheath nematodes are about 1/24 inch (1 mm) long. You need a microscope (magnification of about 30× or more) to see nematodes clearly. Identifying the species of nematode is difficult. Consult a nematologist or other competent expert and submit properly collected samples to a diagnostic laboratory to help verify the presence of nematodes and to identify the species.

Nematodes typically develop within their egg to the second-stage juvenile (J2). After it hatches, the second stage develops through two more larval stages (J3 and J4) before becoming an adult. Except for a free-living period during the second stage after the larva has emerged from the egg, female citrus nematodes spend most of their life cycle feeding with their head embedded in citrus feeder roots. Second-stage larvae settle and feed in epidermal (outer layer) cells of rootlets. As they mature, females feed more deeply in the rootlet. After the last molt, the adult female becomes immobile. Her feeding causes the surrounding root cortical cells to develop into nurse cells (feeding cells). The posterior of the female remains outside of the root, swells, and excretes gelatinous material into which up to 100 eggs are laid. This life cycle is described in more detail and illustrated in Figure 48.

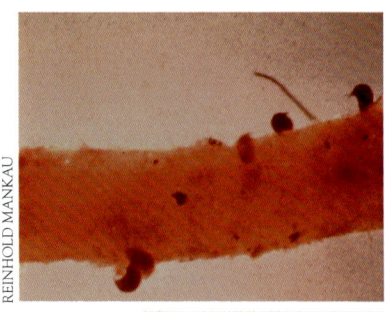

The posterior ends of adult female citrus nematodes are protruding from this citrus root. The root and nematodes are artificially stained red to make them more visible.

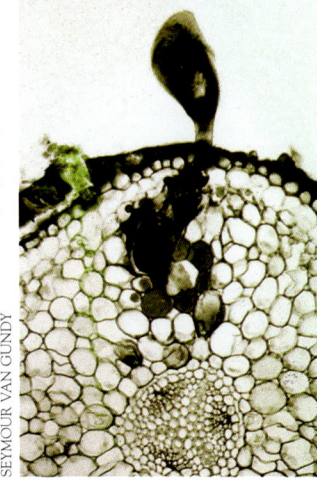

The adult female citrus nematode (shown here in a root cut in cross section) feeds with its anterior (front) end deep inside the root tissue.

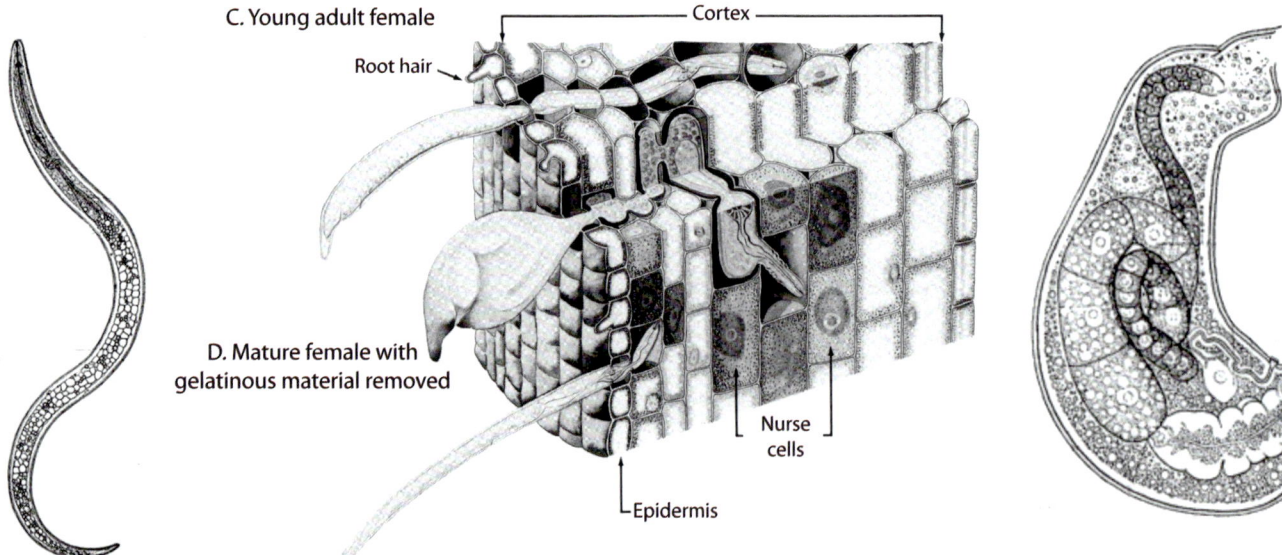

A. Second-stage larva
B. Second- through fourth-stage larvae
C. Young adult female
D. Mature female with gelatinous material removed
E. Mature female reproductive parts

Figure 48. Citrus nematode life cycle and feeding in a citrus root. First-stage larvae complete their development inside the egg case. After molting into the second stage, the larva emerges from its egg. A. The second-stage larva initially is free-living in soil as it seeks a young citrus feeder root. B. The female second-stage larva settles to feed in the root epidermis (outer layer). As females develop through the third- and fourth-larval stages, they feed deeper in cortical cells. C. After the last molt, the young adult female is feeding deep in the root in a cortical cell. D. The female becomes immobile and her head swells. Her feeding causes the surrounding cells to develop into nurse cells (feeding cells). Nurse cells develop denser cytoplasm (the fluid that fills the cell), and their nucleus (the organ containing most genetic material) becomes larger and more visible. The female's posterior remains outside of the root and swells. E. The posterior contains the ovary, eggs, and uterus (reproductive parts shown after removing the body's surface layer). The female excretes sticky, gelatinous material (not shown) into which eggs are laid. Female reproduction occurs with or without mating; males are present but are not required. Nematodes (A., E.) by W. E. Chambers in Cobb 1914.

Males are present but are not required for production of eggs. Females reproduce without mating (parthenogenesis), although mating and sexual reproduction may also occur. Males live about 1 week after leaving the egg and feed very little or not at all on roots. The female life cycle from egg to egg takes 6 to 8 weeks under optimal temperatures of 77° to 86°F (25° to 30°C). Citrus nematode larvae are not active when soil temperatures drop below 60°F (16°C). Without host roots, few citrus nematodes will survive for more than 1 to 2 years. However, low populations of citrus nematodes reportedly can persist for up to 5 years after host removal, perhaps in part because pieces of infested root typically remain in the soil after the host is removed.

Sheath nematode larvae and adult females feed attached to the outside of root tips. Their feeding results in swelling and galling that usually occurs only at the root tips. Fourth-stage larvae and adult females can survive without root hosts for at least 6 months.

Symptoms and Damage

Citrus nematode damage is sometimes called citrus slow decline. Often in combination with root rot pathogens (especially *Phytophthora* spp.), nematode damage to young trees planted where citrus previously grew is one factor in the "citrus replant problem." A nematode infestation also increases the susceptibility of citrus to, and damage from, abiotic problems such as nutritional disorders, salinity, and water stress.

The severity of nematode damage depends on the age and vigor of the tree, race and density of the nematode population, and susceptibility of the rootstock. Mature trees can tolerate a considerable number of citrus nematodes before showing lack of vigor and decline symptoms. Rootstocks with excellent resistance generally are not damaged in the field. In heavily infested soils, young trees may be stunted or fruit production may be reduced on bearing trees that have susceptible rootstocks.

In some cases, nematode infestations can occur without causing any aboveground symptoms. When aboveground symptoms of nematode damage do occur, they include leaves that are yellow and undersized and a reduction in the number and size of fruit. Trees often lack vigorous growth, and twigs may die back. These symptoms do not definitely mean that nematodes are the cause. *Phytophthora* infection and other maladies can produce the same symptoms.

Belowground symptoms of citrus nematode infection are difficult to observe in the field; they include poor growth of feeder roots and a slight thickening of small roots. Infested small roots develop an irregular surface, and the outer portion of the rootlet may separate from the inside core (Figure 49). For the best visual symptoms, collect small roots from healthy and unhealthy trees and wash them for comparison. Uninfested roots appear clean and yellowish after washing. Soil adheres to infested roots, giving them a dirty appearance. This "dirty root" symptom results from soil particles sticking to the females' gelatinous egg masses. Dirty roots are not, however, reliably diagnostic. The presence of citrus nematode can be confirmed only by microscopic examination by an expert. Phytophthora root rot also causes small roots to darken, and *Phytophthora* spp. and citrus nematodes often occur together on the same roots.

There are at least three distinct races or strains of citrus nematode worldwide, with two occurring in California. Susceptible rootstocks differ somewhat in the extent to which they are damaged. Certain rootstocks, including trifoliate and its hybrids, are resistant to the dominant strain of citrus nematode ("Citrus") in California. However, the "Poncirus" strain of citrus nematode was detected in California in 1969, and it reproduces well on trifoliate and

This orange tree is in an advanced stage of citrus slow decline. The cause is numerous citrus nematodes feeding on its roots over a prolonged time period.

A citrus root system from soil infested with citrus nematodes, *Tylenchulus semipenetrans*, next to healthy citrus roots from noninfested soil (left). Soil clings to nematode-damaged roots (right), causing roots from infested soil to appear darker or dirty.

its hybrids. The distribution of "Poncirus" in California is unknown.

Sheath nematode feeding may reduce root growth and the vigor of trees. Below ground, sheath nematode feeding causes swelling and galling of root tips. This appearance differs from root knot nematode swellings, which occur throughout the length of roots. The root knot nematode species in California do not infest citrus.

Monitoring and Diagnosis

To make management decisions, determine which nematode species are present and estimate their population density. Consider the site's cropping history and whether irrigation water, runoff, or drainage may have contaminated the site with nematodes. The soil is almost certainly infested with the citrus nematode if citrus or other susceptible hosts grew there within the last 5 years. Citrus nematodes are likely present if the property floods or receives runoff from other infested citrus orchards, or if the previous crop was irrigated with surface water drawn downstream from infested citrus or other susceptible hosts.

If the history of a site to be planted is unknown, or if an existing orchard is suspected of having nematode problems but the nematode species and abundance have not been identified, collect samples and send them to a diagnostic laboratory.

Flooding left this eroded soil gully. Nematodes readily spread in water and whenever soil is moved. Because this orchard is surrounded by other citrus (not shown), the soil probably is highly infested with citrus nematodes because of extensive movement of soil and surface water.

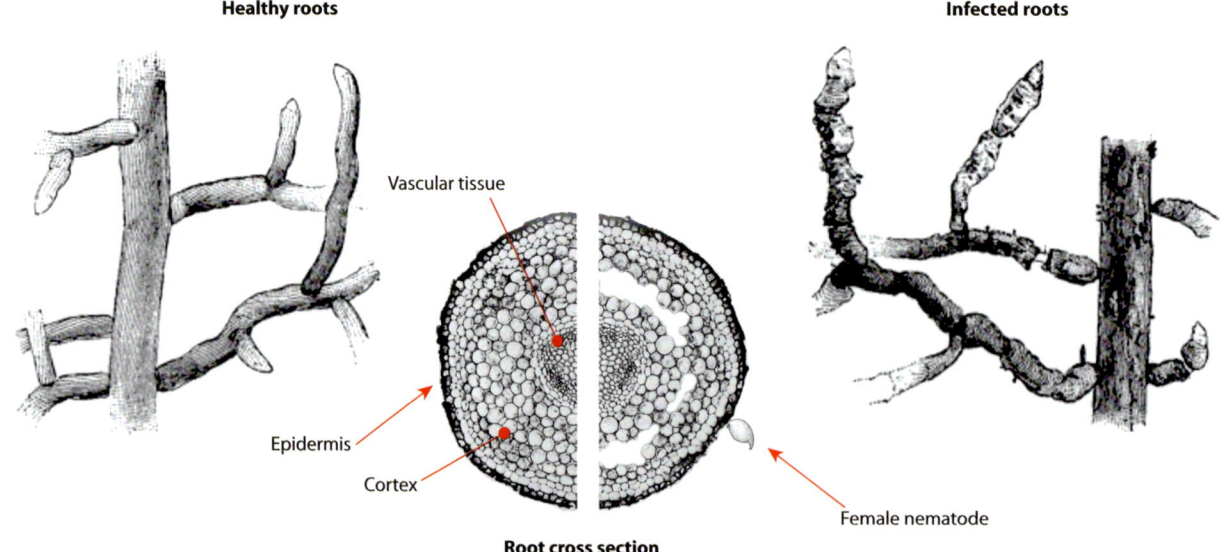

Figure 49. Citrus nematode damage to small roots (rootlets). Healthy citrus roots (left) appear clean and yellowish after washing. When viewed in cross section, healthy roots (center left) have continuous layers of well-connected cells. Roots infected by citrus nematode (right) develop an irregular surface, and portions of the rootlets become slightly thickened. It can be difficult to wash soil from infected roots and clearly observe the tissues. Soil adheres to nematodes' gelatinous egg masses, encrusting infested rootlets and giving roots a dirty appearance. When an infested root is viewed in cross section (center right), the outer layers may be seen to have separated some from the inner layers. Nematode feeding causes destruction of cortical tissue, so that outer tissue may slough off of roots. A female citrus nematode (only the rear portion that extends outside the root) is shown here approximately to scale in relation to the diameter of a rootlet. Her total body length at maturity is about $1/50$ inch (0.4) mm. Adapted from W. E. Chambers in Cobb 1914, Schneider 1968.

Sampling Nematode Larvae in Soil. Take soil samples within the root zone, from 6 to 36 inches (15–90 cm) below the soil surface, depending on soil type as recommended in *UC IPM Pest Management Guidelines: Citrus* (online at www.ipm.ucdavis.edu). Take subsamples from regularly wetted zones at the edge of the tree canopy and include some feeder roots in each sample. In drip-irrigated orchards, take samples around emitters where feeder roots are abundant.

Unless other methods are recommended, divide the field into sampling blocks of about 2 to 4 acres each. The soil submitted for analysis from each block should be representative of that location's cropping history, crop injury, and soil texture. Take 20 or more soil and feeder root subsamples randomly from within each block. Mix the subsamples together gently and thoroughly and make a composite sample of about 1 quart (or 1 liter) for each block. Place the composite sample from each block in a separate plastic bag and seal it. Place a label on the outside of each bag, including your name, address, location, collection date, the current and previous crop, and the crop you intend to grow. Keep samples shaded and cool at about 50°F (10°C, do not freeze them). Transport samples promptly to a diagnostic laboratory.

The number of nematodes in the soil as determined by a soil analysis gives some indication of the damage potential from an infestation. However, many of the nematodes in soil are not pathogenic to citrus. The number of nematodes detected in the soil analysis will vary greatly depending on factors such as soil moisture, soil type, and temperature. Soil samples do not predict yield at the end of the season since factors such as alternate bearing and other pest problems can influence yield.

Sampling Female Nematodes on Roots. The number of female nematodes per unit of feeder roots is a better indicator of tree damage potential than the number of free larvae in soil. Seek out a laboratory that will determine the extent of root infestation, such as the number of citrus nematode females per gram of root. Specifically request this root analysis for females and learn and use the proper root sample collection methods.

Interpreting Sample Results. Laboratories should report nematode counts to at least the genus level. Citrus nematode (*Tylenchulus semipenetrans*) is usually the dominant plant parasitic nematode once soil has been planted with citrus trees for about 3 years or longer. Therefore in established citrus groves, counts reported as *Tylenchulus* species can be assumed to be citrus nematode. For more information on nematode sampling, guidelines on what level of nematodes may reduce tree growth and fruit production, and recommendations on when direct control actions may be warranted, consult the *UC IPM Pest Management Guidelines: Citrus* and "Nematodes in California Citrus" (in preparation). For personal help about sampling methods

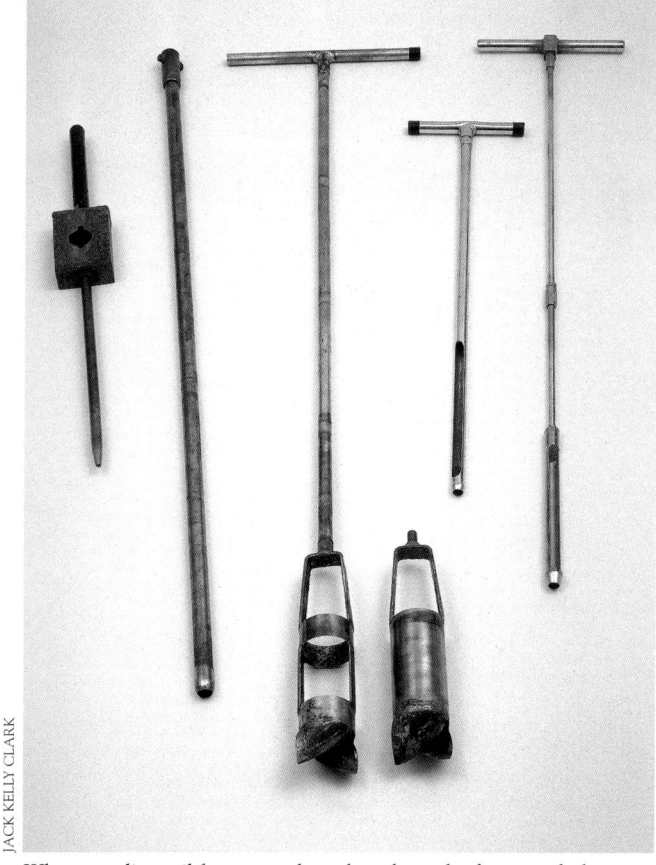

When sampling soil for nematodes, take subsamples from regularly wetted zones at the edge of the tree canopy where feeder roots are likely to be found. Soil augers (center) have a variety of bits for taking a relatively wide-diameter soil core. Soil augers or a shovel (not shown) are good tools for collecting soil and feeder roots for laboratory analysis of nematodes. Soil tubes (right) are useful for monitoring soil type and moisture. The two-piece Veihmeyer tube (left) has a slotted hammer for driving the tube into soil and removing it. Oakfield soil tubes (right) can collect soil cores down to about 2 feet (60 cm).

and interpreting test results, contact your local University of California Cooperative Extension farm advisor or diagnostic laboratory.

Prevention and Management

The best time to control nematodes is before and during planting. Control options are limited once you detect nematode damage in an established orchard. Site preparation, rootstock cultivar selection, sanitation, and other cultural practices are the primary nematode management methods. Consider rotating with annual crops such as cereal grains for at least 1 to 3 years before planting citrus in nematode-infested soil. Rotation will reduce citrus nematode populations and allow the decay of woody roots that harbor nematodes. Preplant soil fumigation or post-

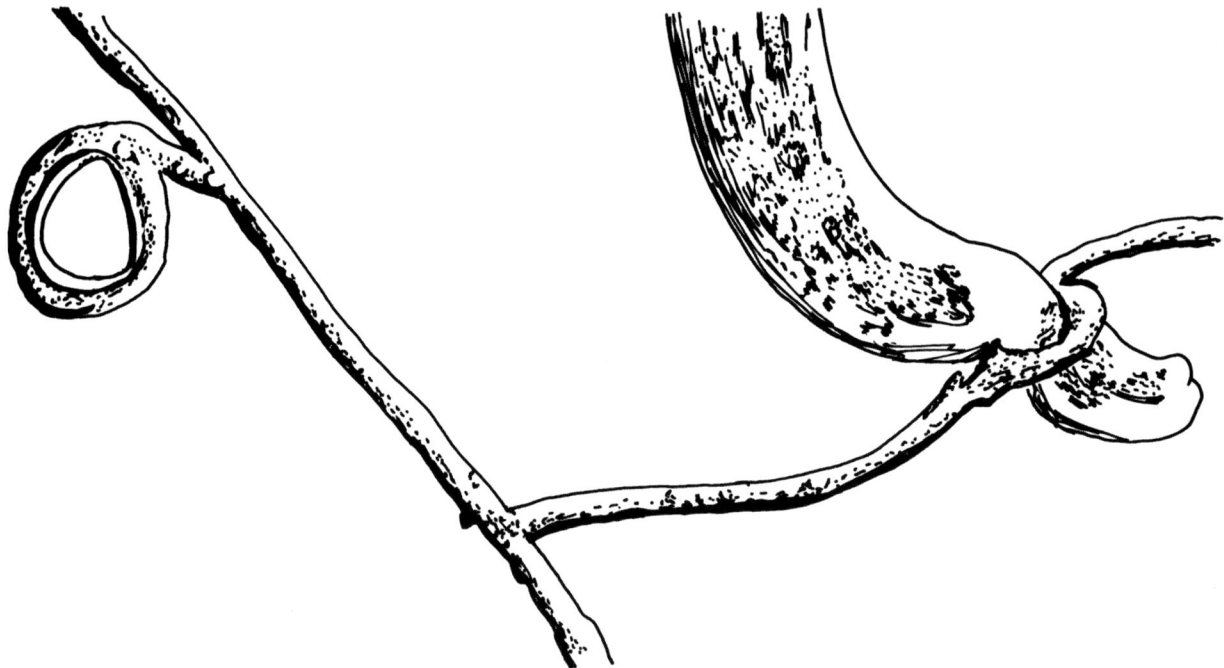

Figure 50. A nematode at right snared by a ring trap fungus (*Arthrobotrys* sp.). Another trap waits for prey at the upper left. When a nematode enters, the loop contracts like a noose. Fungal hyphae then grow into the captured nematode and consume its body. Predaceous nematodes and mites, parasitic fungi, and especially nematode-trapping fungi are common around roots in soil of mature citrus orchards. However, biological control has not been found to be sufficient in controlling citrus nematode on susceptible rootstock cultivars. Illustration by Valerie Winemiller.

plant nematicides may be available, as discussed in the online UC IPM *Pest Management Guidelines: Citrus*.

Rootstock Selection. Plant nematode-resistant rootstocks if they are compatible with other soil conditions and the selected scion. Purchase certified nematode-free nursery stock regardless of site history and whether rotation or fumigation are used. Resistance-breaking nematode biotypes may develop over time, but the most critical period for preventing nematode problems is during the early years of young tree growth. For a list of rootstocks resistant to nematodes and other problems, consult "Rootstocks for Citrus in California" (in preparation), Table 5 in the chapter "Managing Pests In Citrus," the online UC IPM *Pest Management Guidelines: Citrus*, or contact your local University of California Cooperative Extension farm advisor or nursery supplier.

Sanitation and Cultural Practices. Good sanitation practices are essential to avoid nematode infestations and to prevent the introduction of resistance-breaking nematode strains. Do not bring plants, soil, or other amendments to the site unless they are known to be free of plant-parasitic nematodes and other pests. Other tactics you can use to prevent the introduction of pests include

- clean equipment thoroughly before bringing it on-site
- do not allow runoff water to enter the site
- irrigate using water that is not contaminated with citrus nematodes

Practices that favor root growth and reduce tree stress increase tree tolerance to nematode infection. Irrigate and fertilize trees properly, avoid excess salinity, provide good soil drainage, and prevent and control other sources of injury to roots, especially root rot diseases.

Natural Enemies. Parasites, pathogens, and predators of nematodes increase in soils over time. These natural enemies are often abundant around feeder roots in mature citrus orchards. However, biological control apparently is not sufficient to control citrus nematode on susceptible cultivars. Female nematodes feed embedded in roots and at maturity excrete gelatinous material that protects the posterior of the nematode and the eggs. Most known natural enemies attack free-living (soil-dwelling) nematodes, which for the citrus nematode occurs only during the second larval stage.

Beneficial species found in California citrus soils include predaceous nematodes (*Mononchus* spp.), predatory mites (*Lasioseius scapulatus*), and nematode-trapping and parasitic fungi. Nematode-trapping fungi (*Arthrobotrys* and *Dactylaria* spp.) in particular are remarkably widespread in mature citrus orchards. These include those that trap nematodes using constricting rings, a sticky net or mesh, or sticky knobs.

For example, ring-trap fungi produce looplike structures in spaces among soil particles where nematodes commonly travel (Figure 50). When a nematode enters, the loop constricts like a noose, and the snared nematode is consumed by hyphae that grow from the fungi.

Loam soils high in organic matter generally have a greater diversity (more species) and higher populations of natural enemies than sandy soils low in organic matter. However, efforts to enhance biocontrol of citrus nematode, such as by manipulating cultural practices or soil conditions, have been unsuccessful. Nematicides, especially preplant fumigants, may have undesirable side effects on nematode-feeding beneficial organisms.

Chemical Control. If the site was previously infested with citrus nematode, preplant fumigation may be recommended even when a resistant rootstock will be used. Young trees in fumigated orchard sites generally exhibit improved growth and yields compared to those on nonfumigated sites.

Soil fumigation is increasingly restricted because of the fumigants' hazards and contribution to air and water quality problems. Fumigant contamination of groundwater has closed some drinking water wells. Fumigants produce volatile organic compounds (VOCs), and VOCs in the presence of sunlight combine with nitrogen oxides in air to form ozone. Ozone present at ground level can reduce crop yields and result in human respiratory problems. Fumigate soil before planting only as a last resort when other management strategies have not been successful or are not available.

Postplant nematicides may be recommended if citrus nematode densities exceed certain thresholds. Postplant nematicides require precise and repeated application to be effective. Postplant nematicides are expensive and it is difficult to predict whether the increased yield they provide economically justifies their cost. When deciding whether to begin a postplant nematicide application program, weigh the projected increase in fruit yield and sales revenue against the age of the orchard, infestation level, and treatment cost. Compare postplant nematicide costs and benefits with the costs and benefits of replanting the site with new trees on nematode-resistant rootstock. For nematicide recommendations and to compare treatment materials, consult the online *UC IPM Pest Management Guidelines: Citrus*.

Vertebrates

Citrus orchards provide food and shelter for vertebrates. Pest species feed on fruit or chew bark and shoots, which stunts tree growth, allows entry of decay pathogens, and may kill trees. Some vertebrate pests chew or destroy irrigation lines and emitters. Other vertebrates dig holes through the soil surface, which channel irrigation water to undesired areas. Food safety is a concern if animal pathogens contaminate fruit.

The major vertebrate pests in citrus are black-tailed jackrabbit, California ground squirrel, and pocket gophers. Occasional pests include brush and cottontail rabbits, coyote, deer, European starling, tree squirrels, voles, roof rat, and wild hog (wild pig). Deer mice are a rare pest in commercial citrus.

The orchard's location influences whether certain vertebrates are likely to become pests. For example, coyotes, deer, and (most) rabbits do not live in citrus orchards. These vertebrates cause problems when citrus grows adjacent to their preferred habitats, such as forests, rangelands, or unmanaged areas. Rodents (and occasionally rabbits in young orchards) can reside in orchards year-round and are potential pests in all orchards. However, rodents and rabbits generally cause the most damage when citrus grows next to rangeland, riparian areas (vegetation bordering waterways), or unmanaged land where their populations may build up unchecked.

Rabbits chewed the bark off this tree, girdling the trunk. Voles cause similar damage, but vole gnawing occurs no higher than about 2 inches (5 cm) above the ground. Squirrels can chew virtually anywhere on trunks and limbs, while pocket gopher girdling is usually hidden below ground.

Ground squirrels chewed this microsprinkler, but coyotes, dogs, gophers, and rabbits also gnaw irrigation systems. Do not rely solely on recognizing the type of damage caused by the pest because different pests can cause similar damage. Positive identification of the damaging species is important because it allows you to choose effective control actions.

Managing Pest Vertebrates

Manage your orchards to keep pest populations at low levels so that significant damage does not occur. Before planting, remove vertebrate pests and destroy their habitats within the orchard boundaries (such as by plowing burrows). Preventive measures (such as burrow destruction) and direct controls (such as baiting and trapping) usually cost less and are more successful before planting when they can be applied without risk of damaging the crop or irrigation system and one can easily see the pests or their habitats. Be aware of what species are more likely to become serious pests because of the orchard's location. It is often easier to manage vertebrates by implementing controls on the orchard's perimeter versus inside the orchard.

Management programs for vertebrate pests involve four primary steps:

1. Correctly identify the pest species by learning to recognize characteristic damage, burrows, and signs such as feces and tracks.
2. Alter the habitat where feasible to make the area less favorable to the pest species.
3. Implement appropriate control for the orchard and time of year, taking early action and using due consideration for the environment and nontarget species.
4. Establish a monitoring system so you can detect reinfestation and know when additional control measures are needed.

A successful pest management program requires good records and regular monitoring. Some vertebrate pest populations can increase rapidly because of high reproductive rates and abundant food. Keep a record of the management procedures you use and their effectiveness. Good records will help you plan and improve future control strategies.

For most vertebrates, there is more than one control option for reducing populations and damage. Table 11 summarizes the various control measures appropriate for the common vertebrate pests of citrus. Consult the UC IPM *Pest Management Guidelines: Citrus* (online at www.ipm.ucdavis.edu) for more details on materials and latest methods for managing the individual pest species.

Observation and Identification

Positively identify the damaging species so you can learn their biology and choose the effective control actions. Identify vertebrates through direct observation of the pest and evidence of its presence, such as burrows, feces, and tracks. The nature of the damage provides a clue to which species is causing problems, but do not rely solely on this kind of evidence. Many species can cause similar types of damage, such as tree girdling or chewing marks on fruit or irrigation equipment. Management tactics depend on the pest species to be controlled. The descriptions, line drawings, and photographs in this chapter can help you identify the vertebrate pests that are causing problems in your orchards. Consult *Wildlife Pest Control around Gardens and Homes* (Salmon, Whisson, and Marsh 2006), the UC IPM *Pest Management Guidelines: Citrus* and *Pest Notes* (online at www.ipm.ucdavis.edu) on individual species, and other publications listed in "Suggested Reading" for additional information on vertebrate pest identification and biology.

Monitoring

Monitor orchards more often and more carefully if conditions in or near the orchards are especially favorable for vertebrates. For example, certain types of habitat near orchards (unmanaged lands, streams, or other surface water) and conditions within orchards (abundant weeds, ground covers, or trees on or near berms) can increase the likelihood of damage from certain vertebrates. If vertebrate pest populations build up, respond quickly with control actions. After you take a control action, establish a routine monitoring program so you can assess the effectiveness of the control and detect any further problems that may develop. Based on knowledge of their biology, monitor during the time of day and season when your problem species are active (e.g., for ground squirrels, monitor during midmorning in months when squirrels are not hibernating). Record the location and approximate number of pests you see or some measure of their activity (e.g., the number of squirrels and the number of their burrows). Keep a record of the management procedures you use and their effects on vertebrate pest activity. Good records will help you plan future control strategies and improve their effectiveness.

Control Actions

For most vertebrates, you will have more than one option for reducing populations and damage (Table 11). Before you use any of these controls, consult your county agricultural commissioner to find out which procedures work best in your location and what restrictions apply to these techniques. For example, tree squirrels can usually be controlled with a few well-maintained traps; poison baits are not permitted for tree squirrels. If ground squirrels are the problem, you usually need a combination of methods, such as habitat modification, trapping, shooting, poison baits, and burrow fumigation.

The timing of control actions is often critical; timing is determined largely by the season and life cycle of the target pest. Become familiar with the biology of the vertebrates affecting your orchards and learn what factors influence the effectiveness of the available control options so you will be able to plan the most cost-effective management strategy.

When preparing the land and planting the orchard, take steps to prevent or reduce potential vertebrate problems. Baiting, fencing, fumigating burrows, shooting, and trapping

Table 11. Control Methods for Vertebrate Pests of Citrus.

Pest	Control method							
	Habitat modification	Trapping	Baiting	Fencing	Tree guards	Frightening or hazing	Shooting	Fumigating
coyote		◆¹		◆			◆¹	
deer				◆		◆	◆²	
eastern fox squirrel	◆	◆					◆	
ground squirrel	◆	◆	◆				◆	◆
pocket gophers	◆	◆	◆					◆
rabbits	◆	◆³	◆	◆	◆		◆	
rats[4]	◆	◆	◆⁴					
starling		◆				◆	◆	
voles	◆		◆		◆			
wild hog		◆¹					◆¹	

1. Contact local agricultural or wildlife agency officials for assistance before beginning a control program.
2. During hunting season or with a permit.
3. Cottontails are relatively easy to trap. Jackrabbits are difficult to trap; alternatives such as baiting, exclusion, and shooting are more useful.
4. Native kangaroo rats and the riparian woodrat resemble pest rats but are protected by law.

Adapted from Salmon and Lickliter 1984.

are easier and usually more effective if employed before you plant the orchard instead of after.

- Where feasible, deep-plow and disc to destroy burrows, disperse or kill resident populations, and reduce the risk of reinvasion by pocket gophers, voles, and (to a lesser extent) ground squirrels.
- Encircle the growing area around young orchards with properly installed fences to exclude deer and rabbits.
- When you plant the trees, install tree guards to keep rabbits from chewing the bark and to reduce damage from voles.

After the orchard is planted, develop and implement a monitoring and management program to promptly address any vertebrate pest problems that may arise.

Habitat Modification. Changes to the environment in and around orchards affect vertebrate pest problems and their management. Ground covers, mulches, soil berms, and weedy areas in and near orchards tend to increase the population of certain rodents and make their activity harder to observe. Certain ground cover plant species attract pocket gophers and sometimes rabbits. Ground squirrels and pocket gophers often burrow in berms. Thick ground cover is preferred habitat for voles. But because voles mostly travel only a few feet from their burrows, controlling weeds, keeping vegetation short around groves, and maintaining bare soil near trees will greatly reduce vole damage. Brush piles provide shelter and resting places for rabbits, rats, and ground squirrels. Remove any brush piles and debris in or near the orchard to make it easier to observe and control any vertebrate pests that are present. Store materials neatly off the ground.

Entirely eliminating habitat preferred by vertebrates may not be desirable because of other management goals. On heavy clay soils, planting on berms may improve soil aeration and reduce problems associated with waterlogging around trunks. Planting on berms in furrow-irrigated citrus minimizes weed growth near trunks by keeping the berm soil surface dry. Ground covers in row middles (especially on slopes) and vegetation along borders can reduce soil erosion and the off-site movement of potential water contaminants such as fertilizers and pesticides. However, ground covers may increase the risk of frost damage, especially in non-hillside orchards.

If you are not able to modify habitat to discourage vertebrate pests, you must rely more on controls such as burrow fumigation, poison baits, and trapping. Where the habitat favors potential vertebrate pests, monitor more carefully and respond quickly with control actions if populations build up.

Biological Control. Vertebrate populations are affected most by the availability of food and cover. Predators such as coyotes, foxes, hawks, owls, and snakes eat some of the

Debris around orchards provides nesting sites and daytime shelter for vertebrates such as ground squirrels, rabbits, and rats. When you eliminate harborage, you make it easier to monitor vertebrate presence and abundance and may help reduce pest populations.

vertebrates that can become pests, but predators play a relatively minor role in keeping small mammal populations low in orchards. Installation of nest boxes for barn owls and perches for other raptors are common wildlife conservation practices, but they do not appear to reduce vertebrate damage levels in orchards. Even though natural enemies seldom

This Pacific gopher snake (*Pituophis catenifer catenifer*) eats pocket gophers and other pests. Predators play a relatively minor role in keeping small mammal populations low in orchards, but conserve these beneficial species wherever possible.

These heavy cardboard tree wrappers help protect trunks against vole (meadow mouse) and rabbit damage. Correctly installed wire tree guards exclude vertebrates more effectively, but unlike these tree wrappers, wire screen does not protect trunks against sunburn or herbicide spray damage.

This conibear trap is set and placed without bait over the entrance to a ground squirrel burrow, then secured to a stake. Follow restrictions to protect endangered species, such as covering traps with a box that has an entrance no larger than 3 inches (7.6 cm) wide to exclude the San Joaquin kit fox. Placing a tunnel over the burrow entrance, such as with roofing paper about 24 inches (60 cm) long, may increase trap effectiveness by making the trap less visible in the light at the end of the tunnel.

keep vertebrate pests from reaching damaging levels, take precautions to avoid harming predators and other nontarget species when you use toxic baits or traps.

Exclusion. Tree guards, trunk wraps, and fencing exclude deer and rabbits. Tree guards and trunk wraps can also reduce damage from voles. Deer fencing also excludes coyotes and dogs. Although the initial costs can be high, fencing (woven mesh, electric or high-tensile wire), tree guards, and trunk wraps provide long-term control and can be the most effective methods for preventing damage by certain species.

Frightening Devices. Frightening devices can disperse roosting starlings and may temporarily keep away coyotes and wild pigs. For example, several techniques can be used in combination to make noise and visually or physically disturb birds. Frightening devices include shooting, pyrotechnics (such as shellcrackers), recorded distress calls, cannons or noise makers, and forceful sprays of water. Start using frightening tools early after the pests are first observed, and keep using the tools until they leave. For more specific recommendations, consult publications such as the *Bird Hazing Manual* (Gorenzel and Salmon 2008).

Shooting. Where shooting is allowed, coyotes, squirrels, rabbits, and wild hogs can be shot. After obtaining a permit, deer also can be shot. Check both local and wildlife regulations for license requirements and any restrictions on shooting in your area.

Trapping. When direct control is warranted, live or kill traps appropriate for the target species can be effective, especially when populations are relatively low or localized. The need for prebaiting typically varies depending on the pest species. Unlike baits, which often cause vertebrates to die somewhere out of sight, traps provide prompt and reliable evidence as to whether a particular species is present or is being killed.

Contact your county agricultural commissioner, the U.S. Department of Agriculture (USDA), or the California Department of Fish and Game (DFG) to determine whether there are special trapping regulations, such as before beginning a coyote control program.

Baits and Fumigants. Single- and multiple-dose baits (food attractants combined with poison) are available for controlling many types of vertebrates. Anticoagulant baits are often the preferred option because they are fairly economical and very effective, and they do not produce "bait shyness." Also, an antidote is available in case a domestic animal is accidentally poisoned. Anticoagulants that require multiple feedings over a period of several days generally cause death about 2 to 6 days after the bait is first consumed. Certain single-dose baits are also available. They can be more hazardous to humans than multidose anticoagulants, and the use of single-dose baits is more strictly regulated.

Sometimes a bait can be applied openly on the ground, but often it must be contained in bait stations specially designed to keep nontarget animals from getting the bait. Certain species in burrows can be controlled with fumigants that generate poisonous gas. Follow label directions carefully and understand the hazards when using baits and fumigants. For information on baits and fumigants, consult the online *UC IPM Pest Management Guidelines: Citrus*.

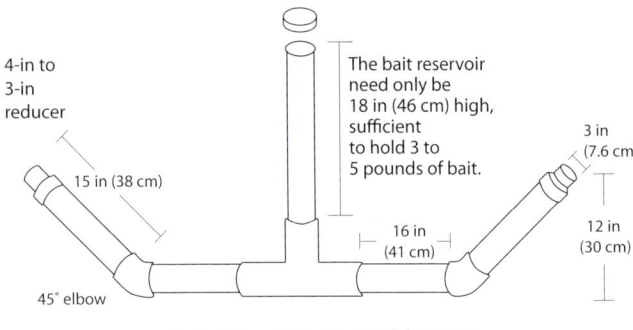

Figure 51. A ground squirrel bait station designed for use within the range of endangered kangaroo rats and the San Joaquin kit fox. End pieces are angled up, and the diameter of their opening is reduced to 3 inches (7.6 cm) to restrict protected species' access to bait. *Source:* Whisson 1997.

Endangered Species Guidelines. Many citrus orchards are located within the range of one or more endangered vertebrate species protected by federal or state regulations. Species likely to be of concern when using traps or poison bait include the San Joaquin kit fox and several species of rare kangaroo rats. If you use burrow fumigants in the San Joaquin Valley and the surrounding foothills, the blunt-nosed leopard lizard is a concern, as this insect feeder seeks shelter in rodent burrows. Special guidelines apply to the use of certain traps, fumigants, and toxic baits for vertebrate pest control in these areas. Modification of ground squirrel bait stations to exclude protected species is one common practice (Figure 51). Other typical guidelines restrict broadcast applications of bait, limit the percentage of active ingredient in baits, prohibit fumigation at certain locations or during certain times of the year, and require that applications be supervised by someone trained to avoid harming endangered species.

Contact your county agricultural commissioner for the latest maps that show the ranges of endangered species and for current information on restrictions that apply to pest control activities in your area. More information on endangered species regulations, including locations of protected species and bait station modifications to protect them, is available online at the California Department of Pesticide Regulation (DPR) Web site (www.cdpr.ca.gov).

Adult pocket gophers are 6 to 8 inches (15–20 cm) long, with stout brown, gray, or yellowish bodies and small ears and eyes. They rarely are seen above ground except when pushing soil from their burrow and sometimes when clipping small plants near a burrow opening.

This bait station constructed of PVC plastic pipe is one of several designs for control of California ground squirrel. When used within the range of the San Joaquin kit fox, entrances must be restricted to 3 inches (7.6 cm) in diameter. Place baffles inside the pipe to keep bait inside the station.

Pocket gophers damage trees by feeding on roots. Some of the roots of this tree have been eaten by gophers, greatly reducing the upper portion of the root system and likely decreasing tree vigor and yield. Note the tunnel several inches deep underground, exposed by excavating soil next to the trunk.

Vertebrate Pests

Pocket Gophers

Thomomys spp.

Pocket gophers can be serious pests, primarily in young orchards and in older orchards that use cover crops or organic production. Gophers feed primarily on the roots of herbaceous plants. Herbaceous cover crops, especially legumes, are their preferred food. They may also come above ground to clip small plants within a few inches of their burrow and pull vegetation into the burrow for feeding. However, pocket gophers also feed on the bark of tree crowns and roots, girdling and killing young trees and reducing the vigor of older trees. Gophers sometimes gnaw on plastic irrigation lines.

Pocket gophers spend most of their time in tunnels they construct 6 to 18 inches (15–46 cm) beneath the soil surface. A single burrow system can cover several hundred square feet. It consists of main tunnels with lateral branches used for feeding or for pushing excavated soil to the surface. Gophers are extremely territorial; except for females with young, you rarely find more than one gopher per burrow system.

Gophers breed throughout the year on irrigated land, with a peak in late winter or early spring. Females bear as many as three litters each year. Once weaned, the young travel to a favorable location to establish their own burrow system. Some take over previously vacated burrows. The buildup of gopher populations in an orchard is encouraged by extensive weed growth or the presence of most cover crops, especially perennial clovers. When cover crops or weeds dry up, gophers may feed extensively on the bark of tree crowns and roots. Damage to trees is always under ground and is usually not evident until the trees show symptoms of stress.

The best times to observe gophers are after irrigation and when mound building peaks in the fall and spring. Likely problem areas are orchard perimeters (where gophers invade the orchard), weedy areas such as roadsides, and extensive weed growth or ground covers in young orchards. Look for darker-colored mounds (Figure 52), which indicate newly removed (more moist) soil. Where management is planned, use a soil probe to locate the gopher tunnels (Figure 53) so you know where to excavate to place traps, as illustrated in Figure 54. For more information, consult the online *UC IPM Pest Management Guidelines: Citrus* and *Pocket Gophers Pest Note* (Salmon and Gorenzel 2002).

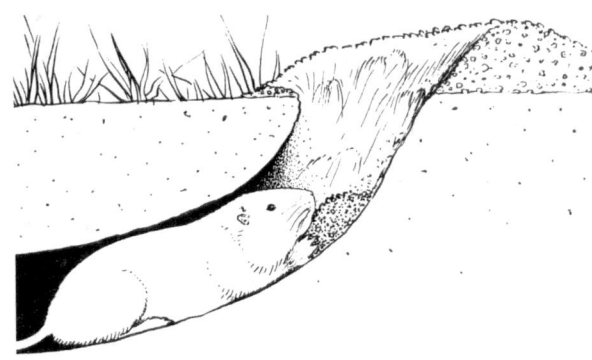

Figure 52. Pocket gophers dig their tunnels by pushing soil out through lateral tunnels. They leave the tunnel surface openings plugged with soil.

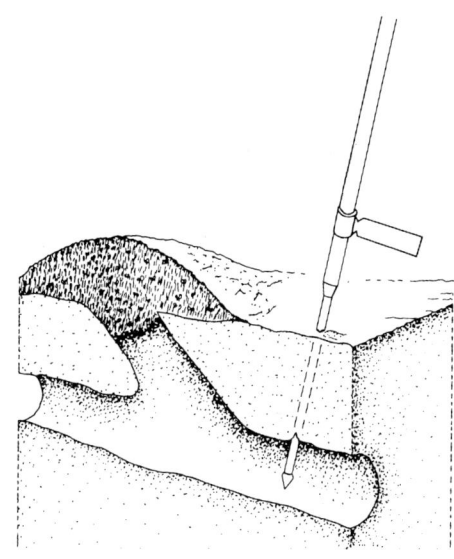

Figure 53. Use a soil probe to locate gopher tunnels (top), such as when determining where to place gopher traps. Probe 8 to 12 inches (20–30 cm) from the plug side of the mound. The probe will suddenly drop a few inches when you hit the main tunnel. Excavate there with a shovel to expose the main tunnel. Consult the online UC IPM *Pest Management Guidelines: Citrus* for instructions on how to place traps. *Source:* Salmon and Lickliter 1984.

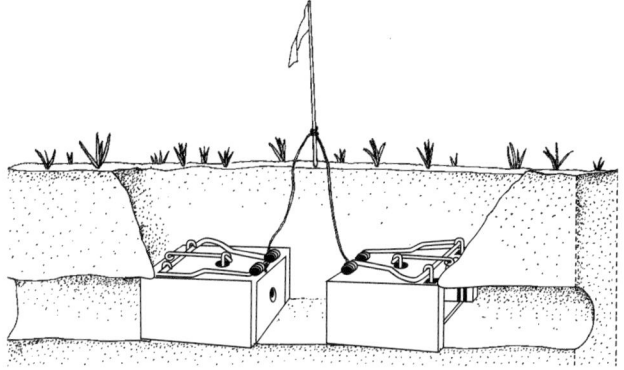

Figure 54. Gophers can be trapped with two 2-pronged pincher traps (such as Macabee traps) or with two box traps (as shown here). After locating and excavating the main tunnel as described in Figure 53, place two traps in the tunnel, one facing in each direction. If using box traps, push each trap tightly against the tunnel opening. Secure traps to a tall stake and carefully cover the excavated hole to ensure that no light enters the tunnel. The traps shown here are set (their triggers are down). Adapted from Valerie Winemiller in Strand 2002.

Mounds of soil in a group, and each mound without openings, is typical of pocket gopher activity. Given the closeness of these mounds to the trunk, it is likely that excavating the root crown would reveal gopher damage to this citrus tree. Identify the cause of mounds before taking control action. Moles also mound soil, but moles feed mostly on earthworms and other invertebrates and are not pests of citrus.

The often conspicuous fan- or crescent-shaped soil mounds over tunnel openings are the most obvious sign of a gopher infestation. Mole mounds appear more circular and, in profile (not shown), are volcano shaped. Unlike gophers, mole feeding tunnels are shallow and just beneath the surface, leaving an elongated raised ridge of soil to mark the path of their activities.

Ground Squirrel

Spermophilus beecheyi

Ground squirrel, officially named California ground squirrel, causes the most damage in orchards adjacent to uncultivated areas where squirrels are not controlled. Ground squirrels gnaw fruit and bark and girdle trunks and scaffold limbs. They occasionally chew plastic irrigation lines. Their burrows can contribute to soil erosion and cause irrigation water to flow where it is not desired.

Ground squirrels are active during cooler times on hot days and are usually most active in the morning and late afternoon. They live in colonies, which may grow very large if left uncontrolled. Each ground squirrel burrow system can have several openings with scattered soil in front. Individual ground squirrel burrows may be 5 to 30 feet (1.5–9 m) long, 2½ to 4 feet (75 cm–1.2 m) below the surface, and about 4 to 6 inches (10–15 cm) in diameter. Burrows provide the ground squirrels a place to retreat, sleep, hibernate, rear their young, and store food. Ground squirrels often dig their burrows along ditches and fence rows and on other uncultivated land. When uncontrolled, they frequently move into orchards and dig burrows beneath trees.

The California ground squirrel can be active throughout the year in coastal areas of southern California. In hot locations, adult ground squirrels become temporarily dormant (estivate), especially when food is scarce or temperatures are extreme, primarily in late summer. Winter hibernation and summer estivation are more typical among ground squirrels in areas away from the coast, where temperature variations are more extreme. Regardless of location, young squirrels tend to be active all summer.

Squirrels that do hibernate generally emerge around January when weather begins to warm. In late winter and

When an irrigation line has been chewed by ground squirrels (as shown here), other rodents, or rabbits, the pipe is scraped or gnawed repeatedly. The hard plastic often retains the impression of the paired incisor tooth marks. When chewed by coyotes or dogs (as pictured later), materials are crushed, compressed, or shredded.

The adult California ground squirrel has a head and body 9 to 11 inches (23–28 cm) long. The fur is mottled or flecked, dark and light brown or gray. The ground squirrel's tail is about as long as its body, and the tail is less bushy than the more-plump tail of a tree squirrel.

Ground squirrels often dig burrows at the base of trees when they invade orchards. Their burrows have large, conspicuous openings that are not plugged.

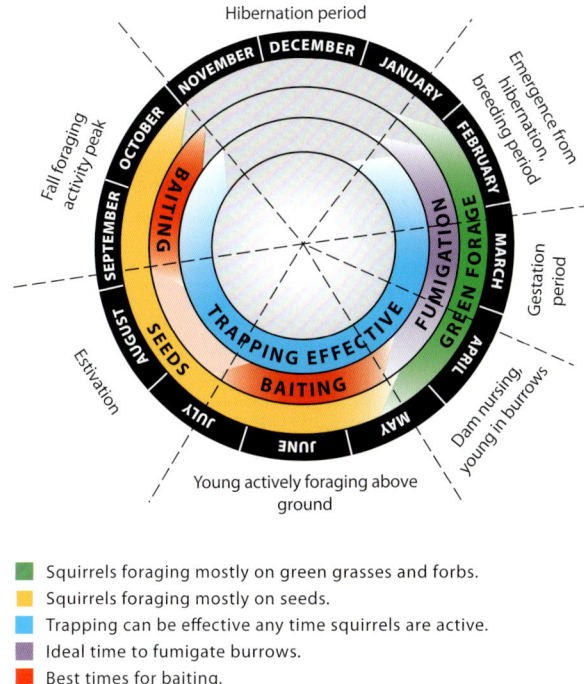

- 🟩 Squirrels foraging mostly on green grasses and forbs.
- 🟨 Squirrels foraging mostly on seeds.
- 🟦 Trapping can be effective any time squirrels are active.
- 🟪 Ideal time to fumigate burrows.
- 🟥 Best times for baiting.
- ⬜ Baiting marginally effective because of estivation.

Figure 55. California ground squirrel reproductive cycle and seasonal food preference, and the best times to take specific control actions. Actual activity varies among growing areas, depending primarily on local climate and weather, and occurs earlier at warmer sites. *Source:* Salmon, Whisson, and Marsh 2006.

spring, the squirrels feed on green vegetation. They switch to seeds and fruit, including citrus, in late spring and early summer as the vegetation dries up. Females have one litter, averaging eight young, in the spring. The young squirrels emerge from their burrow when about 6 weeks old. Young ground squirrels do not estivate their first summer, and most do not hibernate during their first winter.

Figure 55 illustrates the seasonal periods of activity for the California ground squirrel and the times to control ground squirrels. Consult the online *UC IPM Pest Management Guidelines: Citrus, California Ground Squirrels Pest Notes* (Salmon and Gorenzel 2002), and Best Management Practices for California Ground Squirrel Control (http://groups.ucanr.org/GSBMP) for more information.

Tree Squirrels

Two introduced species of tree squirrels are occasional pests in fruit and nut orchards. The eastern fox squirrel (*Sciurus niger*) and the eastern gray squirrel (*Sciurus carolinensis*) occur primarily near wooded, riparian, and suburban areas. The eastern fox squirrel is the more common pest in citrus, primarily in the Central Valley around Fresno and Sacramento and in coastal areas south of San Francisco.

Two species of tree squirrels are native to California. The native western gray squirrel (*Sciurus griseus*) rarely is a problem in orchards; it appears largely to have been displaced by the eastern fox squirrel. The Douglas tree squirrel (*Tamiasciurus douglasii*) lives in conifer forests and is not an agricultural pest.

Tree squirrels prefer to live in wooded areas. In orchards, tree squirrels usually feed only in localized areas of the orchard adjacent to woodlands. Although tree squirrels live mainly in trees, the fox squirrel especially spends considerable time foraging on the ground. The eastern fox squirrel and eastern gray squirrel feed on a variety of green and ripe nuts, seeds, mushrooms, and fruits. The eastern fox squirrel additionally feeds on bird eggs and insects.

Tree squirrels do not hibernate and are active year-round. Tree squirrels can be seen in trees, running on utility lines, and foraging on the ground, primarily in the early morning and late afternoon. In comparison with ground squirrels, tree squirrels are easily distinguished by their long bushy tails and lack of flecklike spots or stripes. Ground squirrels usually run to soil burrows when fleeing danger. Tree squirrels escape by climbing trees and other structures.

Tree squirrels nest in holes in trees or build leaf and twig nests in trees. Females of the eastern gray squirrel bear two

The eastern fox squirrel, sometimes called the red fox squirrel, can be differentiated from other tree squirrels by its brownish red-orange fur.

This pair of box gopher traps is baited with fruit and nuts and set on a tree limb to trap the eastern fox squirrel. Trapping is the primary control method for controlling tree squirrels. If you maintain traps continuously, you need relatively few to control tree squirrels.

A jackrabbit is a hare about the size of a large house cat. It has long ears and long legs. Jackrabbits live in open areas of the Central Valley, coastal valleys, and foothills. They make a depression underneath bushes or other vegetation where they remain secluded during the day. Jackrabbit young are born fully haired, and open their eyes and become active within a few hours.

Cottontail and brush rabbits are smaller than jackrabbits and have shorter ears. They nest where thick shrubs, woods, or rocks and debris provide dense cover. Their young are born hairless and blind and stay in the nest for several weeks.

Droppings help you identify the vertebrate species in your grove. Rabbits scatter coarse, circular fecal pellets. These jackrabbit pellets are about 1/2 inch (13 mm) in diameter. Cottontail pellets average about 1/4 inch (6 mm).

litters per year, one litter in late winter and one in late summer. Each litter has three to five young. Young females of the eastern fox squirrel bear one litter. Older female fox squirrels have two litters per year, between January and April and between July and September. For more information, consult the online *UC IPM Pest Management Guidelines: Citrus* and *Tree Squirrels Pest Notes* (Salmon, Whisson, and Marsh 2005).

Rabbits

Rabbits and hares (collectively called rabbits) can kill young trees or severely stunt young tree growth by chewing bark off trunks and eating buds and shoots. Once the trees are 4 or 5 years old, rabbits usually do not present a serious problem. Rabbits may also gnaw irrigation lines. Cottontail and brush rabbits (*Sylvilagus* spp.) prefer orchards near brushy habitats, ravines, riparian areas, and woodlands. Jackrabbits (*Lepus californicus*) prefer trees bordering open areas, such as grassy fields and rangeland. Rabbits typically live in these more preferred habitats near orchards and feed in orchards from early evening through early morning.

Rabbits are active year-round. They damage trees primarily in winter and early spring, when their other sources of food are limited. Jackrabbits frequently cause more damage than cottontails. If you find damage, look for droppings and tracks that indicate rabbits as the cause. Voles also chew trunk bark, but rabbits damage bark higher on the tree and their tooth marks are distinctly larger. For more information, consult the online *UC IPM Pest Management Guidelines: Citrus* and *Rabbits Pest Notes* (Salmon and Gorenzel 2002).

A cylinder of 1/4- to 1-inch (6–25 mm) mesh wire protects trees from rabbits. Tree guards should be at least 24 inches tall and wide enough to allow several years' growth without crowding the tree. Bury the bottom of the cylinder 2 to 3 inches deep and stake the cylinder away from the trunk so rabbits cannot press it against foliage.

Voles (Meadow Mice)

Microtus spp.

Voles, also called meadow voles or meadow mice, can be a particular problem where dense weeds or cover crops grow near tree trunks. Voles feed on bark around the root crown, and they sometimes chew holes in irrigation lines. Small trees are most susceptible to being completely girdled and killed by voles. Large trees can be damaged, for instance after severe pruning allows sufficient light penetration through the tree canopy for vegetation to grow near trunks, but vole damage to large citrus trees is uncommon and rarely kills the tree.

Voles are active both day and night year-round. Females bear several litters each year, with peaks of reproduction in spring and fall. Populations go through cycles, climaxing every 4 to 7 years and then declining fairly rapidly. Grasses and other dense ground covers provide food and cover that favor the buildup of vole populations. You can recognize vole activity by the presence of narrow runways in grass or other ground cover, connecting numerous shallow burrows with openings about 1½ inches (4 cm) in diameter around the bases of orchard trees. Look in runways for fresh vole droppings and short pieces of clipped vegetation, especially grass stems. If you find burrows, remove the soil from around the trunk and look for bark damage on the lower trunk and base of roots. If you do not check carefully, you may not notice damage until late spring or summer, when it may be too late to prevent significant injury to the trees. Make sure to monitor ditch banks, fence rows, roadsides, and other areas around the orchard where permanent vegetation favors the buildup of vole populations. For more information, consult the online *UC IPM Pest Management Guidelines: Citrus* and *Voles (Meadow Mice) Pest Notes* (Salmon and Gorenzel 2002).

Deer Mice

Peromyscus spp.

Deer mice occasionally strip bark from young trees, chew irrigation lines, and feed on fruit. Large membranous ears and a moderately to well-furred tail more than 70% the length of the head and body distinguish *Peromyscus* spp. from other small rodents also called mice. In contrast to their darker upper side, deer mice have distinctly lighter-colored fur on the underside of their body and tail. "Deer mouse" refers to any of seven *Peromyscus* spp. that occur in California and to the prevalent species *P. maniculatus*. At most locations, deer mice are much less likely than voles (meadow mice) to be a problem in groves. Damage, the situations where they become problems, and management methods for deer mice are generally the same as for voles.

Adult voles are larger than house mice but smaller than rats. Compared to deer mice, voles have a more robust body, less obvious ears, and a relatively shorter tail. Voles' ears are at least partly obscured by the hair in front of them, and their tails are about one-half to one-quarter the length of their head and body combined.

An exposed root crown showing vole damage to bark on the lower trunk and a large root. Voles usually start chewing on bark about 2 inches (5 cm) below the soil line and then move upward to about 2 to 4 inches (5–10 cm) above ground. Because voles seldom travel more than a few feet from their burrows and runways, abundant weeds or ground cover (not shown) probably have been growing near this tree.

Deer mice (*Peromyscus* spp.) sometimes strip bark from young trees, chew irrigation lines, and feed on fruit. But deer mice are less likely than voles to be a problem in orchards. Deer mice have relatively large, prominent ears and a long tail, more than 70% the length of their head and body. The hair on the underside of their body and tail is distinctly lighter than hair on their upper body.

Ground covers and grassy weeds can harbor voles, as evidenced by these abundant runways and shallow burrows. Because of the shallow, open nature of their burrows, fumigation is not effective for controlling voles.

Roof Rat

Rattus rattus

Rats gnaw on electrical wires, wooden structures, branch bark, and fruit on trees. Roof rats can hollow navel fruit, eating just the juice sacks while leaving the rind mostly intact. On lemon, they eat the peel, leaving just the juice sacs hanging on the tree. After harvest, rats damage fruit in bins or boxes by chewing it and leaving excrement. Rats are active throughout the year, mostly at night.

The roof rat, sometimes called the black rat, is a common pest in citrus orchards. It builds leaf and twig nests in citrus or nearby trees or nests in debris piles or thick mulch on the ground. This agile, sleek rat has a pointed muzzle and a tail that is longer than the body and head combined. The Norway rat (*Rattus norvegicus*) is an uncommon pest in citrus orchards. Norway rats may cause problems in packinghouses by chewing boxes or fruit and through fecal and urine contamination. A mature Norway rat is larger and more stout than a roof rat. It has a blunt muzzle and a tail that is shorter than its body and head combined.

Be aware that endangered native kangaroo rats (*Dipodomys* spp.) and the riparian woodrat (*Neotoma fuscipes riparia*) resemble pest rats and are protected by law. Unlike the hairless, scale-covered tail of Norway rats and roof rats, the tails of kangaroo rats and the riparian woodrat are covered with fur. The riparian woodrat is active mostly during the day, and its tail is somewhat shorter than the combined length of its body and head. A kangaroo rat's tail is noticeably longer than its body and head combined. Kangaroo rats are nocturnal, but unlike Norway rats and roof rats, which move on all four legs, kangaroo rats hold their front legs off the ground and travel by hopping on their hind legs.

It is important to know which species of rat is present in order to choose the most effective management methods and to avoid killing nontarget or protected species. For more

Rats can chew bark and girdle limbs, sometimes killing small branches on citrus trees. More commonly, rats chew fruit, hollowing out the inside after gnawing a hole in the rind, as pictured earlier in the introduction to this chapter.

Roof rats have a pointed muzzle and long, hairless tail. It is important to distinguish the species of rat present in order to select the most effective types and placements of baits or traps and to avoid killing nontarget species. For example, when using baits, secure it inside bait stations to minimize the risk of nontarget exposure.

Roof rats often build leaf and twig nests in trees. Rat nests may resemble those of tree squirrels. However, because squirrels are active during the day, their presence is usually easy to recognize.

Native kangaroo rats and the riparian woodrat resemble pest rats but are protected by law. Unlike the hairless, scale-covered tail of Norway rats and roof rats, the tails of kangaroo rats (shown here) and the riparian woodrat are covered with fur.

information, consult the online *UC IPM Pest Management Guidelines: Citrus* and publications such as *Norway Rats* (Timm 1994), *Rats Pest Notes* (Salmon, Marsh, and Timm 2003), and *Roof Rats* (Marsh 1994).

Deer

Odocoileus hemionus

Mule deer, including the subspecies called black-tailed deer (*O. hemionus columbianus*), can be serious pests when trees are young. Young trees can be severely stunted, deformed, or killed when deer browse on new shoots. Bucks occasionally break limbs off of smaller trees or injure the bark when they use trees to rub the velvet off their antlers. Deer feeding on older trees seldom causes significant damage.

Deer feed mostly from late evening through early morning, so they can be present without being directly observed. To confirm their presence, look for fecal pellets and tracks in the vicinity of damaged trees. Deer hooves are split, pointed at the front and more rounded at the rear, and are about 2 to 3 inches (5–7.5 cm) long. You may also use spotlights to check for deer at night. For more information, consult the online *UC IPM Pest Management Guidelines: Citrus* and *Deer Pest Notes* (Salmon, Whisson, and Marsh 2004).

Coyote

Canis latrans

Coyotes feed on certain citrus pests, including rodents and rabbits. Damage to orchards is caused when coyotes chew flexible irrigation lines and enlarge the burrows of animals such as ground squirrels. In recent decades, coyotes have

Deer occur in many foothill and coastal orchards and sometimes in the Central Valley near riparian habitats. Young citrus trees can be severely stunted, deformed, or killed when mule deer browse on new shoots. Deer feeding on older trees seldom causes significant damage.

A well-maintained, sturdy fence 7 to 8 feet (2.1–2.4 m) tall encircling trees is the only completely effective deer control. Where deer are abundant, the lush foliage of irrigated trees is highly attractive, and it is best to install fencing before orchards are planted.

Deer droppings are a good indicator of deer presence. The appearance of droppings varies, but commonly each fecal pellet is oblong, somewhat pointed at one or both ends, and ¼ to ½ inch (6–12 mm) long.

Coyotes are medium-sized members of the dog family. They are larger than foxes but smaller than wolves and weigh about 20 to 35 pounds (9–16 kg) when fully grown. Coloration is usually a blend of rust colored to brown to gray. The coyote resembles a small German shepherd dog but with a longer, narrower snout and a bushy, black-tipped tail.

increased in number and expanded their geographical range. Coyotes can live in almost any habitat in California, from arid desert to foggy coastal regions.

Breeding occurs once annually, typically from late January through February, with pups born in March and April. Parents and offspring continue to remain in a family group for about 6 months. Before giving birth, adults excavate one or more dens in the soil, occasionally by expanding the burrows of other animals. Sometimes they den in hollow logs, rock piles, or culverts, typically choosing sites where human activity is minimal. Litter size is normally four to seven pups.

Pups emerge from the den at about 3 weeks of age and grow quickly, relying primarily on their parents to provide them with food for the first few months. By late fall, juveniles may disperse to live independently, although if food resources are adequate, juveniles can remain with their parents through the next year. Coyotes can be heard vocalizing (barking and howling) in the evening and night throughout most of the year. They vocalize less when in the early stages of pup rearing.

Humans do not often see coyote damage as it is happening. Heavy reliance must be placed on evidence of their presence at the damage site. Inspect the area for "coyote sign" such as hair and droppings and for other evidence such as tracks (Figure 56) or their tooth marks on irrigation pipe.

Drip irrigation pipe and similar materials chewed by coyotes appear to have been compressed and shredded as if chewed by dogs. Coyote hair may be found in fence wire, particularly at locations where coyotes have dug a "slide" to crawl under a chain-link or woven-wire fence. A close look at the hair will reveal bands of color on individual hairs (but many other animals also have hair with color bands). Hairs from a coyote's back are often black tipped.

Coyote droppings or "scats" are typically about the diameter of a cigar, sometimes tapering at one end. Scats are deposited along trails and roadsides and vary in appearance depending on the animal's diet and on the age of the scat. The scat may contain hair, feathers, bones, or other animal parts, as well as plant materials such as grass or seeds. Scats are typically black to light gray in color, becoming bleached out as they remain exposed to sunlight and the elements. For more information, consult the online *UC IPM Pest Management Guidelines: Citrus* and *Coyote Pest Notes* (Timm, Baker, and Beckerman 2007).

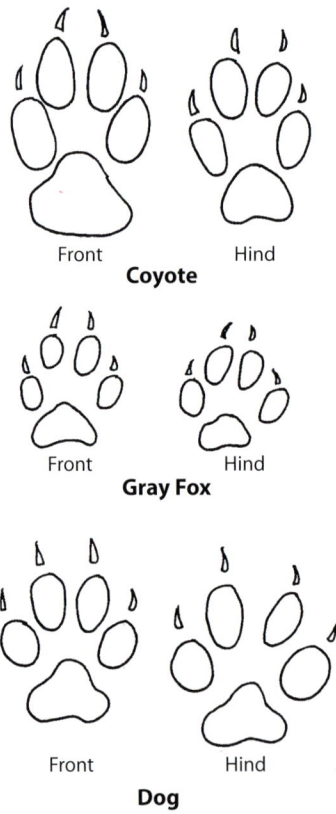

Figure 56. Footprints for distinguishing vertebrate species. Vertebrates often feed at night or hide when people approach, so their damage may not be seen by humans as it is happening. Heavy reliance must be placed on indirect evidence at the damage site, such as animal tracks. For example, coyote tracks can often be distinguished from those of dogs by their shape and appearance. Coyote tracks tend to be more oval shaped and compact than those of common dog breeds. Nail marks of coyotes are less prominent, and the tracks tend to follow a straight line more closely than those of dogs. Tracks of large coyotes can be up to 3 by 1¾ inches (7.5 by 3.4 cm). *Source*: Wade 1973.

Drip irrigation pipe and similar materials chewed by coyotes appear to have been compressed and shredded as if chewed by dogs. This is in contrast to chewing damage done by rodents or rabbits (pictured earlier), where the pipe is scraped or gnawed repeatedly and often retains the impression of the paired incisor tooth marks.

Wild pigs travel in herds, as with the three adult females and piglets shown here. Wild pigs are expanding their range and are commonly found near foothill areas and habitat with dense brush.

Wild Pig

Sus scrofa

Wild pigs or hogs include both domestic swine that escaped captivity and wild boars. They feed on insects and underground vegetation. Their "rooting" behavior to expose food disturbs the orchard floor. Pigs also create large uneven basins or wallows in moist soil where during hot weather they roll their body ("wallow") in water or moist soil, such as in irrigated crops. Soil disturbance from rooting and wallowing interferes with the application of irrigation water. Pigs can get tangled up in flexible irrigation equipment and pull it out. Pigs will also eat lower fruit on the tree. Consult the online *UC IPM Pest Management Guidelines: Citrus* for management recommendations.

Starlings are about the size of a robin. They are dark-colored birds with light speckling on the feathers.

European Starling

Sturnus vulgaris

Large flocks of starlings sometimes roost in citrus orchards. Their excrement or droppings cause unsightly blemishes on fruit and may transmit diseases. Starlings range widely to feed but usually seek closely grouped trees for their communal roost. Consult the online *UC IPM Pest Management Guidelines: Citrus* for management recommendations.

Starlings often travel in flocks and prefer habitat with nearby trees in which to roost. Where they roost in large numbers, surfaces underneath quickly become fouled with excrement.

Weeds

Plants growing in citrus orchards become weeds when they compete with trees for nutrients, water, and light. Competition from weeds is especially damaging to trees during their first 5 years of growth. Weed competition slows the growth of young trees, making the orchards take longer to become productive. Stress from competition may increase young trees' susceptibility to insect and disease damage.

As trees grow older, their canopies increasingly shade much of the orchard floor, reducing weed growth. Thus, direct competition from weeds is less of a concern in mature orchards. Exceptions include when trees are severely pruned or become unhealthy and prematurely drop their leaves, allowing extensive light penetration to the orchard floor. Weed competition can also be significant for older trees in drip- or microsprinkler-irrigated orchards because tree roots are concentrated in a smaller area than in furrow irrigation.

Weeds can be problems for trees of any age by interfering with cultural operations. Weeds and ground covers increase frost hazard by reducing heat absorption and radiation from soil. Frost injury then makes citrus more susceptible to disorders and pathogens affecting fruit, leaves, trunks, and twigs. Weeds also increase populations of certain species of invertebrate pests by providing cover and an alternate food source for snails and various insect pests such as orange tortrix. Weeds around tree trunks may create a favorable environment for pathogens that infect the trunk and roots as well as providing shelter for gophers, voles (meadow mice), and perhaps other vertebrates.

The presence of orchard vegetation, however, can provide some benefits, such as improving water infiltration. Cover crops or managed weeds in row middles and vegetative filter strips along borders and roads can reduce soil erosion and off-site movement of fertilizers and pesticides into surface or groundwater. Where soil becomes waterlogged around young trees, it is beneficial to allow weeds to grow: their transpiration will reduce soil moisture and the likelihood of resultant damage from asphyxiation and root rots. Vegetation provides alternative hosts and shelter for parasites and predators of certain invertebrate pests and reduces dust that contributes to outbreaks of mites and certain insects.

Life Cycles

A plant's life cycle affects both its importance as a weed and the strategies used for its control (Figure 57). Weeds are grouped as annuals, biennials, or perennials based on their life cycles. An annual completes its life cycle of germination, growth, flowering, and seed production within 1 year. Winter annuals germinate in the fall, grow through the winter, and flower in late winter and early spring, They produce seed in the spring and die by early summer. Common winter annuals in citrus orchards include annual bluegrass, chickweeds, common groundsel, foxtails, henbit, miners lettuce, fiddlenecks, filaree, little mallow (cheeseweed), mustards, shepherd's-purse, and wild barley. Winter annuals are the least troublesome species because they grow when competition for water is less of a concern. However, if abundant, weeds during winter may increase crop damage during freeze conditions. Many winter annuals can be managed as desirable cover crops on slopes, in orchard row middles, and along borders to reduce soil erosion and runoff.

Summer annuals germinate in late winter, spring, or early summer. They produce seed in summer or fall and die in fall or early winter. Major summer annuals include barnyardgrass, crabgrasses, common lambsquarters, flax-leaved fleabane, marestail, lovegrasses, pigweeds, puncturevine, spotted spurge, purslane, sprangletop, nightshades, turkey mullein, and witchgrass. Summer annuals are more problematic than winter annuals because summer annuals compete for moisture during the dry season and may interfere with the distribution of irrigation water particularly from emitters in low-volume irrigation systems.

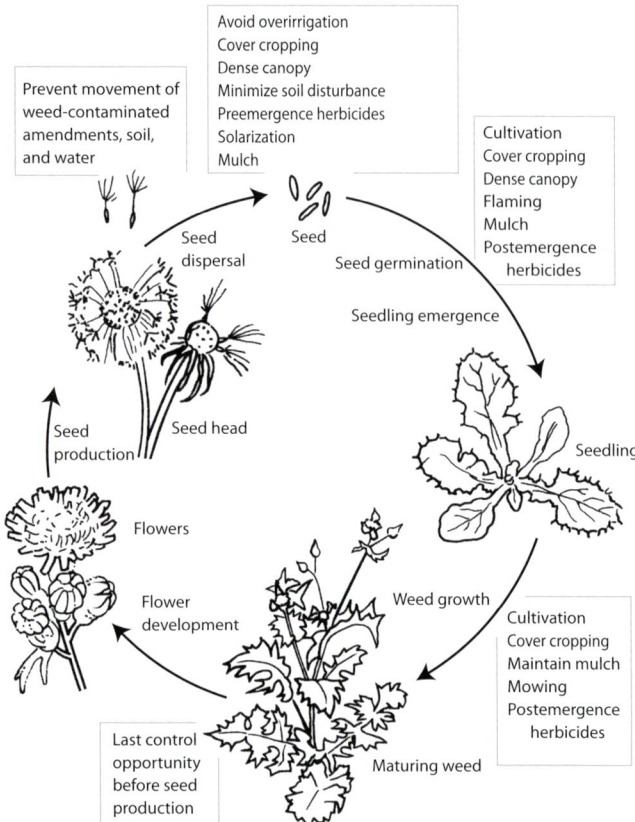

Figure 57. Weed control methods vary depending on when the method is applied relative to the stage of weed growth, as illustrated here with sowthistle, an annual weed. The best times for effective management are before planting, before weeds emerge, and before weeds mature.

Weeds compete with trees for water and nutrients and provide cover for vertebrate feeding, which may go unnoticed. Weeds and ground covers can increase frost hazard. However, orchard vegetation can also be beneficial. Cover crops can help exclude problem weeds, improve water infiltration, and reduce dust, soil erosion, and off-site movement of potential contaminants.

Biennials complete their life cycle in 2 years. Their first year's growth is vegetative; flowering occurs in the second year. In California's mild climate, certain annuals may behave as biennials or short-lived perennials, for example, marestail, little mallow, and sweet clovers.

Perennials live for 2 years or more. Many perennials infesting citrus are herbaceous species with aboveground parts that die back or become dormant during part of the year. The season and vigor of perennial growth above ground varies depending on the weed species and growing location. Important perennials in citrus include field bindweed, johnsongrass, and nutsedges that die back above ground in winter and grow above ground during spring and summer. Bermudagrass becomes seasonally dormant during the winter, when leaves and shoots turn brown. Perennial fescues grow above ground during winter but in hot areas die back during summer. Dallisgrass can grow well above ground throughout the year.

Perennials are the most difficult weeds to control. They form extensive rhizomes, stolons, tubers, or taproots from which they regrow after aboveground parts are killed (Figure 58). Because management of perennials can require repeated cultivation and herbicide application to destroy

This severe infestation of yellow nutsedge is likely to stunt and stress these young trees by competing for water and nutrients. Established perennials can be especially difficult to control around young trees. This grower will likely regret the failure to obtain good weed control before planting.

underground structures, established perennials should be controlled before planting the orchard. Even after perennial and annual weeds are destroyed, the seeds of many species remain viable in the soil for years and produce new plants that require ongoing management.

Weed Control Strategies

Key strategies for citrus weed management include

- obtaining excellent control of perennials and reducing the soil seed bank before planting
- preventative actions, such as using good sanitation and applying cultural practices such as irrigation and fertilization in ways that discourage weed growth and favor the development of healthy tree canopies
- controlling annual species before they produce seeds
- minimizing the opportunity for perennials to produce vegetative propagules and replenish carbohydrate storage in rhizomes, tubers, and stolons

The extent to which orchards are kept free of other vegetation varies greatly depending on factors such as crop age and grower preference. Strip weed control, basal control, and total control are three general strategies for weed management.

With strip weed control, a weed-free area is maintained only in the tree row. By allowing vegetation to grow in the row middles, strip control improves soil structure, reduces erosion, and requires less effort than trying to control weeds over the entire orchard floor. However, perennials can establish quickly in the tree rows because there is no competition from other weeds, so monitor them regularly and spot-treat as required.

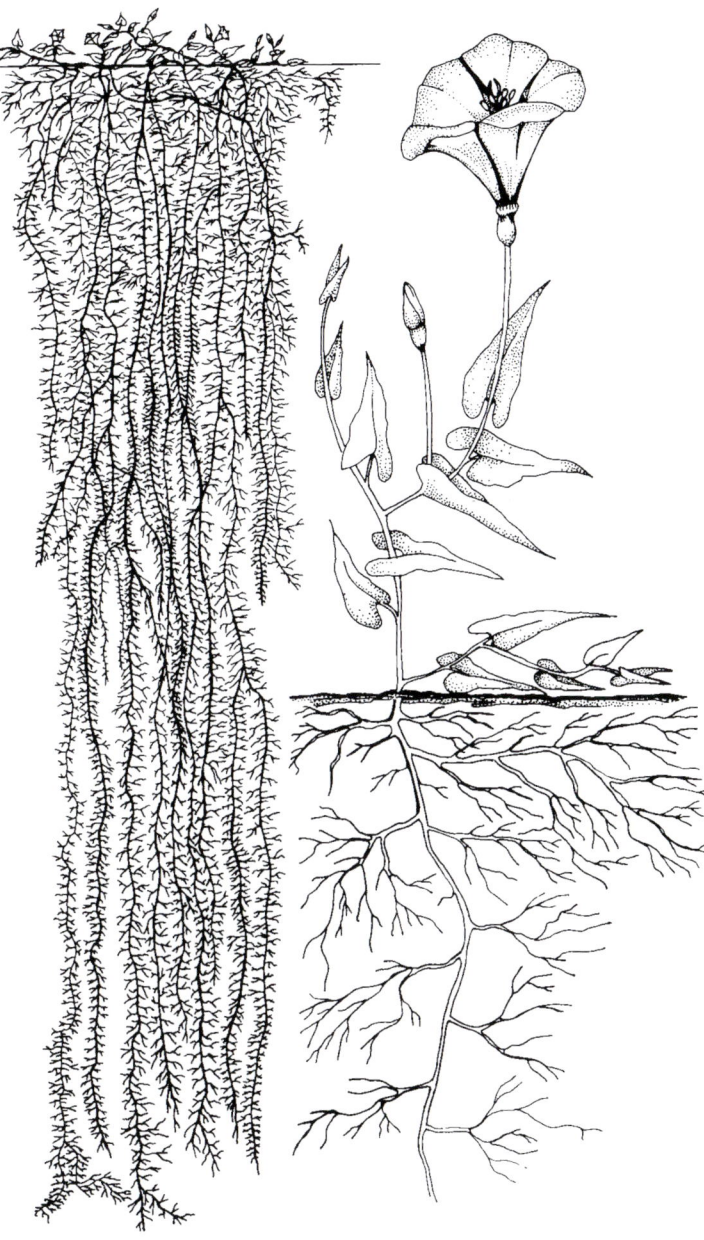

Figure 58. Field bindweed has deep rhizomes from which plants repeatedly regrow. Control established perennials before planting the orchard, since their extensive rhizomes, stolons, taproots, or tubers make control even more difficult after planting. Redrawn from Klitz 1930.

Basal control eliminates weeds in an area several feet in diameter around each tree trunk. If vegetative cover in middles and in rows (some distance away from trunks) is allowed, basal control requires control actions in a smaller area than with total weed control. Basal control also requires a compatible irrigation system with emitters placed in the areas around trees, where weeds are controlled.

Total weed control attempts to maintain the entire orchard floor free of vegetation. This generally requires the most extensive use of herbicides. With total control, herbicides should be used in combination with other methods, such as cultivation and good sanitation. Total weed control can create serious problems, such as soil compaction, erosion, reduced water infiltration, off-site movement of contaminants, and injury to trees. However, cultivation may damage feeder roots that often grow close to the soil surface. Because no single chemical controls all weed species, combinations of herbicides or sequential treatments are generally required. Weeds must be carefully monitored to determine which herbicides to apply and when to apply them. With total weed control, herbicide rotation is especially important to avoid promoting the buildup of weeds that are tolerant to certain herbicides or that become resistant to the herbicides due to the repeated application of chemicals with the same mode of action.

The generally weed-free area around trunks (basal weed control) shown here may be sufficient to minimize weed competition with young citrus trees. Unless nearby weeds are controlled before they produce abundant seed, many weeds may later emerge near trunks, where frequently wet soil reduces preemergent persistence and contact herbicide application or mechanical controls can injure the young trees.

Prevention

Careful management and good sanitation help limit weed infestations. Provide trees with good cultural care to encourage development of a healthy canopy that shades out weeds on the orchard floor. To avoid introducing new weed species or spreading infestations, clean equipment before moving it into orchards. Do not introduce or move compost, mulch, organic fertilizers, soil, or other amendments unless the material is known to be relatively weed-free. To prevent the spread of weeds, make sure that irrigation canals and ditch banks are free of weeds and weed seeds. Do not allow weeds around the orchard perimeter to mature and produce seeds. Provide good drainage, because high moisture in areas such as furrow bottoms, at furrow ends, and around standpipes favors weed growth. Where furrow irrigation is used on slow-draining soils, use shorter furrows or establish lateral furrows halfway into the tree rows to reduce the time water stands in the furrows. Discourage weed seedling establishment by letting the top 2 or 3 inches (5–7.6 cm) of soil dry completely between furrow or sprinkler irrigation cycles.

Strip weed control provides competing vegetation to help exclude less desirable weed species. Cereal grains or resident grasses grown in row middles can be killed with a selective grass herbicide that will not damage citrus. Special mowers can blow clippings into the tree row, providing mulch that helps suppress weeds near tree trunks. On flat land, where the risk of cold damage is higher, ground covers are less common in citrus because vegetation increases the risk of frost damage.

Monitoring

Identify the weed species that are present in your orchard and know their likely emergence dates. Monitoring will help you select effective control methods and tell you when to take action. Begin monitoring at least 1 year before you plant the site and then at regular intervals after planting. Monitor for weeds at least twice per year, in midwinter and late spring. Additional monitoring in midsummer and late fall is desirable, especially before planting and during the first several years of orchard growth, to give you a picture of the full spectrum of weed species that are present. When monitoring, pay special attention to perennials.

Keep records of when and where you monitored, what species were present, and some measure of the weeds' relative abundance so you will be able to make sound decisions on cultural and chemical controls. Information collected over several years tells you how weed populations may be changing and how effective your control operations have been. It is a good idea to draw a map of the orchard and mark a copy with the monitoring date and the location of any perennials that you find. That way you can quickly return to see how well your control actions are working and know whether further control action is needed in that area. A handheld GPS (global positioning system) device is useful during monitoring: it allows you to accurately describe specific locations in the orchard and then find them again on a later inspection. It may be helpful for decision making to record separately weeds near trunks (where management is focused) and weeds in borders or row middles (where control may be less intensive). For suggested record-keeping forms and detailed monitoring recommendations, consult the latest *UC IPM Pest Management Guidelines: Citrus* and *Citrus Year-Round IPM Program* (online at www.ipm.ucdavis.edu).

This scout is monitoring weeds that should be controlled around young trees. Recording what weed species are present and their relative abundance will help you select effective management actions.

Control Methods

In addition to prevention through sanitation and promoting a vigorous citrus canopy, weed control methods include good irrigation management, cultivation (discing or tillage), cover cropping, mulching, heating (solarizing bare soil and flaming emerged weeds), and biological control. Herbicide applications are the most common method in most orchards. Consult the *UC IPM Pest Management Guidelines: Citrus* and publications such as "Integrated Weed Management in Citrus" (in preparation) for practical discussions of weed management.

Irrigation

The location and abundance of weeds are greatly affected by irrigation, such as the irrigation amount, method, and system maintenance. For example, low-volume irrigation generally reduces the percentage of orchard soil surface between trees that is wet and exposed to sunlight, reducing weed growth between trees. However, weed control near to trunks must be adjusted because the continuously moist zone around the emitters favors weed growth and the rapid breakdown of herbicides. See the "Soil and Water Management" section in the chapter "Managing Pests in Citrus" for more discussion.

Cultivation

Except between recently planted trees, weeds in citrus are rarely managed with regular cultivation. Tillage destroys the feeder roots of the trees, which absorb nutrients, water, and oxygen in the topsoil. Disease organisms may enter through root systems or trunks that have been injured by tillage. Discing contributes to soil erosion, especially on sloping land, and to soil compaction, which promotes water-related diseases. If the soil is dry, cultivation creates dust, which causes mite outbreaks, interferes with biological control of insects, and may cause air quality problems. Discing may also increase the weed population by bringing buried seeds to the surface or spreading rhizomes, stolons, or tubers throughout the orchard.

Ground Cover

A ground cover is maintained in some citrus orchards, mainly in northern California, on hilly terrain, and in organic production. A ground cover of resident vegetation, a sown cover crop, or a mix of both prevents soil erosion and improves water penetration and soil structure. Some ground covers can be managed by complete mowing or by mowing the row middles while keeping a strip along the tree rows free of weeds with herbicides. Repeated mowing alone favors the establishment of perennial weeds, which are deep rooted and more competitive with citrus than annuals.

Organic mulch is applied on the soil surface in some orchards to help control weeds. Mulch can moderate the root-growing environment, conserve soil moisture, and reduce dust. Irrigation method and application cycles may need to be modified to be compatible with mulching and to avoid contributing to problems such as root rot disease.

Mulch

Organic mulch can be applied to suppress weeds, moderate the root-growing environment, and conserve moisture. Synthetic fabric mulches that are resistant to UV degradation can be placed around the base of the tree. These fabrics allow moisture to penetrate but prevent weeds from emerging. Install fabric in a 4-by-4- foot square centered around the tree. Use anchoring staples or nails to hold fabric in place. Woven fabrics last at least 5 years. Use in conjunction with tree wraps to prevent the fabric from contacting tree trunks because dark material exposed to intense sunlight can become hot and cause heat injury to the bark, as discussed in the "Sunburn" sections of the chapter "Diseases." Irrigation method and application cycles may need to be modified to be compatible with mulching.

Biological Control

Biological control of weeds includes plant pathogens, insects, and small vertebrates such as birds that destroy or consume seeds. Two small weevils (*Microlarinus* spp.) often provide good biocontrol of puncturevine by feeding inside and killing the weed's vines and seed. Generally, however, biological control does not provide substantial weed control in citrus orchards.

Herbicides

Herbicides can provide effective and relatively economical control of most weeds, which facilitates irrigation and other cultural operations. Herbicides create a relatively weed-free orchard floor, which minimizes other pest problems and

Two introduced weevils (*Microlarinus* spp.) provide good biological control of puncturevine in some areas of California. Weevil presence can be recognized by feeding scars (lighter patches on the stem and brownish areas on the green seed capsule) or by larvae, pupae, or frass in plant crowns, stems, or seed capsules (it may be necessary to dissect the plants to find these). An adult weevil emerged from the hole in this stem (bottom left) after feeding inside.

reduces frost hazard during winter because of the warming influence of the bare ground.

Postemergence herbicides are applied after weeds emerge; preemergence herbicides are applied before weeds emerge. Growers sometimes apply a preemergence herbicide and a nonselective postemergence herbicide in combination if some weeds have already emerged. Translocated (systemic) herbicides such as glyphosate are transported via the plant's vascular system from contacted foliage to other parts of the plant and are the most effective materials for the control of perennials because the herbicide can move to and kill underground portions of plants (Figure 59). Contact herbicides kill only the exposed plant parts on which spray is deposited and are most effective on seedlings and young weeds. Combinations of materials or sequential treatments with different materials are often needed, since no single herbicide controls all weed species.

When using herbicides, choose the best materials, rates, and time of application for use against the weed species prevalent in your orchard. You can customize tables in the

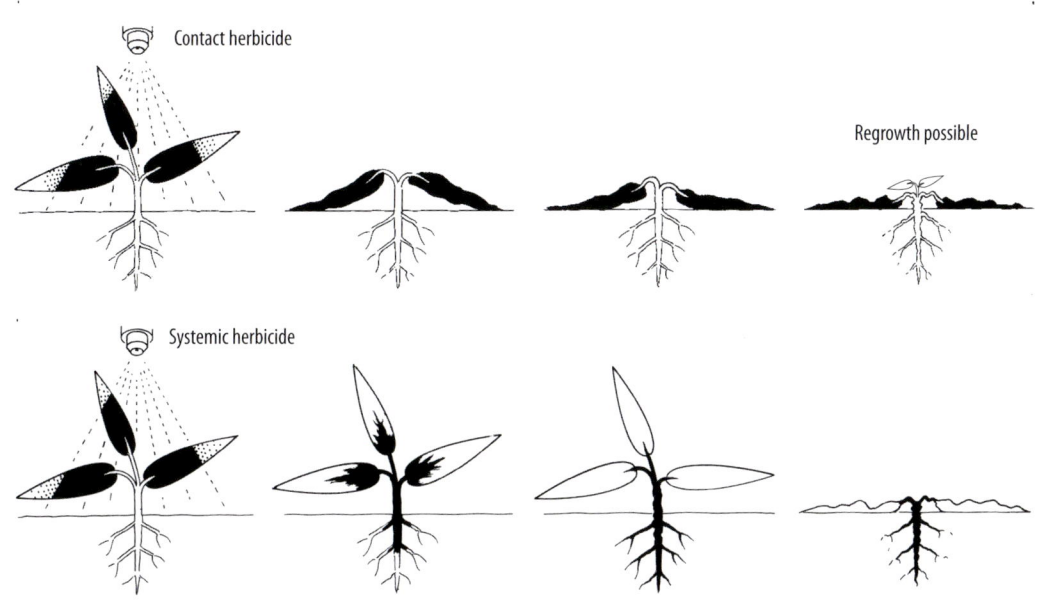

Figure 59. When treating weeds, make sure they are at the growth stage recommended on the pesticide label. Contact herbicides usually kill only the green plant parts on which spray is deposited. Thorough coverage is vital for good control, and even then certain weeds (e.g., established perennials) may regrow from underground parts. Systemic herbicides are taken up by green plant tissue and are transported to the growing tips of roots and shoots. Because they are translocated, systemic herbicides more effectively control perennial weeds, as illustrated in the bottom figure. The herbicide moves within the plant (black color) from sprayed leaves to roots. Unless they are selective for grasses, systemic herbicides pose a greater risk of injury to citrus trees. Adapted from David Kidd in Marer 1991.

online *UC IPM Pest Management Guidelines: Citrus* (illustrated in Figure 60) to identify the herbicides that will be most effective, depending on the application timing and season and on which weed species are most common in your orchard. Selectivity, mobility, mode of action, method and timing of application, the potential for phytotoxicity, and after-application issues such as persistence and environmental fate are among the important considerations when choosing and applying herbicides. Other herbicide selection considerations may include your soil type, irrigation method, the age of your orchard, and Ground Water Protection Area restrictions.

Make sure your spray equipment is calibrated accurately and is functioning properly, as discussed in *The Safe and Effective Use of Pesticides* (O'Connor-Marer 2000) and *Pesticide Safety: A Reference Manual for Private Applicators* (O'Connor-Marer and Cohen 2006), listed in "Suggested Reading." Consider the advantages and disadvantages of the alternative control measures suitable for your situation and consider using these whenever feasible to help minimize problems associated with overreliance on herbicides. Consult publications such as "Integrated Weed Management in Citrus" (in preparation) for a detailed discussion of tillage, cover cropping, mulching, heating, and preventive measures for weeds in citrus orchards.

Apply herbicides and other materials carefully and as directed on the label to avoid phytotoxicity. See the "Phytotoxicity" section in the chapter "Managing Pests in Citrus" for more detailed recommendations, and see the chapter "Diseases" for photographs of injury to fruit, leaves, and shoots to help you diagnose some common causes of phytotoxicity.

Trunk wraps help protect young trees from herbicide spray, sunburn, and vertebrate chewing. But cardboard or plastic collars do not guarantee that trees are completely protected from drift or misdirected herbicide applications. Use care when choosing and applying materials like glyphosate (Roundup) that can damage citrus.

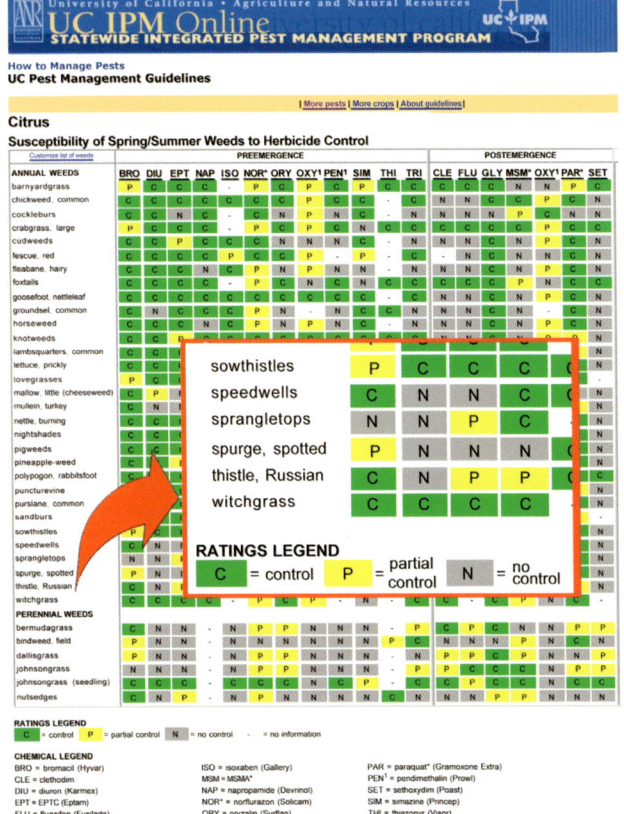

Figure 60. Herbicide susceptibility tables in the online *UC IPM Pest Management Guidelines: Citrus* identify herbicides that will be most effective. Because no single chemical controls all weed species, you can customize these online tables, depending on which weeds are most common in your orchard and depending on the application timing and season. Most growers rely extensively on herbicides, so the most common weed species in an orchard are generally those most tolerant of the herbicides applied there.

Preemergence Herbicides. Preemergence herbicides kill susceptible plants as they germinate. Most are effective only against germinating seeds. To be effective, the chemical must be moved into about the upper 2 inches (5 cm) of soil by rain, light irrigation of about $1/2$ inch (1 cm), or cultivation. Some preemergence herbicides must be moved into the soil immediately, while others may remain on the surface for a short time before incorporation. Certain materials lose effectiveness if the soil is cultivated after application. Follow label directions regarding incorporation. Efficacy typically persists for 2 to 5 months after application but sometimes may persist for more than a year. Persistence is affected by application rate, soil conditions, amount of rainfall or frequency of irrigation, and whether the soil is disturbed. For example, herbicide activity dissipates more quickly during prolonged wet weather and in areas that remain wet, such as around low-volume emitters.

Postemergence Herbicides. Foliar-applied or postemergence herbicides are sprayed on the leaves of weeds that have emerged. Postemergents can be either contact or systemic, as discussed above. Compared to contact herbicides, systemic herbicides pose a greater risk of phytotoxicity to trees. Special sprayers have been designed to apply postemergence herbicides only where they detect the presence of weed foliage. Often called smart sprayers, they reduce the amount of herbicide applied by 40 to 80% when weed populations are low. They are especially useful for strip weed control and spot-treating perennial weeds such as bermudagrass, dallisgrass, field bindweed, and nutsedges.

Herbicide-Resistant Weeds. Tolerance and resistance prevent some herbicides from controlling certain weeds. Tolerant plant species have a natural lack of susceptibility to certain herbicides. Tolerance can be desirable when it allows you to use selective herbicides, such as those that control grasses but do not damage citrus trees or other broadleaves. Resistance occurs when a pest population is no longer controlled by pesticides that previously provided control, as illustrated in Figure 13 in the chapter "Managing Pests in Citrus." After a weed population is repeatedly exposed to the same herbicide or group of herbicides with the same mode of action, the weed population may become dominated by plant biotypes that are resistant to that class of herbicide. For example, hairy fleabane, horseweed, and rigid ryegrass that are resistant to glyphosate (Roundup) have developed in California where glyphosate has been repeatedly applied.

To avoid promoting the development of herbicide-resistant weeds, rotate applications among herbicides with different modes of action (mechanisms or sites of action), which categorize materials according to the way they kill weeds. Avoid repeated application of a single herbicide or

Horseweed (shown here) has become resistant to glyphosate (Roundup) in some California locations where glyphosate has been repeatedly applied. To minimize the development of resistance, rotate applications among herbicides with different mechanisms (or sites) of action. Where feasible, instead of total reliance on herbicides, use cover crops, cultivation, mulch, and preventive measures such as sanitation.

of herbicides that kill weeds in the same way. Use alternative methods to control weeds such as cultivation, mulch, and preventive measures such as sanitation where feasible. Scout growing areas and note weed escapees and species shifts over time, which may indicate that resistance is developing and different materials or methods need to be used. Avoid spreading weed seeds and propagules from infested areas as this can introduce resistant weeds to new locations. For example, control runoff water and clean equipment before you move it to another site. Also see the general "Pesticide Resistance" discussion in the chapter "Managing Pests in Citrus."

Environmental Considerations. Air and water quality are below state and federal standards in much of California. Agricultural practices can contribute to this pollution. Certain herbicides are very soluble in water and pose a risk of leaching or moving in water and contaminating surface water or groundwater. Other herbicides adsorb tightly to organic matter and can move off-site in runoff if soil erodes. Many citrus orchards in California are within regulated Ground Water Protection Areas, where there are special restrictions to minimize herbicide leaching and runoff. Contact the county agricultural commissioner to learn whether the property you are treating is subject to Ground Water Protection Area regulations.

Use of fumigants and emulsifiable concentrate (EC) formulation pesticides are being restricted in part because they produce volatile organic compounds (VOCs) that contribute to ozone formation. If possible, choose formulations that are effective but lower in VOC production. Take steps to avoid pollution where feasible or required by law, such as by employing alternative management practices. Also be aware of and minimize environmental problems that can occur from practices such as tillage that can contribute to soil erosion or generate very small airborne particulates that damage lungs. See *Tillage and Crop Management Effects on Air, Water, and Soil Quality in California* (Horwath, Mitchell, and Six 2008) for more information.

Identifying Major Weed Species

Flowers are used to reliably identify the species of most plants, but weeds should be identified and controlled before they flower. For management decision making, use seedling appearance and the shape and arrangement of vegetative parts such as leaves, stems, and roots or other underground parts (for perennials) to identify weeds (Figure 61).

Seasonal growth patterns also help identify the species of a weed. Weeds are classified as annuals, biennials, or perennials based on their growth habit. As discussed earlier in the "Life Cycles" section, certain annuals such as annual bluegrass and mustards typically germinate in the fall or early winter, so they are classified as winter annuals.

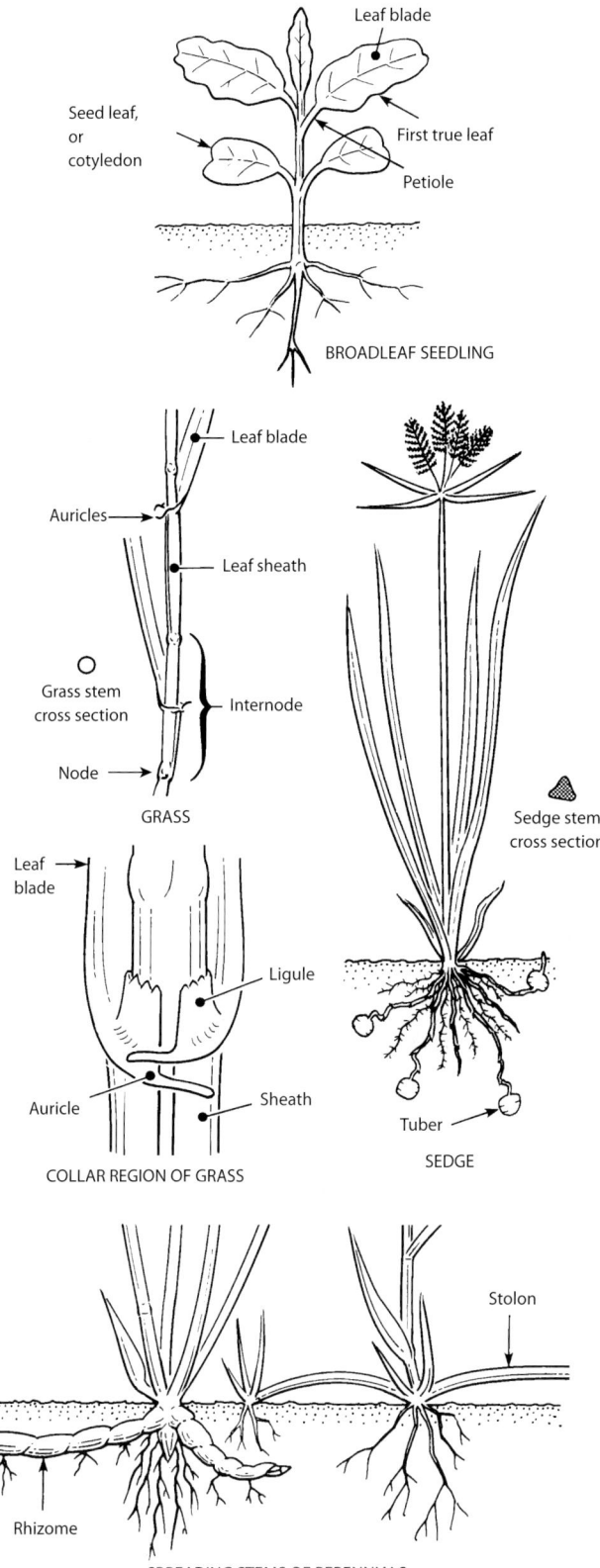

Figure 61. Vegetative parts of weeds and terms used in identification. Grasses and sedges superficially resemble each other, but their biology and management often differ. For the most effective control, the species should be identified and control action taken before weeds mature beyond the seedling stage.

Barnyardgrass and crabgrasses germinate in the spring or early summer, so they are summer annuals.

Weeds are also classified as broadleaves, grasses, or sedges based on botanical characteristics. Broadleaf seedlings have two seed leaves (cotyledons). The seed leaves (and sometimes also the first true leaves) of each broadleaf species have a characteristic color, shape, size, texture, and presence or absence of features such as hairs or spines that make the seedlings relatively easy to identify. Seed leaves (and sometimes also the first true leaves) are different from true leaves that form later. If newly emerged broadleaves lack cotyledons, they are not developing from seed; they are perennials developing from rhizomes, tubers, or other vegetation structures. Most broadleaf weeds have true leaves with netlike veins, and most develop showy flowers as they mature.

Grass leaves are longer than they are wide and have parallel veins. Grass seedlings are more difficult to identify than broadleaves because the seed leaves of different species are similar and grass seed leaves differ little from true leaves. Characteristics of the collar region (where the leaf blade joins the leaf sheath) help distinguish grasses. These distinguishing features include presence or absence of hairs, projections called auricles, and a collarlike ligule. Perennial grasses may be distinguished by the appearance of their rhizomes or stolons.

Grass and sedge seedlings have one seed leaf. Grasses and sedges resemble one another but are very different in many ways. The true leaves of grasses alternate from one side of the stem to the other; sedge leaves are joined to the stem in groups of three. Grass stems are hollow, rounded, and have nodes (joints) that are hard and closed. Sedge species have three-sided, solid stems that are triangular in cross section. The most important sedges are yellow and purple nutsedge. They are perennials that form characteristic tubers on their rhizomes.

Plants in approximately 60 different genera are considered to be weeds in California citrus (Table 12). However, within any one orchard usually only a few species cause significant problems. Which weeds are problematic depends on growing location, soil type, and the cropping history and management practices at that site and on adjacent properties. Because most growers rely extensively on herbicides, the predominant weed species are commonly those most tolerant of the herbicides applied there.

Identification Resources. Several excellent printed and online publications can help you identify the particular weed species that are problems in your orchard. Compare your problem species to the photographs and descriptions in the Citrus Weed Photo Gallery online at www.ipm.ucdavis.edu and illustrated in Figure 62 and the Weed Research and Information Center Web site (online at http://wric.ucdavis.edu). Consult the highly illustrated printed publications *Weeds of California and Other*

Table 12. Weeds in California Citrus.

Common name	Scientific name
barley, hare	*Hordeum leporinum*
barnyardgrass	*Echinochloa crus-galli*
bermudagrass	*Cynodon dactylon*
bindweed, field	*Convolvulus arvensis*
bluegrass, annual	*Poa annua*
bromegrasses	*Bromus* spp.
burclover, California	*Medicago polymorpha*
canarygrass	*Phalaris canariensis*
chickweed, common	*Stellaria media*
cockleburs	*Xanthium* spp.
crabgrass, large	*Digitaria sanguinalis*
cudweeds	*Gnaphalium* spp.
dallisgrass	*Paspalum dilatatum*
eveningprimrose, cutleaf	*Oenothera laciniata*
fescue, red	*Festuca rubra*
fiddlenecks	*Amsinckia* spp.
filarees	*Erodium* spp.
fleabane, hairy	*Conyza bonariensis*
foxtails	*Setaria* spp.
goosefoot, nettleleaf	*Chenopodium murale*
groundcherries	*Physalis* spp.
groundsel, common	*Senecio vulgaris*
henbit	*Lamium amplexicaule*
horseweed	*Conyza canadensis*
johnsongrass	*Sorghum halepense*
knotweeds	*Polygonum* spp.
lambsquarters, common	*Chenopodium album*
lettuce, prickly	*Lactuca serriola*
lovegrasses	*Eragrostis* spp.
mallow, little (cheeseweed)	*Malva parviflora*
miner's lettuce	*Claytonia perfoliata*
morningglory, annual	*Ipomoea* spp.
mullein, turkey	*Croton* (=*Eremocarpus*) *setigerus*
mustards	*Brassica* spp.
nettle, burning	*Urtica urens*
nightshades	*Solanum* spp.
nutsedge, purple	*Cyperus rotundus*
nutsedge, yellow	*Cyperus esculentus*
oat, wild	*Avena fatua*
pigweeds	*Amaranthus* spp.
pineapple-weed	*Chamomilla suaveolens*
polypogon, rabbitsfoot	*Polypogon monspeliensis*
puncturevine	*Tribulus terrestris*
purslane, common	*Portulaca oleracea*
radish, wild	*Raphanus raphanistrum*
redmaids (desert rockpurslane)	*Calandrinia ciliata*
rocket, London	*Sisymbrium irio*
ryegrass, Italian	*Lolium multiflorum*
sandburs	*Cenchrus* spp.
shepherd's-purse	*Capsella bursa-pastoris*
sowthistles	*Sonchus* spp.
speedwells	*Veronica* spp.
sprangletops	*Leptochloa* spp.
spurge, spotted	*Euphorbia* (=*Chamaesyce*) *maculata*
sweetclovers	*Melilotus* spp.
thistle, Russian	*Salsola tragus*
tobacco, tree	*Nicotiana glauca*
wild cucumber	*Marah* spp.
witchgrass	*Panicum capillare*

Figure 62. The Citrus Weed Photo Gallery is an example of pest identification information accessible through the Internet at www.ipm.ucdavis.edu. Also online are seasonal monitoring and decision-making guides, weed survey record-keeping forms, and herbicide recommendations.

UC IPM Online - University of California Statewide Integrated Pest Management Program

How to Manage Pests
Identification: Weed Photo Gallery

| More weeds | More ID helpers |

Johnsongrass

Scientific name: *Sorghum halepense* (Grass Family: Poaceae)

Click on image to enlarge

DESCRIPTION:

Johnsongrass is one of the most troublesome of perennial grasses. It reproduces from underground stems and seeds. The seedling resembles a young corn seedling, but can be distinguished by examining the attached seed after careful removal from the soil. The seed is football to egg-shaped and dark reddish-brown to black. The first blade is parallel to the ground. Blades are hairless with smooth margins and the midvein is whitish at the base. Auricles are lacking and ligules are membranous to hairy. The mature plant grows in spreading, leafy patches that may be as tall as 6 to 7 feet (1.8 - 2.1 m). Leaves have a prominent whitish midvein, which snaps readily when folded over. The flower head is large, open, well-branched, and often reddish tinged. Underground stems are thick, fleshy, and segmented. Roots and shoots can rise from each segment. The ligule consists of a fringe of dense, fine hairs.

Grass ID illustration.

Table 13. Weed Identification and Management Resources.

Resource	Format (e.g., printed publication)	ID	Monitoring	Decision making	Management recommendations
UC farm advisors (and PCAs) www.ucanr.org/ce.cfm		•	•	•	•
County agricultural commissioner www.cdfa.ca.gov/exec/county/county_contacts.html		•	•	•	•
Cover Crops SAREP www.sarep.ucdavis.edu/ccrop/index.htm	computer	•		•	•
Cover Cropping in Vineyards. UC ANR Publication 3338. anrcatalog.ucdavis.edu/SustainableandOrganic/3338.aspx	printed	•		•	•
DPR groundwater protection www.cdpr.ca.gov/docs/gwp/	computer			•	
Integrated Pest Management for Citrus manual www.ipm.ucdavis.edu/IPMPROJECT/ADS/manual_citrus.html	printed		•	•	•
Covercrops for California Agriculture anrcatalog.ucdavis.edu/VegetableCrops/21471.aspx	printed			•	•
Pest Management Guidelines (PMG) www.ipm.ucdavis.edu/PMG/selectnewpest.citrus.html	both	•	•	•	•
Pesticide labels MSDS CDMS www.cdms.net/LabelsMsds/LMDefault.aspx?t=	both			•	•
Citrus Production Manual	printed		•	•	•
Soil Solarization anrcatalog.ucdavis.edu/InOrder/Shop/ItemDetails.asp?ItemNo=21377	printed			•	•
Weed Photo Gallery www.ipm.ucdavis.edu/PMG/weeds_common.html	computer	•	•		
Weeds of California and Other Western States, with CD-ROM weed gallery anrcatalog.ucdavis.edu/InOrder/Shop/ItemDetails.asp?ItemNo=3488	printed	•	•		
Year-Round IPM Program www.ipm.ucdavis.edu/PMG/C107/m107yi01.html	both	•	•	•	

Western States (DiTomaso and Healy 2007) and *Weeds of the West* (Whitson et al. 1991). Expert computer-based identification guides to grasses and broadleaf weeds of California are available for confidently determining weed species. A wide variety of identification and decision-making resources are available to you, including those illustrated in Figure 63 and listed in Table 13. Which are most helpful depends on your initial level of knowledge, the type of information you need, and whether you prefer to use, and have access to, printed versus computerized information.

Special Weed Problems in Citrus

A few example weeds are discussed below. For more comprehensive information, consult other University of California publications (Table 13). These include the online *UC IPM Pest Management Guidelines: Citrus*, which are regularly revised and updated and include species-specific management recommendations.

Johnsongrass
Sorghum halepense

Johnsongrass is one of the most troublesome of perennial grasses. It can grow from either seed or rhizomes (underground stems). Rhizomes are thick, fleshy, and segmented. Roots and shoots can rise from each segment. The seeds have a red to purple tint and remain viable in the soil for at least 5 years.

A. Johnsongrass seedling.

- **A.** Johnsongrass seedlings have broad, light green leaves with smooth leaf sheaths that may have a reddish tinge. The midvein appears as a broad, white line at the base of the first leaf.

- **B.** Mature johnsongrass plants grow in spreading, leafy patches that may be as tall as 6 to 7 feet (1.8–2.1 m). The flower head is large, open, well branched, and often reddish tinged.

- **C.** The collar region (where a leaf blade meets the stem) often provides key features for distinguishing among species of grasses. There may be a ligule (an outgrowth) at the top of the sheath. In johnsongrass (shown here), the ligule is a fringe of fine hairs (at tips of arrows). Barnyardgrass superficially resembles johnsongrass, but the barnyardgrass ligule is long, thin, and delicate, not hairy. See Figure 61 for an illustration of these diagnostic plant parts.

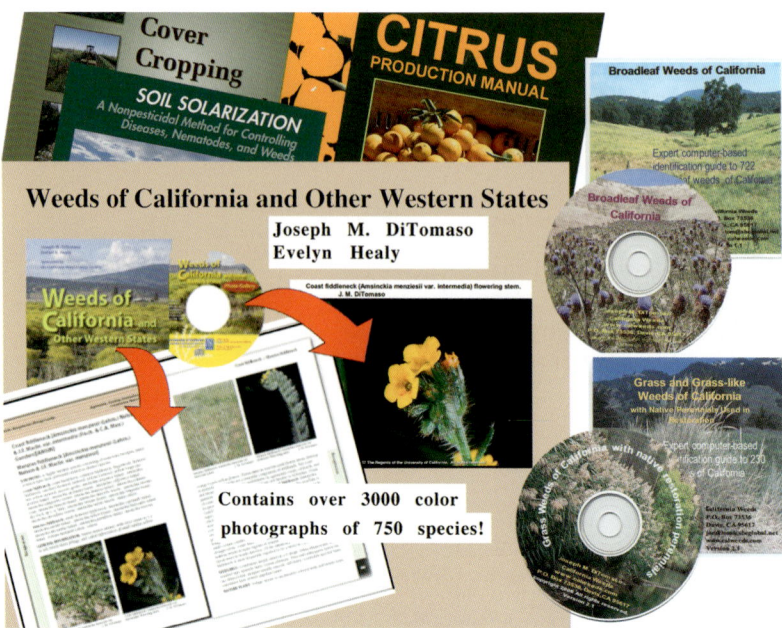

Figure 63. Well-illustrated printed and computerized publications on weed identification and management useful for growers and pest managers with different needs and levels of expertise are available from the University of California.

B. Mature johnsongrass.

C. Johnsongrass collar.

D. Johnsongrass rhizomes are exposed here after digging in soil. Perennials are the hardest weeds to control because they form extensive rhizomes, stolons, taproots, or tubers from which plants can regrow after treatment. A new johnsongrass plant can grow from each of the many segments shown on these rhizomes.

D. Johnsongrass rhizomes.

Nutsedges

Cyperus spp.

Nutsedges resemble grasses but have leaves that are triangular in cross section, whereas grass leaves are round. Purple nutsedge and yellow nutsedge are the two most common weedy species. Their flowers resemble those of grasses: yellow nutsedge flowers are yellow, while purple nutsedge flowers are purple. Nutsedges can grow from seed, but most infestations develop from rhizomes or tubers that persist and spread in soil.

Yellow and purple nutsedge are easily distinguished from each other by looking at their tubers. Yellow nutsedge tubers are nearly round and somewhat smooth. Purple nutsedge tubers are oblong and very rough and scaly. Purple nutsedge tubers are linked together by rhizomes (underground stems), whereas yellow nutsedge tubers are found singly only at the ends of rhizomes. Tubers of both species have three to seven buds, each capable of forming a new plant. Nutsedge plants develop from sprouts on a tuber; the sprout forms a bulb just under the soil surface. Leaves then grow from the basal bulb.

E. This yellow nutsedge seedling resembles grass, but nutsedge leaves are thicker and stiffer than most grasses. Nutsedge leaves are V-shaped in cross section and grow from the base in sets of three. The true leaves of grasses alternate from one side of the stem to the other.

E. Yellow nutsedge seedling.

F. Mature nutsedge plants.

F. Mature nutsedge plants are 1 to 2 feet (30–60 cm) tall with a flower head at the tip of each stem. Flowering stems are triangular in cross section. This yellow nutsedge has three long, leaflike bracts at the base of each flower head. These bracts are short in purple nutsedge.

G. The tubers of this yellow nutsedge are produced singly on the end of a rhizome. The tubers have a pleasant nutlike flavor. Rhizomes and tubers can develop into new plants, and their spread is the main source of new nutsedge infestations.

H. In contrast with the single tuber per rhizome on yellow nutsedge, these purple nutsedge tubers are linked together in chains with several tubers on a single rhizome. Each purple nutsedge tuber is oblong, rough, and scaly. The tubers have a bitter flavor.

H. Purple nutsedge tubers.

G. Yellow nutsedge tubers.

Harvesting and Pest Management

Fruit is susceptible to physical injury and fruit rot diseases as well as pathogen contamination during harvest and handling, up until it arrives at the packinghouse. Good harvest and handling practices can minimize injuries that damage rinds and promote the development of postharvest fruit rots. Fruit decays such as Alternaria rot, blue and green mold (caused by *Penicillium* spp.), Botrytis disease (gray mold), brown rot (Phytophthora fruit rot), and Septoria spot can develop after harvest as a result of improper crop management practices or the mishandling of fruit in the orchard.

Pest management practices become important at harvest when they include the use of pesticides. Laws limit the level

Decay organisms readily colonize split fruit and produce abundant spores that disperse and contaminate fruit throughout the tree. Provide a good growing environment and appropriate cultural care to minimize tree stress and reduce fruit splits and rots.

Most fruit rots begin in the orchard. However, this clear rot (blue and green mold) and other fruit decays often do not become apparent until after harvest when fruit arrives at the packinghouse or market. Fruit rot prevention requires good orchard management practices and proper fruit handling during harvest.

of pesticide residue that is allowable after fruit is picked. Using the pesticide preharvest interval to harvest at the correct time will help ensure that pesticide residues on fruit are below regulatory standards. Fruit destined for export could have additional constraints, as some countries have more stringent maximum pesticide residue levels.

Prevent Fruit Rot Diseases

Preharvest fungicide applications can reduce the incidence of certain fruit rot diseases. For certain other fruit rots, no effective materials are available. Even when preharvest treatment can be effective, it may not be economical to make the frequent in-orchard applications required under disease-favoring conditions, such as prolonged cool, wet weather. Rely primarily on modifying the growing environment and providing good cultural practices to prevent fruit rot diseases. See the chapter "Diseases" for photographs of fruit rots and discussion of their biology and prevention. For preharvest treatment recommendations, consult the latest *UC IPM Pest Management Guidelines: Citrus* (online at www.ipm.ucdavis.edu).

Pick a Good Harvest Time

Harvesting under the proper conditions can decrease the risk of postharvest disease. The risks of contamination by decay pathogens in the orchard and fruit rot development after harvest are especially high if fruit is handled when it is wet. If conditions have been wet, delay the harvest until foliage and fruit are dry.

Selecting the correct date of harvest is also important to minimize pesticide residues on fruit. Consult orchard treatment records before you schedule picking and comply with any preharvest intervals for pesticides (the minimum time, required by law, after the last application before harvest is permitted). If your fruit may be exported, learn whether your export markets have established maximum residue levels (MRLs) for the materials you use or have more stringent MRLs than U.S. pesticide tolerance levels. Be aware that certain countries also have special pest-exclusion quarantines and phytosanitary requirements. Export information resources include the California Citrus

Brown rot primarily affects low-growing fruit. The subtle rind darkening caused during early stages of this disease's development can easily be overlooked in the orchard. Pruning tree skirts to 24 inches (60 cm) or more above the ground significantly reduces this disease.

Fruit is susceptible to injury, diseases, and contamination until it arrives at the packinghouse. Use proper production, harvest, and handling techniques and industry standard good agricultural practices (GAPs) to avoid the fate of this fruit culled in the packhouse, which will feed livestock or go to the landfill.

Quality Council's *Export Manual* (available online at www.citrusresearch.org) and the U.S. Department of Agriculture, Foreign Agricultural Service's online database *International Maximum Residue Limits* (http://mrldatabase.com). Because standards and requirements frequently change, communicate regularly with your handler or packer to be certain you know and comply with requirements needed for your fruit to qualify to be shipped under a variety of export standards.

When Picking Fruit

Conditions during harvest affect fruit quality after harvest. Guidelines to follow include:

- Stop picking if fruit or trees become wet, such as during foggy or rainy weather.
- Instruct workers not to pick fruit that show vertebrate chewing damage or bird droppings on the surface. Tell workers not to place rotted or wounded fruit (such as from snail feeding) into bins. Fruit that is decayed or injured may constitute a food safety hazard, and it promotes decay in other fruit that comes in contact with it.
- Do not pick fruit that is on the ground, on downed branches, or on lower limbs that touch the ground. Prevent fruit from touching the ground.
- Pick fruit using clippers or a blade to cut the stem to $1/8$ inch (3 mm) or less to minimize fruit handling injury from protruding stems.
- Keep clippers, bags, and bins clean and dry during harvest. Wash and dry picking bags periodically. Direct workers to remove all plant debris each time the picking bag is emptied by moving away from the field bins, then shaking any debris or dust out of their empty bags before they begin to pick again. Keep twigs and leaves out of the bins and remove them if you see them.
- Keep gloves as clean as practical and free from contamination during the workday. Wash gloves thoroughly and frequently and do not allow them to contact soil or other potential sources of pathogens.
- Do not allow workers to stand in bins. Shoes are a common source of soilborne pathogens. Clean bins with steam or a high-pressure wash, rinse, and sanitizing treatment when you bring them in from off site, before you move them into orchards. Cover clean bins that you are not using to keep them from being contaminated by birds and other animals.

Do not allow fruit that drops or touches the ground to go into bins for human consumption. Soil is often contaminated with pathogens that cause postharvest fruit rots or human illness. Once produce is contaminated, it is very difficult to kill or remove pathogens.

The long pedicel stems visible here can directly injure fruit and allow pathogen entry. Direct workers to cut fruit stems to $1/8$ inch (3 mm) or less when harvesting citrus.

Transport fruit promptly to the packinghouse, where it usually will be washed, waxed, and chilled to remove pathogens and preserve fruit quality. Fruit is also sorted and culled in the packinghouse (shown here) to ensure that only high-quality citrus is shipped to consumers.

After Fruit Are Picked

Even after fruit are picked, damage can occur that makes fruit more susceptible to disease pathogens. Ensure that bins are rigid and do not flex or cause rubbing injury when they are moved. Disinfect and/or steam clean bins between loads and keep them free of any protrusions that might injure fruit. Load fruit carefully to minimize abrasion and transport it promptly to the packinghouse, where fruit usually will be washed, waxed, and chilled to preserve fruit quality. Do not leave bins with fruit in the orchard overnight: fruit quality will decline, and citrus is more susceptible to vertebrate pest damage, theft, and potential tampering.

Good Agricultural Practices

Good agricultural practices (GAPs) are industry standards that help growers provide safe and nutritious produce while sustaining agricultural productivity. As illustrated in Figure 64, GAPs include eight key elements:

1. Communicate regularly with the packer to ensure that the grower and packer understand, agree upon, and follow current fruit quality standards.
2. Follow a written GAP program plan that allows the source and recipients of fruit and potential contamination to be traced. Document all production inputs and operations, the locations of all orchard facilities and nearby sources of potential contaminants, tests of amendments and water quality, and activities such as employee training.
3. Comply with all pesticide regulations and document that compliance was achieved. For exported fruit, there may be additional standards beyond those of the U.S. Environmental Protection Agency, California Department of Pesticide Regulation, and local county agricultural commissioner.
4. Ensure good employee hygiene, training, and supervision, and comply with all Cal-OSHA requirements. This includes providing employees with clean water, hand-washing facilities, and properly maintained and well-stocked toilet facilities.
5. Adopt field sanitation and harvest practices that prevent fruit contamination. Provide clean bins, harvest tools, and equipment, including gloves.
6. Use pathogen-free water for all production practices, including irrigation, foliar sprays, and cleaning hands and equipment. Prevent water contamination and regularly test and document water quality.
7. Avoid fruit contamination by organic amendments and

Dirty bins and equipment are a source of fruit contaminants. Clean or disinfect picking bins before bringing them into the orchard, such as by using high-pressure water or steam cleaning.

soils, such as by testing amendments before use for pathogens. Reduce potential contaminant risks based on investigation of your orchard's soil type, drainage, and land history. For example, avoid planting citrus on land previously used for livestock production or biological or chemical waste disposal. When using amendments, properly composted manure is a safer choice than fresh or aged manure. Shortly after harvest is a suitable time for applying soil amendments.

8. Prevent fruit and orchard contamination by wildlife and domestic animals. Regularly monitor for and manage any vertebrate pest problems. Learn about their biology and prevention in the chapter "Vertebrates." Also consult the latest *UC IPM Pest Management Guidelines: Citrus* for management information.

For more information, consult the most recent *Food Safety Good Agricultural Practices (GAP) for California Citrus Growers* (available online at www.citrusresearch.org). Other resources include *Guidance for Industry: Guide to Minimize Microbial Food Safety Hazards for Fresh Fruits and Vegetables* (FDA 1998), *Key Points of Control and Management of Microbial Food Safety: Information for Growers, Packers, and Handlers of Fresh-Consumed Horticultural Products* (Suslow 2003), "Postharvest Handling Systems: Subtropical Fruits"

Shortly after harvest is a suitable time for applying soil amendments. Properly composted manure is a safer choice than fresh or aged manure. Consider testing amendments for pathogens. More information is online at the CalRecycle Web site, www.calrecycle.ca.gov.

(Kader and Arpaia 2002), and *UC GAP Self-Audits* (online at http://groups.ucanr.org/UC_GAPs).

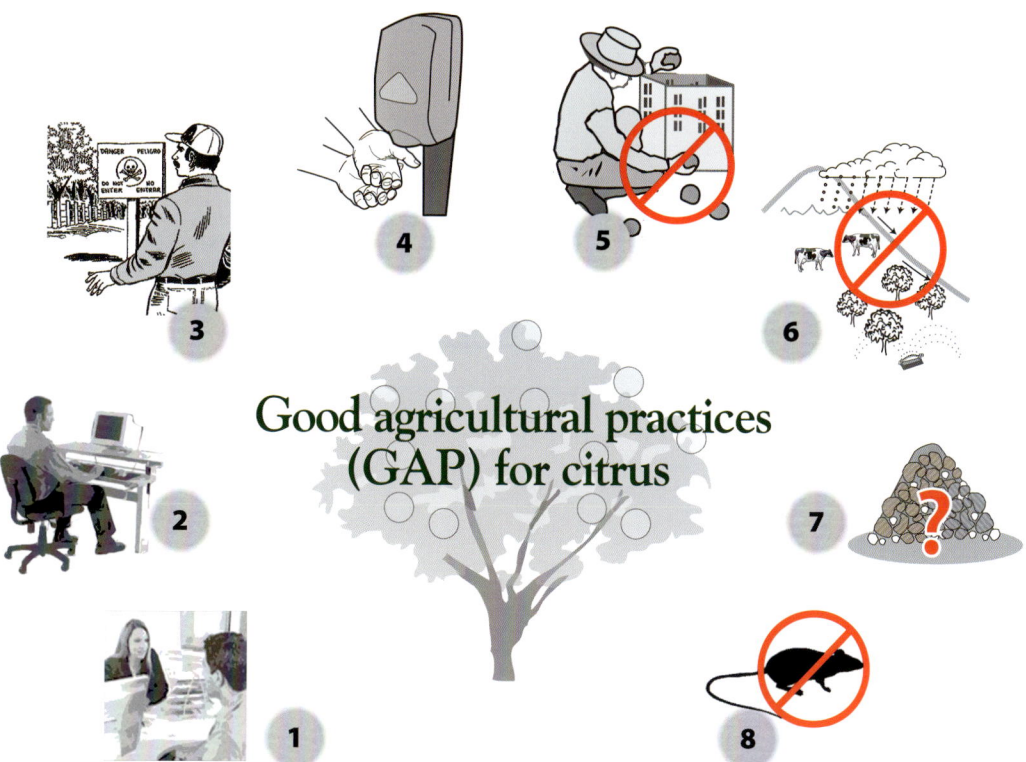

Figure 64. Good agricultural practices (GAP) for citrus include several key elements:
1. Communicate regularly with the packer regarding fruit quality standards.
2. Follow a written GAP program plan.
3. Comply with all pesticide regulations and phytosanitary standards.
4. Ensure good employee hygiene.
5. Adopt field sanitation and harvest practices that prevent fruit contamination. This includes providing clean bins, harvest tools, and equipment.
6. Use pathogen-free water for all production practices.
7. Avoid fruit contamination by organic amendments and soils.
8. Prevent fruit and orchard contamination by vertebrate pests and domestic animals.

World Wide Web Sites

The Internet's World Wide Web is a vast source of information on crop production and pest management. Among the many resources available online through the Web are color photographs of crop damage, pests, and natural enemies; information on biology, management recommendations, and decision-making models; and communication channels for pest control experts and practitioners. This section contains selected relevant Web sites and their online addresses (URLs).

Agricultural Research Service. U.S. Department of Agriculture. www.ars.usda.gov.

Association of Natural Biocontrol Producers. www.anbp.org.

California Invasive Plant Council. www.cal-ipc.org.

California Irrigation Management Information System (CIMIS). www.cimis.water.ca.gov/cimis/welcome.jsp.

Center for Invasive Species Research. University of California. www.cisr.ucr.edu.

Center for Irrigation Technology. California State University. www.wateright.org.

Citrus Clonal Protection Program. http://ccpp.ucr.edu.

Citrus Entomology. University of California Kearney Agricultural Center. http://ucanr.org/sites/KACCitrusEntomology.

Citrus Research Board. www.citrusresearch.org.

Cover Crops Database. UC Sustainable Agriculture Research and Education Program. www.sarep.ucdavis.edu.

Endangered Species Project. California Department of Pesticide Regulation. www.cdpr.ca.gov/docs/es.

Exotic Pests. California Department of Food and Agriculture. www.cdfa.ca.gov/phpps.

Fungicide Resistance Action Committee (FRAC). www.frac.info.

Good Agricultural Practices. University of California. http://ucgaps.ucdavis.edu.

Groundwater Protection. California Department of Pesticide Regulation. www.cdpr.ca.gov/docs/gwp.

Herbicide Resistance Action Committee (HRAC). www.hracglobal.com.

Insecticide Resistance Action Committee (IRAC). www.irac-online.org.

National Invasive Species Information Center. USDA Agricultural Library. www.invasivespeciesinfo.gov/unitedstates/ca.shtml.

National Weather Service. www.nws.noaa.gov.

Natural Enemies Gallery. www.ipm.ucdavis.edu/PMG/NE/index.html.

Natural Resources Conservation Service. www.nrcs.usda.gov.

Nematodes, Insect Parasitic. http://oardc.osu.edu/nematodes.

Pesticide Information, Extension Toxicology Network (EXTOXNET). http://ace.orst.edu/info/extoxnet.

Pesticide Labels and Material Safety Data Sheets (MSDS). Crop Data Management Systems. www.cdms.net/LabelsMsds/LMDefault.aspx?t.

Pesticide Regulation, California Department of. www.cdpr.ca.gov.

Red Imported Fire Ant (RIFA) Program. California Department of Food and Agriculture. www.fireant.ca.gov.

Statewide Integrated Pest Management Program. University of California. www.ipm.ucdavis.edu.

Sustainable Agriculture Research and Education. U.S. Department of Agriculture. www.sare.org.

Sustainable Agriculture Research and Education Program (SAREP). University of California. http://sarep.ucdavis.edu.

Water Resources, California Department of. www.dwr.water.ca.gov.

Water Resources Control Board, California State. www.swrcb.ca.gov.

Weed Research and Information Center. University of California. http://wric.ucdavis.edu.

Suggested Reading[1]

The publications listed below are referenced in text as sources for further information and are organized here by general topic. For a list of publications that are cited in figures and tables as the sources of data and illustrations, see the "Literature Cited," following this list.

Cultural Practices, Environmental Quality, Horticulture, and Water

An Annotated Checklist of Woody Ornamental Plants of California, Oregon, & Washington. 1979. E. McClintock and T. Leiser. Oakland: UC ANR Publication 4091.

Citrus physiology and phenology. In preparation. In *Citrus Production Manual*, L. Ferguson et al., eds. UC ANR Publications, Oakland.

Citrus Production Manual. In preparation. L. Ferguson et al., eds. UC ANR Publications, Oakland.

Determining Daily Reference Evapotranspiration (Et_o). 1987. R. L. Snyder, W. O. Pruitt, and D. A. Shaw. Oakland: UC ANR Publication 21426.

Establishing the citrus grove. In preparation. In *Citrus Production Manual*, L. Ferguson et al., eds. UC ANR Publications, Oakland.

Evapotranspiration and Irrigation Water Requirements. 1990. M. E. Jensen, R. D. Burman, and R. G. Allen, eds. ASCE Manuals and Reports on Engineering Practices No. 70. American Society of Civil Engineers, New York.

Fertigation. In preparation. In *Citrus Production Manual*, L. Ferguson et al., eds. UC ANR Publications, Oakland.

Ground cover height affects pre-dawn orchard floor temperature. 1993. R. L. Snyder and J. H. Connell. *California Agriculture* 47(1): 9–12.

Irrigation. In preparation. In *Citrus Production Manual*, L. Ferguson et al., eds. UC ANR Publications, Oakland.

Irrigation Scheduling: A Guide for Efficient On-Farm Management. 1989. D. A. Goldhamer and R. L. Snyder, eds. Oakland: UC ANR Publication 21454.

Leaf analysis as a guide to citrus fertilization. 1978. Embleton, T. W., W. W. Jones, and R. G. Platt. In *The Citrus Industry.* Vol. 3. W. L. Ruether, ed. UC ANR Publications, Berkeley and Oakland.

---

1. Many University of California Publications are available for free download or purchase via the World Wide Web. Look online at http://anrcatalog.ucdavis.edu or www.ipm.ucdavis.edu.

Nutrient deficiency and correction. In preparation. In *Citrus Production Manual*, L. Ferguson et al., eds. UC ANR Publications, Oakland.

Ozone air pollution. In preparation. In *Citrus Production Manual*, L. Ferguson et al., eds. UC ANR Publications, Oakland.

Protecting Groundwater Quality in Citrus Production. 1994. C. Ingels. Oakland: UC ANR Publication 21521.

Pruning. In preparation. In *Citrus Production Manual*, L. Ferguson et al., eds. UC ANR Publications, Oakland.

Questions and Answers About Tensiometers. 1981. A. W. Marsh. Oakland: UC ANR Leaflet 2264.

Rootstocks for citrus in California. In preparation. In *Citrus Production Manual*, L. Ferguson et al., eds. UC ANR Publications, Oakland.

Soil/water analysis and amendment strategies for citrus. In preparation. In *Citrus Production Manual*, L. Ferguson et al., eds. UC ANR Publications, Oakland.

The California citrus clonal protection program. In preparation. In *Citrus Production Manual*, L. Ferguson et al., eds. UC ANR Publications, Oakland.

Tillage and Crop Management Effects on Air, Water, and Soil Quality in California. 2008. W. R. Horwath, J. P. Mitchell, and J. W. Six. Oakland: UC ANR Publication 8331.

Water management. 2002. J. Hartin and B. Faber. In *California Master Gardener Handbook*, D. R. Pittenger, ed. Oakland: UC ANR Publication 3382.

Water Quality for Agriculture. 1994. R. S. Ayers and D. W. Westcot. Irrigation and Drainage Paper 29. Food and Agriculture Organization. Rome. Online at www.fao.org/docrep/003/T0234E/T0234E00.HTM.

Watershed Function. 2002. M. Padgett-Johnson and T. Bedell. Farm Water Quality Planning Reference Sheet 10.1. Oakland: UC ANR Publication 8064.

Watertight. 2005. Center for Irrigation Technology, California State University, Fresno. Online at www.wateright.org.

Western Fertilizer Handbook: 9th Edition. 2001. California Plant Health Association. Interstate, Sacramento.

Diseases and Disorders

Bacterial diseases of citrus. In preparation. In *Citrus Production Manual*, L. Ferguson et al., eds. UC ANR Publications, Oakland.

Citrus Bacterial Canker Disease and Huanglongbing (Citrus Greening). 2007. M. L. Polek, G. Vidalakis, and K. E. Godfrey. Oakland: UC ANR Publication 8218.

Citrus frost protection. In preparation. In *Citrus Production Manual*, L. Ferguson et al., eds. UC ANR Publications, Oakland.

Citrus: UC IPM Pest Management Guidelines: Diseases. 2009. J. E. Adaskaveg, J. A. Menge, and H. D. Ohr. University of California Statewide Integrated Pest Management Program. Oakland: UC ANR Publication 3441.

Environmental, physiological, and cultural injuries and genetic disorders. In preparation. In *Citrus Production Manual*, L. Ferguson et al., eds. UC ANR Publications, Oakland.

Fungal diseases of citrus. In preparation. In *Citrus Production Manual*, L. Ferguson et al., eds. UC ANR Publications, Oakland.

Fungi on Plants and Plant Products in the United States. 1989. D. F. Farr, G. F. Bills, G. P. Chamuris, and A. Y. Rossman. American Phytopathological Society, St. Paul, MN.

Invasive pests—Exotic plant pathogens. In preparation. In *Citrus Production Manual*, L. Ferguson et al., eds. UC ANR Publications, Oakland.

Plant Pathology. 4th ed. 1997. G. N. Agrios. Academic Press. San Diego.

Plants Resistant or Susceptible to Verticillium Wilt. 1981. A. H. McCain, R. D. Raabe, and S. Wilhelm. Oakland: UC ANR Publication 2703.

Soil solarization: A natural mechanism of integrated pest management. 1995. J. J. Stapleton and J. E. DeVay. In *Novel Approaches to Integrated Pest Management*, R. Reuveni, ed. Lewis Publishers. Boca Raton, FL.

Sooty Molds Pest Notes. 2003. F. F. Laemmlen. University of California Statewide Integrated Pest Management Program. Oakland: UC ANR Publication 74108.

Virus and viroid diseases of citrus. In preparation. In *Citrus Production Manual*, L. Ferguson et al., eds. UC ANR Publications, Oakland.

Harvesting and Pest Management

Compost Use in Orchards. 2001. California Integrated Waste Management Board (CIWMB) Publication 442-01-031. Online at www.calrecycle.ca.gov.

Composting Reduces Growers' Concerns About Pathogens. Undated. D. Crohn, C. P. Humpert, and P. Paswater. California Integrated Waste Management Board Publication 442-00-014. Online at www.calrecycle.ca.gov.

Export Manual. California Citrus Quality Council. Online at www.citrusresearch.org.

Food Safety Good Agricultural Practices (GAP) for Cafornia Citrus Growers. 2010. California Citrus Research Board, Visalia. Online at www.citrusquality.org.

Guidance for Industry: Guide to Minimize Microbial Food Safety Hazards for Fresh Fruits and Vegetables. 1998. FDA. Food Safety Initiative Staff. Washington, DC: U.S. Food and Drug Administration HFS-32. Online at www.fda.gov.

International Maximum Residue Limits (database). Undated. USDA Foreign Agricultural Service. Online at www.mrldatabase.com.

Key Points of Control and Management of Microbial Food Safety: Information for Growers, Packers, and Handlers of Fresh-Consumed Horticultural Products. 2003. T. V. Suslow. Oakland: UC ANR Publication 8102.

Postharvest handling systems: Subtropical fruits. 2002. A. A. Kader and M. L. Arpaia. In *Postharvest Technology of Horticultural Crops*, A. A. Kader, ed. Oakland: UC ANR Publication 3311.

UC GAP Self-Audits. Updated. University of California. Online at http:// groups.ucanr.org/UC_GAPs.

Integrated Pest Management

Citrus IPM. In preparation. In *Citrus Production Manual*, L. Ferguson et al., eds. UC ANR Publications, Oakland.

Citrus: UC IPM Pest Management Guidelines. 2009. E. E. Grafton-Cardwell, J. G. Morse, N. V. O'Connell, P. A. Phillips, C. E. Kallsen, and D. R. Haviland (General Pesticide Information, and Insects, Mites, and Snails); J. E. Adaskaveg, J. A. Menge, and H. D. Ohr (Diseases); J. O. Becker and B. B. Westerdahl (Nematodes); C. J. Lovatt (Plant Growth Regulators); M. W. Freeman, R. E. Marsh, and T. P. Salmon (Vertebrates); A. Shrestha and N. V. O'Connell (Weeds). University of California Statewide Integrated Pest Management Program. Oakland: UC ANR Publication 3441.

Citrus Year-Round IPM Program. 2008. E. E. Grafton-Cardwell, B. A. Faber, D. R. Haviland, C. E. Kallsen, J. G. Morse, N. V. O'Connell, P. A. Phillips, A. Shrestha, and S. H. Dreistadt. University of California Statewide Integrated Pest Management Program. Online at www.ipm.ucdavis.edu.

Integrated Pest Management for Citrus. 2nd ed. 1991. University of California Statewide Integrated Pest Management Program. Oakland: UC ANR Publication 3303.

IPM in Practice: Principles and Methods of Integrated Pest Management. 2001. M. L. Flint and P. Gouveia. University of California Statewide Integrated Pest Management Program. Oakland: UC ANR Publication 3418.

Insects, Mites, Nematodes, and Snails

A Key to the Most Common and/or Economically Important Ants of California, with Color Photographs. 1983. P. Haney, P. A. Philips, and R. Wagner. Oakland: UC ANR Publication 21433.

A Photographic Guide to Citrus Fruit Scarring. E. E. Grafton-Cardwell, N, V. O'Connell, C. E. Kallsen, and J. G. Morse. 2003. Oakland: UC ANR Publication 8090.

Amorbia: A California Avocado Insect Pest. 1980. J. B. Bailey and M. P. Hoffmann. Oakland: UC ANR Publication 21156.

Ants Pest Notes. 2000. M. Rust and J. Klotz. University of California Statewide Integrated Pest Management Program. Oakland: UC ANR Publication 7411.

Asian Citrus Psyllid. 2006. E. E. Grafton-Cardwell, K. E. Godfrey, M. E. Rogers, C. C. Childers, and P. A. Stansly. Oakland: UC ANR Publication 8205.

California Insects. 1979. J. A. Powell and C. L. Hogue. Univ. Calif. Press, Berkeley.

Citrus Leafminer and Citrus Peelminer. 2008. E. E. Grafton-Cardwell, K. E. Godfrey, D. H. Headrick, P. A. Mauk, and J. E. Peña. Oakland: UC ANR Publication 8321.

Citrus: UC IPM Pest Management Guidelines: Insects, Mites, and Snails. 2009. E. E. Grafton-Cardwell, J. G. Morse, N. V. O'Connell, P. A. Phillips, C. E. Kallsen, and D. R. Haviland. University of California Statewide Integrated Pest Management Program. Oakland: UC ANR Publication 3441.

Color-Photo and Host Keys to California Whiteflies. 1982. R. J. Gill. Scale and Whitefly Key #2. California Department of Food and Agriculture, Sacramento.[2]

Color-Photo and Host Keys to the Armored Scales of California. 1982. R. J. Gill. Scale and Whitefly Key #5. California Department of Food and Agriculture, Sacramento.[2]

Color-Photo and Host Keys to the Mealybugs of California. 1982. R. J. Gill. Scale and Whitefly Key #3. California Department of Food and Agriculture, Sacramento.[2]

Color-Photo and Host Keys to the Soft Scales of California. 1982. R. J. Gill. Scale and Whitefly Key #4. California Department of Food and Agriculture, Sacramento.[2]

Common Names of Arachnids. 1995. R. G. Breene. American Tarantula Society, South Padre Island, TX.

---

2. Publication is out of print. You may be able to find a reference copy at a library.

Common Names of Insects & Related Organisms 1997. 1997. J. J. Bosik, ed. Entomological Society of America, Lanham, MD.

Destructive and Useful Insects. 5th ed. 1993. R. L. Metcalf and R. A. Metcalf. McGraw-Hill, New York.

Diaprepes Root Weevil. 2004. E. E. Grafton-Cardwell, K. E. Godfrey, J. E. Peña, C. W. McCoy, and R. F. Luck. Oakland: UC ANR Publication 8131.

Fruittree Leafroller on Ornamental and Fruit Trees Pest Notes. 2000. W. J. Bentley, C. Pickel, R. E. Rice, and R. Van Steenwyk. University of California Statewide Integrated Pest Management Program. Oakland: UC ANR Publication 7473.

Giant Whitefly Pest Notes. 2006. T. S. Bellows, J. N. Kabashima, and K. L. Robb. University of California Statewide Integrated Pest Management Program. Oakland: UC ANR Publication 7400.

Glassy-Winged Sharpshooter Pest Notes. 2007. L. G. Varela, J. M. Hashim-Bucky, C. A. Wilen., and P. A. Phillips. University of California Statewide Integrated Pest Management Program. Oakland: UC ANR Publication 7492.

Insects and Mites of Western North America. 1958. E. O. Essig. MacMillan, New York.[2]

Key to Identifying Common Household Ants. 2004. C. A. Reynolds. University of California Statewide Integrated Pest Management Program. Online at www.ipm.ucdavis.edu/TOOLS/ANTKEY.

Life Stages of California Red Scale and Its Parasitoids. 1995. L. D. Forster, R. F. Luck, and E. E. Grafton-Cardwell. Oakland: UC ANR Publication 21529. Online at http://ucanr.org/sites/KACCitrusEntomology/files/4457.pdf

Light Brown Apple Moth in California: Quarantine, Management, and Potential Impacts. 2007. M. W. Johnson, C. Pickel, L. L. Strand, L. G. Varela, C. A. Wilen, M. P. Bolda, M. L. Flint, W. K. F. Lam, and F. G. Zalom. University of California Statewide Integrated Pest Management Program. Online at www.ipm.ucdavis.edu/PDF/PUBS/lbam070717.pdf

Managing Insects and Mites with Spray Oils. 1991. N. A. Davidson, J. E. Dibble, M. L. Flint, P. J. Marer, and A. Guye. University of California Statewide Integrated Pest Management Program. Oakland: UC ANR Publication 3347.

Mealybugs in California Vineyards. 2002. K. E. Godfrey, K. M. Daane, W. J. Bentley, R. J. Gill, and R. Malakar-Kuenen. Oakland: UC ANR Publication 21612.

Natural Enemies Are Your Allies! 1990. M. L. Flint and J. K. Clark. Poster. University of California Statewide Integrated Pest Management Program. UC ANR Publication 21497.

Natural Enemies Handbook: The Illustrated Guide to Biological Pest Control. 1998. M. L. Flint and S. H. Dreistadt. University of California Statewide Integrated Pest Management Program. Oakland: UC ANR Publication 3386.

Nematode diseases of citrus. 1999. L. W. Duncan. In *Citrus Health Management.* L. W. Timmer and L. W. Duncan, eds. APS Press, St. Paul, MN.

Nematodes in California citrus. In preparation. In *Citrus Production Manual,* L. Ferguson et al., eds. UC ANR Publications, Oakland.

Pest Thrips of North America. Undated. C. O'Donnell, G. Moritz, L. Mound, S. Nakahara, and M. Parrella. Online at http://entomology.ucdavis.edu/thrips/about00.html.

Red Imported Fire Ant Pest Notes. 2007. L. Greenberg, J. H. Klotz, and J. N. Kabashima. University of California Statewide Integrated Pest Management Program. Oakland: UC ANR Publication 7487.

Scale Insects of California, Part 1: The Soft Scales. R. J. Gill. 1988. California Department of Food and Agriculture, Sacramento.

Scale Insects of California, Part 2: The Minor Families. R. J. Gill. 1993. California Department of Food and Agriculture, Sacramento.

Scale Insects of California, Part 3: The Armored Scales. R. J. Gill. 1997. California Department of Food and Agriculture, Sacramento.

Snails and Slugs Pest Notes. 2009. M. L. Flint and C. A. Wilen. University of California Statewide Integrated Pest Management Program. Oakland: UC ANR Publication 7427.

Stages of the Cottony Cushion Scale (Icerya purchasi) *and its Natural Enemy, the Vedalia Beetle* (Rodolia cardinalis). 2002. B. Grafton-Cardwell. Oakland: UC ANR Publication 8051.

Sticky Trap Monitoring of Insect Pests. 1998. S. H. Dreistadt, J. P. Newman, and K. L. Robb. Oakland: UC ANR Publication 21572.

Suppliers of Beneficial Organisms in North America. C. D. Hunter. 1997. California Department of Food and Agriculture, Sacramento. Online at www.cdpr.ca.gov.

2. Publication is out of print. You may be able to find a reference copy at a library.

Thrips of California: Distinguishing Pest Species Among California's Rich Native Thrips Fauna. 2008. M. S. Hoddle, L. A. Mound, and D. Paris. Online at www.lucidcentral.org.

Thrips Pest Notes. 2007. S. H. Dreistadt and P. A. Phillips. University of California Statewide Integrated Pest Management Program. Oakland: UC ANR Publication 7429.

Pesticides

Bordeaux Mixture Pest Notes. 2010. J. C. Broome and D. R. Donaldson. University of California Statewide Integrated Pest Management Program. Oakland: UC ANR Publication 7481.

Classification of Herbicides According to Mode of Action. 2005. Herbicide Resistance Action Committee (HRAC). Online at www.hracglobal.com.

Fungicides Sorted by Modes of Action. 2007. Fungicide Resistance Action Committee (FRAC). Online at www.frac.info.

How to Reduce Bee Poisoning from Pesticides. 1999. D. F. Mayer, C. A. Johansen, and C. R. Baird. Pullman: Washington State University Pacific Northwest Extension Publication PNW591. Online at http://cru.cahe.wsu.edu/CEPublications/pnw0518/pnw0518.pdf

IRAC Mode of Action Classification. 2008. Insecticide Resistance Action Committee (IRAC). Online at www.irac-online.org.

Managing Insects and Mites with Spray Oils. 1991. N. A. Davidson, J. E. Dibble, M. L. Flint, P. J. Marer, and A. Guye. University of California Statewide Integrated Pest Management Program. Oakland: UC ANR Publication 3347.

Managing pesticide resistance in insects, mites, weeds, and fungi infesting California citrus. In preparation. In *Citrus Production Manual*, L. Ferguson et al., eds. UC ANR Publications, Oakland.

Pesticide principles. In preparation. In *Citrus Production Manual*, L. Ferguson et al., eds. UC ANR Publications, Oakland.

Pesticide Safety: A Reference Manual for Growers. 1998. P. J. O'Connor-Marer. University of California Statewide Integrated Pest Management Program. Oakland: UC ANR Publication 3383.

Pesticide Safety: A Reference Manual for Private Applicators. 2nd. ed. 2006. P. J. O'Connor-Marer and S. Cohen. University of California Statewide Integrated Pest Management Program. Oakland: UC ANR Publication 3383.

Pesticides: Theory and Application. G. W. Ware. 1983. W. H. Freeman, San Francisco.

The Safe and Effective Use of Pesticides. 2nd ed. 2000. P. J. O'Connor-Marer. University of California Statewide Integrated Pest Management Program. Oakland: UC ANR Publication 3324.

Vertebrates

Bird Hazing Manual: Techniques and Strategies for Dispersing Birds from Spill Sites. 2008. W. P. Gorenzel and T. P. Salmon. Oakland: UC ANR Publication 21638.

California Ground Squirrel Pest Notes. 2010. T. P. Salmon and W. P. Gorenzel. University of California Statewide Integrated Pest Management Program. Oakland: UC ANR Publication 7438.

Coyote Pest Notes. 2007. R. M. Timm, C. C. Coolahan, R. O. Baker, and S. F. Beckerman. Statewide Integrated Pest Management Program. Oakland: UC ANR Publication 74135.

Deer Pest Notes. 2004. T. P. Salmon, D. A. Whisson, and R. E. Marsh. University of California Statewide Integrated Pest Management Program. Oakland: UC ANR Publication 74117.

Norway Rats. 1994. R. M. Timm. In *Prevention and Control of Wildlife Damage*, Vol. 1. S. E. Hygnstrom, R. M. Timm, and G. E. Larson, eds. Lincoln: University of Nebraska Cooperative Extension. B.105–120.

Pocket Gophers Pest Notes. 2009. T. P. Salmon and R. A. Baldwin. University of California Statewide Integrated Pest Management Program. Oakland: UC ANR Publication 7433.

Rabbits Pest Notes. 2010. T. P. Salmon and W. P. Gorenzel. University of California Statewide Integrated Pest Management Program. Oakland: UC ANR Publication 7447.

Rats Pest Notes. 2003. T. P. Salmon, R. W. Marsh, and R. M. Timm. University of California Statewide Integrated Pest Management Program. Oakland: UC ANR Publication 74106.

Roof rats. 1994. R. E. Marsh. In *Prevention and Control of Wildlife Damage*. Vol. 1. S. E. Hygnstrom, R. M. Timm, and G. E. Larson, eds. Lincoln: University of Nebraska Cooperative Extension. B.125–132.

Tree Squirrels Pest Notes. 2005. T. P. Salmon, D. A. Whisson, and R. W. Marsh. University of California Statewide Integrated Pest Management Program. Oakland: UC ANR Publication 74122.

Voles (Meadow Mice) Pest Notes. 2010. T. P. Salmon and W. P. Gorenzel. University of California Statewide Integrated Pest Management Program. Oakland: UC ANR Publication 7439.

Wildlife Pest Control around Gardens and Homes. 2nd ed. 2006. T. P. Salmon, D. A. Whisson, and R. E. Marsh. Oakland: UC ANR Publication 21385.

Weeds

Annual Bluegrass Pest Notes. 2003. D. W. Cudney, C. L. Elmore, and V. A. Gibeault. University of California Statewide Integrated Pest Management Program. Oakland: UC ANR Publication 7464.

Bermudagrass Pest Notes. 2007. C. L. Elmore and D. W. Cudney. University of California Statewide Integrated Pest Management Program. Oakland: UC ANR Publication 7453.

Broadleaf Weeds of California. 2006. J. M. DiTomaso. California Weeds, Davis. CD-ROM.

Common Purslane Pest Notes. 2007. D. W. Cudney, C. L. Elmore, and R. H. Molinar. University of California Statewide Integrated Pest Management Program. Oakland: UC ANR Publication 7461.

Composite List of Weeds. 1989. J. F. Alex, G. A. Bozarth, C. T. Bryson, J. W. Everest, E. P. Flint, F. Forcella, D. W. Hall, H. F. Harrison Jr., L. W. Hendrick, L. G. Holm, D. E. Seaman, V. Sorensen, H. V. Strek, R. H. Walker, and D. T. Patterson. Weed Science Society of America, Champaign, IL.

Cover Cropping in Vineyards. 1998. C. A. Ingels, R. L. Bugg, G. T. McGourty, and L. P. Christensen, eds. Oakland: UC ANR Publication 3338.

Covercrops for California Agriculture. 1989. P. R. Miller, W. L. Graves, W. A. Williams, and B. A. Madison. Oakland: UC ANR Publication 21471.

Field Bindweed Pest Notes. 2003. C. L. Elmore and D. W. Cudney. University of California Statewide Integrated Pest Management Program. Oakland: UC ANR Publication 7462.

Grass and Grass-like Weeds of California. 2006. J. M. DiTomaso. California Weeds, Davis. CD-ROM.

Integrated weed management in citrus. In preparation. In *Citrus Production Manual*, L. Ferguson et al., eds. UC ANR Publications, Oakland.

Kikuyugrass Pest Notes. 2003. D. W. Cudney, C. L. Elmore, and V. A. Gibeault. University of California Statewide Integrated Pest Management Program. Oakland: UC ANR Publication 7458.

Managing Cover Crops Profitably. 2nd ed. 1998. Sustainable Agriculture Network. USDA Sustainable Agriculture Research & Education, Beltsville, MD. Online at www.sare.org.

Nutsedge Pest Notes. 2010. C. A. Wilen, M. E. McGiffen, and C. L. Elmore. University of California Statewide Integrated Pest Management Program. Oakland: UC ANR Publication 7432.

Soil solarization: A natural mechanism of integrated pest management. 1995. J. J. Stapleton and J. E. DeVay. In *Novel Approaches to Integrated Pest Management*, R. Reuveni, ed. Lewis Publ. Boca Raton, FL. 309–322.

Soil Solarization: A Nonpesticidal Method for Controlling Diseases, Nematodes, and Weeds. 1997. C. L. Elmore, J. J. Stapleton, C. E. Bell, and J. E. DeVay. Oakland: UC ANR Publication 21377.

The Jepson Manual: Higher Plants of California. 1993. J. C. Hickman, ed. Oakland: Univ. Calif. Press. Berkeley.

Weeds of California and Other Western States. 2007. J. M. DiTomaso and E. A. Healy. Oakland: UC ANR Publication 3488.

Weeds of the West. 1991. T. D. Whitson, L. C. Burrill, S. A. Dewey, D. W. Cudney, B. E. Nelson, R. D. Lee, and R. Parker. Wyoming Agricultural Extension, Jackson. Available as UC ANR Publication 3350.

Literature Cited[1]

The publications listed here are cited in this book's figures and tables as the sources of information and illustrations. For publications referenced in the text as recommended sources for further information, see the "Suggested Reading."

Bailey, D., and T. Bilderback. 1998. *Alkalinity Control for Irrigation Water Used in Nurseries and Greenhouses*. Raleigh: North Carolina Cooperative Extension Service Horticulture Information Leaflet HIL #558. Online at www.ces.ncsu.edu/depts/hort/hil/hil-558.html.

Bailey, L. H. 1941. *The Standard Cyclopedia of Horticulture*. Vol. 3. New York: MacMillan.

Bosik, J. J. 1997. *Common Names of Insects & Related Organisms*. Lanham, MD: Entomological Society of America.

CCQC. 2003. *Pest Management Strategic Plan For Citrus Production in California*. Auburn, CA: California Citrus Quality Council.

Childers, C. C., and T. R. Fasulo. 1995. *Citrus Red Mite*. Gainesville: University of Florida Cooperative Extension Fact Sheet ENY-817.

Cobb, N. A. 1914. Citrus-root nematode. *Journal of Agricultural Research* 3:217–230.

Davenport, T. L. 1990. Citrus flowering. *Horticulture Reviews* 12:349–408.

Davidson, N. A., J. E. Dibble, M. L. Flint, P. J. Marer, and A. Guye. 1991. *Managing Insects and Mites with Spray Oils*. University of California Statewide Integrated Pest Management Program. Oakland: UC ANR Publication 3347.

DeBach, P., and J. Landi. 1961. The introduced purple scale parasite, *Aphytis lepidosaphes* Compere, and a method of integrating chemical with biological control. *Hilgardia* 31(14): 459–497.

DeBach, P., R. M. Hendrickson, and M. Rose. 1978. Competitive displacement: Extinction of the yellow scale, *Aonidiella citrina* (Coq.) (Homoptera: Diaspididae), by its ecological homologue, the California red scale, *Aonidiella aurantii* (Mask.) in Southern California. *Hilgardia* 46:1–35.

Dreistadt, S. H. 2002. *Integrated Pest Management for Floriculture and Nurseries*. University of California Statewide Integrated Pest Management Program. Oakland: UC ANR Publication 3402.

Ebeling, W. 1959. *Subtropical Fruit Pests*. Los Angeles: University of California Division of Agricultural Science.

———. 1978. *Urban Entomology*. Berkeley: University of California Division of Agricultural Science.

Essig, E. O. 1949. Aphids in relation to quick decline and *Tristeza* of citrus. *Pan-Pacific Entomologist* 25:13–23.

European Plant Protection Organization 2005. Web Fig. 2, Morphological differences between *Aonidiella citrina* and *Aonidiella aurantii*. Online at www.eppo.org/QUARANTINE/insects/Aonidiella_citrina/pm7-51(1)%20AONDCI%20web.pdf

Ewing, H. E. 1939. *A Revision of the Mites of the Subfamily Tarsoneminae of North America, the West Indies, and the Hawaiian Islands*. U.S. Department of Agriculture Technical Bulletin 653.

FAO. 2007. FAOSTAT. The United Nations Food and Agriculture Organization. Online at http://faostat.fao.org.

Fasulo, T. R. 2007. *Broad Mite*, Polyphagotarsonemus latus (Banks) (Arachnida: Acari: Tarsonemidae). Gainesville: University of Florida Publication EENY-183.

Flint, M. L. 1995. *Whiteflies in California: A Resource for Cooperative Extension*. Davis: University of California Statewide Integrated Pest Management Program Publication 19.

Forster, L. D., R. F. Luck, and E. E. Grafton-Cardwell. 1995. *Life Stages of California Red Scale and Its Parasitoids*. Oakland: UC ANR Publication 21529. Online at http://ucanr.org/sites/KACCitrusEntomology/files/4457.pdf

FRAC. 2007. *Fungicides Sorted by Modes of Action*. Fungicide Resistance Action Committee. Online at www.frac.info.

Gill, R. J. 1997a. *Scale Insects of California, Part 3: The Armored Scales*. Sacramento: California Department of Food and Agriculture.

———. 1997b. Thrips. *California Plant Pest Disease Report* 16(3–6):33.

———. 1982. *Color-Photo and Host Keys to the Mealybugs of California*. Sacramento: California Department of Food and Agriculture Scale and Whitefly Key #3.

Godfrey, K. E., K. M. Daane, W. J. Bentley, R. J. Gill, and R. Malakar-Kuenen. 2002. *Mealybugs in California Vineyards*. Oakland: UC ANR Publication 21612.

Gorham, J. R., ed. 1991. *Insect and Mite Pests in Food: An Illustrated Key*. Vol. 2. Washington, DC: U.S. Department of Agriculture Handbook 655.

Grafton-Cardwell, E. E. Undated. *Interactive Key for Differentiating Brown Citrus Aphid from Aphids that Infest California Citrus*. University of California Kearney Agricultural Center.

Grafton-Cardwell, E. E., D. H. Headrick, P. A. Mauk, and J. G. Morse. In preparation. Citrus IPM. In *Citrus Production Manual*. L. Ferguson et al., eds. UC ANR Publications, Oakland.

Grafton-Cardwell, E. E., J. G. Morse, N. V. O'Connell, P. A. Phillips, and D. R. Haviland (General Pesticide Information, and Insects, Mites, and Snails); Adaskaveg, J. E., H. D. Ohr, and J. A. Menge (Diseases); Becker, J. O. and B. B. Westerdahl (Nematodes); Lovatt, C. J. (Plant Growth Regulators); Freeman, M. W., R. E. Marsh, and T. P. Salmon (Vertebrates); Shrestha, A. and N. V. O'Connell (Weeds). 2009. *UC IPM Pest Management Guidelines: Citrus*. University of California Statewide Integrated Pest Management Program. Oakland: UC ANR Publication 3441.

Grafton-Cardwell, E. E., N, V. O'Connell, C. E. Kallsen, and J. G. Morse. 2003. *A Photographic Guide to Citrus Fruit Scarring*. Oakland: UC ANR Publication 8090.

HRAC. 2005. *Classification of Herbicides According to Mode of Action*. Herbicide Resistance Action Committee. Online at www.hracglobal.com.

IRAC. 2008. *Classification of Insecticides According to Mode of Action*. Insecticide Resistance Action Committee. Online at www.irac-online.org.

Kahn, T. L. 2007. Birds do it; bees do it, even citrus with seeds do (did) it: Part 1—The biology behind seedlessness in mandarins. *Topics in Subtropics* 5(1):3–5.

Keifer, H. H. 1952. The Eriophyid mites of California. *Bulletin of the California Insect Survey* 2(1):1–123.

Klitz, B. F. 1930. Perennial weeds which spread vegetatively. *Journal of the American Society of Agronomy* 22:216–234.

Kobbe, B. 1984. *Integrated Pest Management for Citrus*. 1st ed. Oakland: UC ANR Publication 3303.

Kono, T., and C. S. Papp. 1977. *Handbook of Agricultural Pests: Aphids, Thrips, Mites, Snails, and Slugs*. Sacramento: California Department of Food and Agriculture.

Lovatt, C. J. 1999. Timing citrus and avocado foliar nutrient applications to increase fruit set and size. *HortTechnology* 9(4): 607–612.

———. In preparation. Citrus physiology and phenology. In *Citrus Production Manual*, L. Ferguson et al., eds. UC ANR Publications, Oakland.

Marer, P. J. 1991. *Residential, Industrial, and Institutional Pest Control*. University of California Statewide Integrated Pest Management Program. Oakland: UC ANR Publication 3334.

McGregor, S. E. 1976. *Insect Pollination of Cultivated Crop Plants*. Washington, DC: U.S. Department of Agriculture Handbook 496.

McKenzie. H. L. 1935. Life history and control of the gladiolus thrips in California. Calif. Agriculture Exp. Sta. Circ. 337:1–16.

———. 1956. The armored scale insects of California. *Bulletin of the California Insect Survey* 5:1–209.

Moore, I., E. F. Legner, and M. E. Badgley. 1975. Description of the developmental stages of the mite predator, *Oligota oviformis* Casey, with notes on the osmeterium and its glands (Coleoptera: Staphylinidae). *Psyche* 82:181–188.

Morse, J. G., P. A. Phillips, P. B. Goodell, D. L. Flaherty, C. J. Adams, and S. I. Frommer. 1987. Monitoring Fuller rose beetle populations in citrus groves and egg mass levels on fruit. *The Pest Control Circular* 547:1–8.

Mound, L. A., and G. Kibby. 1998. *Thysanoptera: An Identification Guide*. New York: CAB International.

Nechols, J. R., L. A. Andres, J. W. Beardsley, R. D. Goeden, and C. G. Jackson, eds. 1995. *Biological Control in the Western United States*. Oakland: UC ANR Publication 3361.

Newcomer, E. J., and M. A. Yothers. 1929. *Biology of the European Red Mite in the Pacific Northwest*. Washington, DC: U.S. Department of Agriculture Technical Bulletin 89.

O'Donnell, C. A., L. A. Mound, and M. P. Parrella. Undated. *Multilevel Identification System for Thrips Associated with Flower Crops in North America*. Online at http://entomology.ucdavis.edu/thrips/Thesis_Section_I_III.pdf

Padgett-Johnson, M., and T. Bedell. 2002. *Watershed Function*. Farm Water Quality Planning Reference Sheet 10.1. Oakland: UC ANR Publication 8064.

Peterson, A. 1960. *Larvae of Insects*. Part 2. Ann Arbor, MI: Edwards Brothers.

———. 1962. *Larvae of Insects*. Part 1. Ann Arbor, MI: Edwards Brothers.

Quayle, H. J. 1912. *Red Spiders and Mites of Citrus Trees*. Berkeley: University of California Agricultural Experiment Station Bulletin 234.

———. 1938. *Insects of Citrus and other Subtropical Fruits*. Ithaca, NY: Comstock.

Reynolds, C. A., M. L. Flint, M. K. Rust, P. S. Ward, R. L. Coviello, and J. H. Klotz. 2001. *Key to Identifying Common Household Ants*. University of California Statewide Integrated Pest Management Program. Online at www.ipm.ucdavis.edu.

Roose, M. L. In preparation. Rootstocks for citrus in California. In *Citrus Production Manual*, L. Ferguson et al., eds. UC ANR Publications, Oakland.

Rose, M., and G. Zolnerowich. 1997. *The Genus Eretmocerus (Hymenoptera: Aphelinidae) Parasites of Whitefly (Homoptera: Aleyrodidae)*. College Station: Texas A&M University Department of Entomology.

Salmon, T. P., and R. E. Lickliter. 1984. *Wildlife Pest Control around Gardens and Homes*. Oakland: UC ANR Publication 21385.

Salmon, T. P., D. A. Whisson, and R. E. Marsh. 2006. *Wildlife Pest Control around Gardens and Homes*. 2nd ed. Oakland: UC ANR Publication 21385.

Sanderson, E. D., and C. F. Jackson. 1912. *Elementary Entomology*. Boston: Ginn.

Schauff, M. E., G. A. Evans, and J. M. Heraty. 1996. A pictorial guide to the species of *Encarsia* (Hymenoptera: Aphelinidae) parasitic on whiteflies (Homoptera: Aleyrodidae) in North America. *Proceedings of the Entomological Society of Washington* 98:1–35.

Schneider, H. 1968. The anatomy of citrus. In *The Citrus Industry*. Vol. 2. W. L. Ruether, D. Batchelor, and H. J. Webber, eds. Berkeley and Oakland: UC ANR Publications.

Silva, D., C. Lovatt, and M. L. Arpaia. 2002. Citrus. In *California Master Gardener Handbook*, D. R. Pittenger, ed. Oakland: UC ANR Publication 3382.

Simanton, F. L. 1916. *Hyperaspis binotata*, a predatory enemy of the terrapin scale. *Journal of Agricultural Research* 6:197–204.

Smith, M. H. 1965. *House-Infesting Ants of the Eastern United States*. Washington, DC: U.S. Department of Agriculture Technical Bulletin 1326.

Smith, H. S., and H. M. Armitage. 1931. *The Biological Control of Mealybugs Attacking Citrus.* Oakland: University of California Bulletin 509.

Smith, R. F., and K. S. Hagen. 1956. Enemies of spotted alfalfa aphid. *California Agriculture* 10(4): 8–10.

Smith, D., G. A. C. Beattie, and R. H. Broadley. 1997. *Citrus Pests and Their Natural Enemies.* Brisbane: Queensland Department of Primary Industries.

Strand, L. L. *Integrated Pest Management for Almonds.* 2nd ed. 2002. University of California Statewide Integrated Pest Management Program. Oakland: UC ANR Publication 3308.

Timmer, W., and L. W. Duncan. 1999. *Citrus Health Management.* St. Paul, MN: The American Phytopathological Society.

Truog, E. 1948. Lime in relation to availability of plant nutrients. *Soil Science* 65:1–7.

USDA. 2007. Situation and outlook for citrus: Citrus—Special feature article. Washington, DC: U.S. Department of Agriculture, Foreign Agricultural Service.

———. 2008. *2008 California Citrus Acreage Report.* Sacramento: U.S. Department of Agriculture, National Agricultural Statistics Service. Online at www.nass.usda.gov.

Wade, D.A. 1973. *Control of Damage by Coyotes and some other Carnivores.* Fort Collins: Colorado State University Cooperative Extension Service Publication WRP-11.

Ward, P. S. 2005. A synoptic review of the ants of California (Hymenoptera: Formicidae). *Zootaxa* 936:1–68.

Ware, G. W. 1983. *Pesticides: Theory and Application.* San Francisco: W. H. Freeman.

Watson, G. W. 2005. *Neohydatothrips* sp. (probably *N. burungae* (Hood)) (Thripidae) A thrips. *California Plant Pest Disease Report* 22(1): 21–23.

Weller, S. C., and F. D. Hess. 1997. Herbicide use and mode of action. In *Weed Management in Horticultural Crops,* E. McGriffen, ed. Alexandria, VA: American Society for Horticultural Science. 74–115.

Whisson, D. 1997. *Modifying a "T" Bait Station for Ground Squirrel Control in Kangaroo Rat Habitat.* Sacramento: California Department of Pesticide Regulation Endangered Species Project. Online at www.cdpr.ca.gov/docs/es/espdfs/baitsta2.pdf

Whitson, T. D., L. C. Burrill, S. A. Dewey, D. W. Cudney, B. E. Nelson, R. D. Lee, and R. Parker. 1991. *Weeds of the West.* Jackson: University of Wyoming. Available as UC ANR Publication 3350.

Wilcox, W. F. 1992. *Phytophthora Root and Crown Rots,* Phytophthora *spp. (deBary).* Geneva: New York State Agricultural Experiment Station Tree Fruit Crops IPM Disease Identification Sheet No. 7.

Tables and Figures

Tables

Table 1. Seedless and Seeded Cultivars of Mandarin, Tangelo, and Tangor Citrus 8
Table 2. Cultural Practices and Growing Conditions That Growers Can Manipulate (Y) in an IPM Program 14
Table 3. Citrus Fruit Scarring Causes 16
Table 4. Approximating Degree-Days (°D) Manually 19
Table 5. Effects of Citrus Rootstocks on Tree Performance, Disease Tolerance, and Fruit Quality 23
Table 6. Examples of the Relative Selectivity of Selected Pesticides Used in Citrus 30
Table 7. Biological Control of Citrus Pests in Major Growing Areas 89
Table 8. Lady Beetles and Their Prey in Citrus 93
Table 9. Selected Thrips in Citrus and How to Distinguish Them 117
Table 10. Pest Mites in California Citrus 182
Table 11. Control Methods for Vertebrate Pests of Citrus 211
Table 12. Weeds in California Citrus 234
Table 13. Weed Identification and Management Resources 235

Figures

Figure 1. Citrus production areas of California 1
Figure 2. Acres of fruit-bearing citrus (and approximate percentage of total acreage) for the citrus-growing regions in California 1
Figure 3. Fruit-bearing acreage (and percentage of total acres) for major citrus cultivars grown in California 3
Figure 4. Citrus is grown in most countries with a tropical, subtropical, or Mediterranean climate 3
Figure 5. Citrus development stages 6
Figure 6. Average seasonal development for navel orange grown in the San Joaquin Valley 7
Figure 7. Citrus flower and fruit parts and stages of their development 7
Figure 8. Photosynthesis is a chemical process within green plant tissue 9
Figure 9. The basic anatomy and functions of key vegetative parts of a tree 10
Figure 10. Approximate monitoring times for diseases, nematodes, vertebrates, and weeds in citrus 21
Figure 11. Approximate monitoring times for invertebrate pests and key natural enemies in citrus 21
Figure 12. Approximate times for citrus-growing practices in California's Central Valley 22
Figure 13. Resistance to pesticides develops through genetic selection in populations of pests 33
Figure 14. Disrupting biological control often results in secondary outbreaks of insects and mites 34
Figure 15. Pesticides and fertilizers applied in orchards can contaminate groundwater or surface water by moving ... 37
Figure 16. The disease triangle 39
Figure 17. The disease cycle of *Phytophthora* species infecting citrus 44
Figure 18. Armillaria root rot development cycle and spread 47
Figure 19. Green lacewing life cycle and stages 90
Figure 20. Minute pirate bug (*Orius* spp.) life cycle and stages 92
Figure 21. Convergent lady beetle life cycle and stages 93
Figure 22. California red scale and yellow scale life stages and their development cycle 97
Figure 23. Features for distinguishing California red scale (*Aonidiella aurantii*) from yellow scale 99
Figure 24. Photomicrographs for distinguishing California red scale from yellow scale 99

Figure 25. California red scale can often be well-controlled by natural enemies 101
Figure 26. Life cycle and stages of *Aphytis lepidosaphes*, the primary parasite of purple scale 107
Figure 27. Understanding the seasonal cycle of citricola scale 108
Figure 28. Lady beetle life cycle and stages, illustrated here with a predator of scales 109
Figure 29. Cottony cushion scale seasonal development 114
Figure 30. Citrus thrips life cycle and stages 116
Figure 31. Seasonal development of citrus thrips on navel orange in the San Joaquin Valley 120
Figure 32. How to distinguish citrus thrips from flower thrips 121
Figure 33. Citrus thrips and western flower thrips compared 122
Figure 34. How to distinguish adults of citrus thrips, *Neohydatothrips burungae*, and avocado thrips 126
Figure 35. Distinguishing characters for adults of four species of thrips found in California citrus 129
Figure 36. Key to caterpillars (Lepidoptera larvae) in California citrus 132
Figure 37. Identification key to the species of mealybugs in California citrus 156
Figure 38. Whitefly species that occur on citrus in California 159
Figure 39a. Key to winged-aphids in California citrus 162
Figure 39b. Key to nonwinged aphids potentially infesting California citrus 164
Figure 40. Fuller rose beetle adults can be distinguished from two other nondamaging species 175
Figure 41. Identification key to ant species 177
Figure 42. Citrus red mite life cycle and stages 184
Figure 43. Citrus broad mite life cycle and stages 190
Figure 44. Citrus broad mite adult male 191
Figure 45. Eriophyid mite life cycle and stages 192
Figure 46. Predaceous midge (*Feltiella occidentalis*) life cycle and stages 196
Figure 47. Predatory rove beetle life cycle and stages 197
Figure 48. Citrus nematode life cycle and feeding in a citrus root 202
Figure 49. Citrus nematode damage to small roots 204
Figure 50. A nematode at right snared by a ring trap fungus 206
Figure 51. A ground squirrel bait station designed for use within the range of endangered kangaroo rats and the San Joaquin kit fox 214
Figure 52. Pocket gophers dig their tunnels by pushing soil out through lateral tunnels 215
Figure 53. Use a soil probe to locate gopher tunnels 215
Figure 54. Gophers can be trapped with 2-pronged pincher traps 215
Figure 55. California ground squirrel reproductive cycle and seasonal food preference 217
Figure 56. Footprints for distinguishing vertebrate species 222
Figure 57. Weed control methods vary depending on when the method is applied 226
Figure 58. Field bindweed has deep rhizomes from which plants repeatedly regrow 227
Figure 59. When treating weeds, make sure they are at the growth stage recommended on the pesticide label 231
Figure 60. Herbicide susceptibility tables in the online *UC IPM Pest Management Guidelines: Citrus* 232
Figure 61. Vegetative parts of weeds and terms used in identification 233
Figure 62. The Citrus Weed Photo Gallery 235
Figure 63. Well-illustrated printed and computerized publications on weed identification 236
Figure 64. Good agricultural practices (GAP) for citrus 243

Glossary

abdomen. the posterior of the three main body divisions of an insect.

abiotic disorder. a disease caused by factors other than a pathogen, such as adverse environmental conditions or inappropriate cultural practices.

aestivation (also estivation). a state of inactivity during the summer months.

albedo. white, spongy inner part of citrus fruit rind (see Figure 7).

alkaline. basic, having a high pH or pH greater than 7.

annual. a plant that normally completes its life cycle of seed germination, vegetative growth, reproduction, and death in a single year.

antenna (plural, antennae). the paired segmented sensory organs on each side of the head.

anther. the pollen-producing organ of a flower (see Figure 7).

anticoagulant. a substance that prevents blood clotting, resulting in internal hemorrhaging; may be used as a rodenticide.

auricles. the earlike projections at the base of leaves of some grasses; used to identify species (see Figure 61).

axil. the upper (narrow) angle between a twig or leaf petiole and the stem from which it is growing.

axillary bud. undeveloped branch or flower tissue (a bud) formed in an axil.

bacterium (plural, bacteria). a single-celled microscopic organism that lacks a nucleus. Some bacteria cause plant or animal diseases.

biofix. an identifiable event that signals when to begin degree-day accumulation or take a management action.

biotic disease. a disease caused by a pathogen such as a bacterium, fungus, phytoplasma, or virus.

biotype. a strain of a species that has certain biological characters distinguishing it from other individuals of that species.

broad-spectrum pesticide. a pesticide that kills a large number of unrelated species.

button. the doughnut-shaped structure where the stem connects to the fruit (see Figure 7).

calcareous soils. soils containing high levels of calcium carbonate.

calibrate. to standardize or correct measuring devices on instruments; to properly adjust a sprayer's output.

calyx. an outer structure of the flower consisting of the whorl of sepals (see Figure 7).

cambium. thin layer of undifferentiated, actively dividing cells that produces new bark (phloem) on the outside and new wood (xylem) on the inside (see Figure 9).

canker. a dead, discolored, and often sunken area (lesion) on a root, trunk, stem, or branch.

caterpillar. the larva of a butterfly or moth.

chlorophyll. the green pigment of plant cells, necessary for photosynthesis.

chlorosis. yellowing or bleaching of normally green plant tissue.

chorion. the outer membrane of an insect egg.

cocoon. a sheath, such as a covering of silk, formed by an insect larva as a chamber for pupation.

collar region. in grasses, the region where the leaf blade and sheath meet; used in identifying species (see Figure 61).

conidium (plural, conidia). an asexual fungal spore formed by fragmentation or budding at the tip of a specialized hypha.

cortex. in stems or roots, the tissue between the epidermis and conducting tissue.

cotyledons. the first leaves of the embryo formed within a seed and present on seedlings immediately after germination; seed leaves (see Figure 61).

crawler. the mobile first instar of certain types of insects, such as scales and mealybugs.

cross-resistance. in pest management, resistance of a pest population to a pesticide to which it has not been exposed that accompanies the development of resistance to a pesticide to which it has been exposed.

crown. the point at or just above the soil surface where the main stem (trunk) and roots join.

cultivar. an identifiable strain within a plant species that is specifically bred for particular properties; sometimes used synonymously with *variety*.

degree-day (°D, or DD). a measurement unit that combines temperature and time; used in calculating growth rates.

developmental threshold. the lowest temperature at which growth occurs in a given species.

diapause. a period of physiologically controlled inactivity or dormancy in insects.

disease. any disturbance of a plant that interferes with its normal structure, function, or economic value.

economic threshold. a level of pest population or damage at which the cost of a control action equals the crop value gained from that control action.

ectoparasite. a parasite that lives on the outside of its host.

endoparasite. a parasite that lives inside its host.

epidermis. the outermost layer of cells on the bodies of animals or on plant surfaces.

estivation. see *aestivation*.

evapotranspiration. the loss of soil moisture by the combined mechanisms of soil surface evaporation and transpiration by plants.

feeder roots. the youngest roots with root hairs, most important in absorption of water and minerals.

flavedo. outer part of the rind of citrus fruit, bearing oil glands and pigments (see Figure 7).

frass. a mixture of feces and food fragments produced by an insect in feeding.

fumigation. treatment with a pesticide active ingredient that is in gaseous form under treatment conditions.

girdle (n.). damage that completely encircles a stem or root, often resulting in death of plant parts above or below the girdle.

girdle (v.). to kill or damage a ring of bark tissue around a stem or root; such damage interrupts the transport of water and nutrients.

honeydew. an excretion from insects, such as aphids, mealybugs, whiteflies, and soft scales, consisting of modified plant sap.

hypha (plural, hyphae). a filament that is the structural unit of a fungus.

indexing. testing a plant for a virus infection, usually by grafting tissue from it onto an indicator plant that develops characteristic symptoms if certain viruses are present.

infection. the entry of a pathogen into a host and establishment of the pathogen as a parasite in the host.

inflorescence. a flower cluster.

inoculum. any part or stage of a pathogen, such as a spore or virus particle, that can infect a host.

instar. an insect between successive molts; the first instar is between hatching and the first molt.

internode. the area of a stem between nodes (see Figure 61).

invertebrate. an animal having no internal skeleton.

larva (plural, larvae). the immature form of an insect that hatches from an egg, feeds, and then enters a pupal stage.

lesion. a well-defined area of diseased tissue, such as a canker or leaf spot.

ligule. in many grasses, a short, membranous projection on the inner side of the leaf blade at the junction where the leaf blade and leaf sheath meet (see Figure 61).

meconium. fecal pellet excreted by a larva before pupation.

metamorphosis. a change in form during development.

microorganism. an organism of microscopic or tiny size.

molt. in insects and other arthropods, the shedding of skin before entering another stage of growth.

mummy. the crusty skin of certain insects (such as an aphid) whose inside has been consumed by a parasite.

mutation. the abrupt appearance of a new, heritable characteristic as the result of a change in the genetic material of one individual cell.

mycelium (plural, mycelia). the vegetative body of a fungus, consisting of a mass of slender filaments called hyphae.

mycoplasma. living organisms smaller than bacteria, now called phytoplasmas; they have a unit membrane but no cell wall as do bacteria.

natural enemies. predators, parasites, or pathogens that are considered beneficial to crops because they attack and kill organisms that we normally consider to be pests.

necrosis. the death of tissue accompanied by dark brown discoloration, usually occurring in a well-defined part of a plant, such as the portion of a leaf between leaf veins or in the xylem or phloem in a stem or tuber.

node. the slightly enlarged part of a stem where buds are formed and where leaves, stems, and flowers originate (see Figure 61).

nymph. the immature stage of insects such as aphids and grasshoppers that hatch from eggs and gradually acquire adult form through a series of molts without passing through a pupal stage.

oviposit. to lay or deposit eggs.

parasite. an organism that lives in or on the body of another organism (the host) without killing the host directly; in this manual the term is used to refer to insect parasitoids, which spend their immature stages on or within the body of a single host that dies before the parasite emerges (see Figure 26).

pathogen. a disease-causing organism.

perennial. a plant that may live three or more seasons and flower at least twice.

pest resurgence. the increase of a pest population following a pesticide treatment to levels higher than before the treatment as a result of the pesticide having killed natural enemies of the pest.

petiole. the stalk connecting the leaf to a stem (see Figure 61).

pH. a value used to express the relative degree of acidic or basic conditions; the hydrogen ion concentration as expressed in a negative logarithmic scale ranging from 0 to 14.

pheromone. a chemical produced by an animal to affect the behavior or development of other members of the same species. Sex pheromones that attract the opposite sex for mating are used in monitoring or management of certain insects.

phloem. the food-conducting tissue of a plant's vascular system (see Figure 9).

photosynthesis. the process whereby plants use light energy to form sugars and other compounds needed to support growth and development (see Figure 8).

phytoplasma. living organisms smaller than bacteria, formerly called mycoplasmas or mycoplasmalike organisms; they have a unit membrane but no cell wall, as do bacteria.

phytotoxic. a material such as a pesticide or fertilizer causing injury to plants.

pistil. the female part of a flower, usually consisting of ovules, ovary, style, and stigma (see Figure 7).

predator. an animal that attacks and feeds on other animals (its prey), usually consuming many prey during its lifetime.

preemergence herbicide. an herbicide applied before target weeds emerge.

presence/absence sampling. a sampling method that involves recording only whether members of the population being sampled (such as an insect pest) are present or absent on a sample unit (such as a leaf), rather than counting the numbers of individuals. Also called binomial sampling.

propagule. any part of a plant from which a new plant can grow, including seeds, bulbs, rootstocks, etc.

prothorax. the anterior of the three thoracic segments of an insect.

pulp. juice-containing vesicles underneath the rind of citrus fruit (see Figure 7).

pupa (plural, pupae). the nonfeeding, inactive stage between larva and adult in insects with complete metamorphosis.

pupate. to develop from the larval stage to the pupa.

resistant. able to tolerate conditions (such as pesticide sprays or pest damage) harmful to other individuals of the same species.

rhizome. a horizontal underground stem, especially one that roots at the nodes to produce new plants (see Figures 58, 61).

rootstock. the lower portion of a graft that grows into the root system.

rosette. a cluster of leaves arranged in a compact circular pattern, often at a shoot tip or on a shortened stem.

scion. the portion above a graft that becomes the trunk, branch, and tree top; the cultivar or variety used for that part of a graft.

sclerotium (plural, sclerotia). a compact mass of hardened mycelium that serves as a dormant stage in some fungi.

secondary pest outbreak. the sudden increase in a pest population that is normally at low or nondamaging levels caused by the destruction of natural enemies by treatment with a nonselective pesticide to control a primary pest (see Figure 14).

sedges. a group of grasslike herbaceous plants that, unlike grasses, have unjointed stems. Stems are usually solid and often triangular in cross section (see Figure 61).

seed leaf. the first leaf (grasses) or first two leaves (broadleaf plants) on a seedling; cotyledons (see Figure 61).

seedling rootstock. a rootstock propagated from seed.

sepal. one of the outermost structures of the flower, one of a whorl of parts collectively called the calyx (Figure 7).

seta (plural, setae). a bristle.

sheath. the part of a grass leaf that encloses the stem below the collar region (see Figure 61).

sieve tubes. see **phloem**.

soil profile. a vertical section of the soil through all its horizontal layers, extending into the parent material.

sooty mold. dark coating on foliage or fruit formed by the mycelia of fungi that live on honeydew secreted by certain insects.

spiracle. an external opening of the system of ducts, or tracheae, that serves as a respiratory system in insects.

sporangium. a structure containing asexual spores.

spore. a reproductive body produced by certain fungi and other organisms and capable of growing into a new individual under proper conditions (see Figure 17).

stamen. a flower structure made up of the pollen-bearing anther and a stalk or filament (see Figure 7).

stele. the central cylinder inside the cortex of many roots and stems.

stigma. the receptive portion of the female flower part to which pollen adheres.

stylar end. the side of a fruit opposite to that with stem, typically the bottom end of a fruit when hanging on the tree; the location of the navel on navel oranges, where the flower style connected to the ovary (see Figure 7).

stolon. a stem that grows horizontally along the surface of the ground, often rooting at the nodes and forming new plants (see Figure 61).

stoma (plural, stomata or stomates). a natural opening in a leaf surface that serves for gas exchange and water evaporation and has the ability to open and close in response to environmental conditions.

systemic herbicide. an herbicide that is able to move throughout a plant (translocated herbicide) after being applied (see Figure 59).

taproot. a large primary root that grows vertically downward, giving off small lateral roots.

tensiometer. a device that measures how tightly water is held by the soil; used for estimating water content of the soil.

thorax. the second of three major divisions in the body of an insect, and the one bearing the legs and wings.

tolerant. able to withstand the effects of a condition without suffering serious injury or death.

translocated herbicide. a systemic herbicide.

transpiration. the evaporation of water from plant tissue, usually through stomata.

treatment threshold. a level of pest population or damage, usually measured by a specified monitoring method, at which a pesticide application is recommended.

topworking. changing the cultivar by pruning off the existing scion and budding or grafting new plant material onto the rootstock or basal scion of an established tree.

true leaf. any leaf produced after the cotyledons (see Figure 61).

tuber. a much-enlarged, fleshy underground stem.

vascular system. the system of plant tissues that conducts water, mineral nutrients, and products of photosynthesis through the plant, consisting of the xylem and phloem.

vector. an organism able to transport and transmit a pathogen to a host.

vegetative growth. growth of stems, roots, and leaves, but not flowers and fruits.

viroid. a portion of infectious nucleic acid, without the protein coat of a virus.

virulent. capable of causing a severe disease; strongly pathogenic.

virus. a small infectious agent, consisting only of nucleic acid and a protein coat, that can reproduce only within the living cells of a host.

xylem. plant tissue that conducts water and nutrients from the roots up through the plant; the woody portion of a tree located inside the cambium (see Figure 9).

zoospore. a motile spore (see Figure 17).

Index

Tables are indicated with *t* and figures (illustrations and photographs) are indicated with *f*.

abiotic disorders, 42, 55–57, 62–66, 69–77
acreage, citrus production, 1*f*
Actia interrupta (tachinid fly), 139
Aeolothrips spp. (banded thrips), 117*t*, 118, 119*f*, 123
aeration deficit, 69
air pollution, 36–37, 72, 233
Aleurodicus dugesii (giant whitefly), 158*f*, 159*f*
Aleurothrixus floccosus (woolly whitefly), 89*t*, 158*f*, 159–161
alternate crop hosts, 81, 151, 152, 169
Alternaria rot (*Alternaria citri*), 58R
amendments, soil, 242–243
Amitus spiniferus (parasitic wasp), 160*f*
amorbia (*Amorbia cuneana*), 16*t*, 135*f*, 142*f*, 143–144
Anacamptodes fragilaria (citrus looper), 133*f*, 148–149
Anagyrus kamali (parasitic wasp), 89*t*, 157
Anastrepha spp. (fruit flies), 198–199
anise swallowtail, black, 147–148
annual weeds, life cycles, 226–227, 233–234
Anthracnose (*Colletotrichum gloeosporioides*), 59, 68
ants
 overview, 30, 175–176
 gummosis relationship, 51
 identification key, 177–179
 scale relationship, 111*f*, 112, 113, 176
 species descriptions, 180–181
Anystis agilis (whirligig mite), 123, 124*f*, 194
Aonidiella spp. (scales), 29*f*, 97–105, 106*f*
Apanteles spp. (braconid wasps), 139, 142*f*, 149
aphids, 81, 83, 93–94, 109, 161–166
 pathogen vector, 68, 80–83
Aphis gossypii (cotton aphid, or melon aphid), 83, 161–166
Aphytis spp. (parasitic wasps) 89*t*
 California red scale control, 29*f*, 100–102, 103, 105, 176
 purple scale control, 107
apple moth, light brown (*Epiphyas postvittana*), 135*f*, 144–145

Archips argyrospila (fruittree leafroller), 135*f*, 139–141
Argentine ant (*Linepithema humile*), 31*f*, 111*f*, 175*f*, 176, 178*f*, 180
Argyrotaenia franciscana (orange tortrix), 131, 135*f*, 138–139
Armillaria root rot (*Armillaria mellea*), 14*t*, 43, 46–48
armored scale
 overview, 96–97
 California red scale, 97–105
 purple scale, 106–107
 yellow scale, 97–105
armyworm, beet (*Spodoptera exigua*), 131, 134*f*, 136, 149–150
Arthrobotrys spp., 206
ashy gray lady beetle (*Olla v-nigrum*), 94, 166
Asian citrus psyllid (*Diaphorina citri*), 15*f*, 166–167
Asian lady beetle, multicolored (*Harmonia axyridis*), 94, 166
assassin bugs, 93, 131
augmentation method, biological control, 29
 brown garden snail, 200
 California red scale, 100–102, 105
 cottony cushion scale, 116
 mealybug destroyer, 158
Australian lady beetle (*Rhyzobius lophanthae*), 100, 104, 107, 111
avocado thrips (*Scirtothrips perseae*), 118, 125–126

Bacillus thuringiensis (Bt), 31
bacterial blast (*Pseudomonas syringae*), 60, 66–67
Bactrocera dorsalis (oriental fruit fly), 198–199
baits
 ant control, 31, 176
 invertebrate monitoring, 87
 vertebrate management, 213–214
Banchus sp. (ichneumonid wasp), 137
banded thrips (*Aeolothrips* spp.), 117*t*, 118, 119*f*, 123
bark disorders, 55
 See also trunk diseases

basal weed control strategy, 228
bayberry whitefly (*Parabemisia myricae*), 89*t*, 158*f*, 159–161
bean thrips (*Caliothrips fasciatus*), 116–119, 128–130
bees, 11, 35
beet armyworm (*Spodoptera exigua*), 131, 134*f*, 136, 149–150
beet leafhopper (*Circulifer tenellus*), 81, 169
Bermudagrass, 226
bicolored pyramid ant (*Dorymyrmex insanus*), 179*f*
bin counts, scale monitoring, 105
bio-fix, degree-days, 19, 87
 scale monitoring, 104–105
biological control methods
 overview, 28–30, 85–86, 89*t*
 nematodes, 206–207
 pesticide cautions, 29–30, 34–35
 regional considerations, 2
 snails, 200
 species descriptions, 88, 90–95
 vertebrates, 212–213
 weeds, 230
 See also augmentation; caterpillars; insects; scale insects; thrips
birds, 172, 223
black citrus aphid (*Toxoptera aurantii*), 161–166
black hunter thrips (*Leptothrips mali*), 118, 119*f*, 123, 127
black pit (*Pseudomonas syringae*), 60, 66–67
black scale (*Saissetia oleae*), 89*t*, 112–113, 175–176
black scavenger caterpillar (*Holocera iceryaeella*), 134*f*, 147
blister beetles, 172
blue mold (*Penicillium italicum*), 58, 61, 239–240
blunt-nosed leopard lizard, 214
boron, 74, 77
Botryosphaeria ribis (Dothiorella gummosis), 51–52
Botrytis rot (*Botrytis cinerea*), 39, 61, 67
braconid wasps (*Apanteles* spp.), 139, 142*f*, 149
Brevipalpus lewisi (flat mite), 80, 182*t*, 193–194

263

broadleaf species, identifying, 234
broad mite (*Polyphagotarsonemus latus*), 182–183, 190–191
brown citrus aphid (*Toxoptera citricida*), 161–166
brown garden snail (*Cornu aspersum*), 28, 199–200
brown lacewings (*Hemerobius* spp.), overview, 90–91
See also lacewings *entries*
brown rot, 39, 53, 57–58, 239–240
brown soft scale (*Coccus hesperidum*), 89t, 110–111, 112f, 176
brush rabbits (*Sylvilagus* spp.), 218
Bt (*Bacillus thuringiensis*), 31
bucket traps, 17f
buds, development stages, 5–8
bud union disorders, 56
budwood, disease management, 41

cabbage looper (*Trichoplusia ni*), 133f, 136, 149
Cachexia, 83–84
Cales noacki (parasitic wasp), 160f
calibration guidelines, pesticide sprayers, 32–33
California ground squirrel (*Spermophilus beecheyi*), 209f, 210–214, 216–217
California harvester ant (*Pogonomyrmex californicus*), 179f
California lady beetle (*Coccinella californica*), 93
California orangedog (*Papilio zelicaon*), 132f, 147–148
California rainbow ant (*Forelius pruinosus*), 178f
California red scale (*Aonidiella aurantii*)
overview, 96–97
biology of, 97–98
damage characteristics, 100
identifying, 98–99
monitoring guidelines, 97f, 104–105, 106f
natural enemies of, 29f, 89t, 100–104
Caliothrips fasciatus (bean thrips), 116–119, 128–130
Candidatus Liberibacter spp. (Huanglongbing), 15f, 57f, 78–79, 166
Canis latrans (coyote), 221–222
cankers, 51–52, 53
Capnodium spp. (sooty molds), 61–62

Caribbean fruit fly (*Anastrepha suspensa*), 198–199
carpenter ant, 177f
caterpillars, management guidelines
overview, 130–131
apple moth, 144–145
beet armyworm, 149–150
cutworms, 136–138
identification key, 132–134
leafrollers, 139–142
loopers, 148–149
orangedogs, 147–148
orange tortrix, 138–139
scavengers, 146–147
tussock moth, 145–146
Ceratitis capitata (Mediterranean fruit fly), 198–199
chewing insects, management guidelines, 86, 170–175
Chilocorus orbus (twicestabbed lady beetle), 107, 109, 111, 112
chimeras, 62, 73f, 74
Chrysopa/Chrysoperla spp. (green lacewings), overview, 88, 90–91
See also lacewings *entries*
Circulifer tenellus (beet leafhopper), 169
Cirrospilus coachellae (eulophid wasp), 151
citricola scale (*Coccus pseudomagnoliarum*), 89t, 107–109, 110
citrophilus mealybug (*Pseudococcus fragilis*), 89t, 154–158
citrus aphids (*Toxoptera* spp.), 161–166
citrus bacterial canker (*Xanthomonas axonopodis*), 77–78
citrus blast (*Pseudomonas syringae*), 39–40, 60, 66–67
citrus bud mite (*Eriophyes sheldoni*), 2, 192
citrus canker (*Xanthomonas axonopodis*), 77–78
Citrus Clonal Protection Program, 41
citrus cutworm, 131, 134f, 136–138
Citrus exocortis viroid, 55
citrus greening (Huanglongbing), 15f, 57f, 78–79, 166
citrus leafminer (*Phyllocnistis citrella*), 150, 152–154
Citrus leprosis (*Citrus leprosis virus*), 80
citrus looper (*Anacamptodes fragilaria*), 133f, 148–149
citrus mealybug (*Planococcus citri*), 89t, 154–158, 176
citrus nematode (*Tylenchulus semipenetrans*), 23t, 201–208

citrus peelminer (*Marmara gulosa*), 150–152
citrus production, overview
regional variations, 1–4
tree development stages, 5–9
tree growth requirements, 9–11
See also integrated pest management (IPM), overview
Citrus psorosis virus, 54
citrus red mite (*Panonychus citri*), 34f, 181f, 182–186, 194, 195f
citrus rust mite (*Phyllocoptruta oleivora*), 193
citrus thrips (*Scirtothrips citri*), 34f, 116–125, 126f, 129f
Citrus tristeza virus (Tristeza), 23t, 40, 82–83, 161, 163
citrus variegated chlorosis (*Xylella fastidiosa*), 79, 167, 168
citrus whitefly (*Dialeurodes citri*), 89t, 158f, 159–161
classical biological control, overview, 29
clear rot, 61
Clementine mandarin, 11
climate, regional variations, 2
See also weather conditions
cobweb weavers, 95
Coccinella spp. (lady beetles), 93
Coccophagus spp. (parasitic wasps), 89t, 109, 111
Coccus spp. (soft scale), 107–111, 112f, 176
Colletotrichum gloeosporioides (Anthracnose), 59, 68
Comperiella bifasciata, 100, 101f, 102–103, 105
compost, 243
Comstock mealybug (*Pseudococcus comstocki*), 89t, 154–158
conibear traps, 213f
Coniopteryx spp. (dustywings), 91–92
conservation guidelines, natural enemies, 29–30
control action guidelines, overview, 19–20
convergent lady beetle (*Hippodamia convergens*), 93
Conwentzia spp. (dustywings), 91–92
Copidosoma truncatellum (parasitic wasp), 149
copper spray, cautions, 66f, 72–73
Cornu aspersum (brown garden snail), 28, 199–200
Cotesia (=*Apanteles*) *medicaginis*, 142

cotton aphid (*Aphis gossypii*), 83, 161–166
cotton cushiony scale (*Icerya purchasi*), 96, 113–116
cottontail rabbits (*Sylvilagus* spp.), 218
cover crops
    overview, 28
    vertebrate management, 212, 215, 219
    weed management, 225, 226f
cowpea aphid (*Aphis craccivora*), 163f, 164f
coyote (*Canis latrans*), 221–222
CPsV (*Citrus psorosis virus*), 54
crease disorder, 65
Cryptochaetum iceryae (parasitic fly), 114, 115
Cryptolaemus montrouzieri (mealybug destroyer), 157f, 158
CTV (Tristeza), 23t, 40, 82–83
cultivar selection, 2, 23t, 41
cultivation, weed control, 228, 229
cultural practices
    for pest control, 14t, 20–28
    nematode management, 23t, 205–206
    weed control, 227, 228, 229–230
    See also specific practices, e.g., fertilizer management; irrigation
cutworms, 131, 133–134t, 136–138
CVC (citrus variegated chlorosis), 79, 167, 168
Cyperus spp. (nutsedges), 237–238

Dactylaria spp., 206
dallisgrass, 226
damage diagnosis, 15–16, 40–41, 203–204
decollate snail (*Rumina decollata*), 200
deer mice (*Peromyscus* spp.), 219
deer (*Odocoileus hemionus*), 221
degree-days, 18–19, 101f, 104–105, 138
Delphastus spp. (lady beetles), 160f
desert valleys, citrus production overview, 1–2
devastating grasshopper (*Melanoplus devastator*), 171–172
developmental stages, pests. *See specific pests, e.g., diseases; insects; weeds*
development stages, citrus trees, 5–9
diagnosing pests, overviews, 15–16, 40–41
    See also specific pests, e.g., diseases; weeds
Dialeurodes citri (citrus whitefly), 89t, 158f, 159–161
Diaphorina citri (Asian citrus psyllid), 15f, 166–167

Diaprepes root weevil (*Diaprepes abbreviatus*), 172–173
Dipodomys spp. (kangaroo rats), 214, 220
dirty root symptom, nematodes, 203
diseases, management guidelines
    overview, 39–42, 239–243
    exotic types, 40, 57, 77–80, 167
    fruit, 57–62, 239f, 240
    habitat/growth types, 80–84
    leaf and twig, 66–68
    root rots, 42–48, 49f
    trunk, 48–55
    See also abiotic disorders
disease triangle, 39f
Dothiorella gummosis (*Botryosphaeria ribis*), 51–52
Douglas tree squirrel (*Tamiasciurus douglasii*), 217
drainage guidelines, 15f, 24, 42, 228
dry bark, 55
dry root rot (*Fusarium solani*), 44–46
Dacus =*Bactrocera cucurbitae* (melon fly), 198–199
dust control, invertebrate management, 30, 183
dustywings (*Conwentzia/Coniopteryx* spp.), 91–92, 123, 185

earwigs, 172
eastern fox squirrel (*Sciurus niger*), 217–218
eastern gray squirrel (*Sciurus carolinensis*), 217–218
Elachertus proteoteratis (eulophid wasp), 142
Empoasca fabae (potato leafhopper), 169
Encarsia pernisiosi, 100, 101f, 103, 158f
endangered species guidelines, 214
Entedononecremnus krauteri (parasitic wasp), 158f
entomopathogenic nematodes, 174
Eotetranychus spp. (mites), 188–190
Epiphyas postvittana (light brown apple moth), 135f, 144–145
Eriophyes sheldoni (citrus bud mite), 2, 192
eulophid wasps, 142, 151
European earwig (*Forficula auricularia*), 172
European starlings (*Sturnus vulgaris*), 223
Euseius spp. (predatory mites)
    overview, 94–95, 194–195
    mite control, 185, 187, 188, 189
    pesticide cautions, 34f
    thrips control, 123–124, 127
Eutetranychus banksi (Texas citrus mite), 187–188

exclusion methods
    invertebrate management, 28, 30, 176
    vertebrate management, 211t, 213, 218f, 221f
Exochus spp., 139
exocortis (*Citrus exocortis viroid*), 55
exotic natural enemies, overview, 29
exotic pests
    diseases, 40, 57, 77–80, 167
    insects, 15f, 89t, 161–167, 176, 181, 198–199
    prevention overview, 15

feeder roots
    disease damage, 41f, 43–44
    nematode damage, 203
Feltiella spp. (predaceous midges)
    overview, 196–197
    mite control, 185, 187, 189
fencing, vertebrate management, 213, 221f
fertilizer management, 10–11, 26–27, 74–77
fescues, 226
Fidiobia citri (parasitic wasp), 175
field bindweed, 227f
fire ants (*Solenopsis* spp.), 179f, 181
fire damage, 71
flat mite (*Brevipalpus lewisi*), 80, 182t, 193–194
flowers, development stages, 5–8
flower thrips, 116, 117f
Forficula auricularia (European earwig), 172
forktailed katydid (*Scudderia furcata*), 170–171
Formica aerata (native gray ant), 176, 178f, 180
fox squirrel, eastern (*Sciurus niger*), 217–218
Franklinothrips spp., 118, 119f, 123, 127
freeze/frost damage, 27, 57, 63, 70–71
frightening devices, vertebrate management, 213
frost/freeze damage, 27, 57, 63, 70–71
fruit, citrus
    development stages, 5–8
    diseases, 57–62, 239–240
    disorders, 62–66
    production growing areas, 1–3
fruit flies, 198–199
fruittree leafroller (*Archips argyrospila*), 131f, 135f, 139–141
Fuller rose beetle (*Pantomorus cervinus*), 174–175

fumigants/fumigation, 73, 207, 213–214, 233
fungi, beneficial, 10–11, 163, 165f
fungicides, 72–73
funnel weaver spiders, 95f
furrow irrigation, 25, 26f
Fusarium solani (dry root rot), 44–46

GAPs (good agricultural practices), 242–243
garden snail, brown (*Cornu aspersum*), 199–200
genetic mutations, chimera production, 62, 73f, 74
giant orangedog (*Papilio cresphontes*), 132f, 148
giant whitefly (*Aleurodicus dugesii*), 158f, 159f
glassy-winged sharpshooter (*Homalodisca vitripennis*), 79, 167–169
glossary, 259–262
glyphosate, 73, 231, 232
Gonatocerus spp. (parasitic wasps), 168
good agricultural practices (GAPs), 242–243
gophers, pocket (*Thomomys* spp.), 209f, 214f, 215, 216f
gopher snake, Pacific (*Pituophis catenifer catenifer*), 212f
grapefruit production, overview, 2–4
grasses, identifying, 234
grasshoppers, 171–172
gray ant, native (*Formica aerata*), 176, 178f, 180
gray lady beetle, ashy (*Olla v-nigrum*), 94, 166
gray mold, 61, 67
gray squirrels (*Sciurus* spp.), 217–218
greenhouse thrips (*Heliothrips haemorrhoidalis*), 116–119, 126–128, 129f
greenhouse whitefly (*Trialeurodes vaporariorum*), 159f
green lacewings (*Chrysopa/Chrysoperla* spp.), overview, 88, 90–91
See also lacewings entries
green mold (*Penicillium digitatum*), 61, 239–240
ground covers. See cover crops
ground squirrel, California (*Spermophilus beecheyi*), 32, 209f, 210–214, 216–217

growth/habitat diseases, 80–84
growth regulators, cautions, 65, 66, 72
growth requirements, overview, 9–11
gummosis diseases, 24f, 40f, 50–52, 53

habitat/growth diseases, 80–84
habitat modification, vertebrate control, 211–212
hail damage, 63, 71
Halmus chalybeus (steelblue lady beetle), 107
hares, 218
Harmonia axyridis (multicolored Asian lady beetle), 94, 166
harvest guidelines, 27, 57, 239–243
Heliothrips haemorrhoidalis (greenhouse thrips), 116–119, 126–128, 129f
Hemerobius spp. (brown lacewings), overview, 90–91
See also lacewings entries
Hemicycliophora arenaria (sheath nematode), 201, 203, 204
Hemipteran insects, 154–158
Henderson tree and branch wilt (*Nattrassia mangiferae*), 52
herbicides, 32, 73, 228, 230–233
See also pesticides
Hippodamia convergens (convergent lady beetle), 93
HLB (Huanglongbing), 15f, 57f, 78–79, 166
Holocera iceryaeella (black scavenger caterpillar), 134f, 147
Homalodisca vitripennis (glassy-winged sharpshooter), 79, 167–169
Homopteran insects, 154–158
honey ant, small (*Prenolepis imparis*), 178f
honey bees, 11, 35
honeydew, 109, 110, 166, 176
Hop stunt viroid, 83
horseweed, 232
Huanglongbing (*Candidatus* Liberibacter spp.), 15f, 57f, 78–79, 166
hydrangea mite (*Tetranychus kanzawai*), 190
Hyphoderma gummosis (*Hyphodontia sambuci*), 53
Hyposoter spp. (parasitic wasps), 146, 147f, 148, 149

Icerya purchasi (cotton cushiony scale), 96, 113–116

ichneumonid wasps, 137, 138f, 139, 146
identifying pests, overview, 15–16
See also specific pests, e.g., diseases; insect management
injury prevention, importance, 41–42
insect growth regulators, 104–105
insecticides
 aphid control, 163
 caterpillar control, 138
 mealybug control, 158
 miner control, 152, 154
 mite outbreaks, 34, 184
 scale control, 106
 thrips control, 125
 See also pesticides
insect management
 overview, 32, 85–88, 89t
 aphids, 161–166
 Asian citrus psyllid, 15f, 166–167
 chewing insects, 170–175
 leafhoppers, 167–169
 mealybugs, 154–158, 176
 miners, 150–154
 sharpshooters, 167–169
 whiteflies, 158f, 159–161, 176
 See also caterpillars; scale insects; thrips
integrated pest management (IPM), overview
 biological control methods, 28–30
 control action decision-making, 19–20
 cultural practices, 14t, 20, 22–28
 defined, 1
 identification and diagnosis, 15–16
 monitoring guidelines, 16–19, 21t
 prevention practices, 13, 15
 regional considerations, 1–4
 seasonal considerations, 20, 21–22f
 See also specific pests, e.g., diseases; insects; nematodes
Internet resources, 15, 235t, 245–246
invertebrate pests. See insect management
IPM. See integrated pest management (IPM), overview
iron, 74, 76–77
irrigation guidelines
 overview, 24–26
 disease management, 41, 42, 43
 disorder prevention, 69–70, 71
 mite control, 183
 nutrient deficiencies, 75, 76–77
 vertebrate damage, 209f, 216f, 222
 weed management, 228, 229
Iseropus stercorator orgyiae (parasitic wasp), 146

jackrabbits *(Lepus californicus)*, 218
johnsongrass *(Sorghum halepense)*, 235f, 236, 237f
jumping spiders, 95

kangaroo rat *(Dipodomys* spp.), 214, 220
kaolin treatments, 30
katydid, forktailed *(Scudderia furcata)*, 170–171

lacewings, insect control
    aphids, 163
    caterpillars, 131
    scale, 111, 112, 123
    thrips, 123, 127
lacewings, mite control, 185, 187, 188, 189
lady beetles, insect control, 89t
    mealybugs, 157f
    psyllid, 166
    species descriptions, 93–94, 109f
    whiteflies, 160f
lady beetles, scale control
    overview, 100
    armored species, 104, 107
    soft species, 109, 111, 112, 113f, 114–116
Lasioseius scapulatus (predatory mites), 181, 183, 206
leaf and twig diseases, 40f, 66–68
leaf and twig disorders, 69–77
leafhoppers, 81, 167–169
leafrollers, 131f, 135f, 139–142
lemon production, overview, 2–4
lemon sieve tube necrosis, 84
Lepidosaphes beckii (purple scale), 106–107
Leptomastix dactylopii (citrus mealybug parasite), 158
Leptothrips mali (black hunter thrips), 118, 119f, 127
Lepus californicus (jackrabbits), 218
Lewis spider mite *(Eotetranychus lewisi)*, 189–190
light brown apple moth *(Epiphyas postvittana)*, 135f, 144–145
lime treatments, 30
Linepithema humile (Argentine ant), 31f, 111f, 175f, 176, 178f, 180
liquid traps, invertebrate monitoring, 87
longtailed mealybug *(Pseudococcus longispinus)*, 89t, 154–158
loopers, 131, 133f, 136, 138, 148–149
low-volume irrigation, 25, 27f

Lysiphlebus sp. (parasitic wasp), 161f, 165f

Maconellicoccus hirsutus (pink hibiscus mealybug), 154–158
Macrosiphum euphorbiae (potato aphid), 162f, 164f
magnesium, 74, 75
management guidelines. *See* integrated pest management (IPM), overview
mandarin
    cultivars, 8t
    rind disorder, 64
    seediness, 11
manganese, 66f, 74, 76
Marietta sp., 104
Marmara gulosa (citrus peelminer), 150–152
meadow mice *(Microtus* spp.), 219
mealybugs, 89t, 154–158, 176
Mediterranean fruit fly *(Ceratitis capitata)*, 198–199
Megaphragma mymaripenne (parasitic wasp), 128
Melanoplus devastator (devastating grasshopper), 171–172
melon aphid, or cotton aphid *(Aphis gossypii)*, 83, 161–166
melon fly *(Dacus =Bactrocera cucurbitae)*, 198–199
mesophyll collapse, 71
Metaphycus spp. (parasitic wasps), 89t, 109, 110, 111, 112–113, 176
Mexican fruit fly *(Anastrepha ludens)*, 198–199
Microlarinus spp., 230
Microtus spp. (voles), 209f, 219
midges. *See Feltiella* spp. (predaceous midges)
mineral deficiencies/toxicities, 74–77
miners, 150–154
minute pirate bugs *(Orius* spp.)
    overview, 92
    aphid control, 163
    caterpillar control, 131
    mite control, 188
    thrips control, 123, 127
mites, predatory
    overview, 94–95, 194–195
    mite control, 185, 187, 188, 189
    pesticide cautions, 34f

thrips control, 123–124, 127
mites, spider
    overview, 85–88, 181–183
    citrus red mite, 183–186
    hydrangea mite, 190
    Lewis spider mite, 189–190
    sixspotted mite, 189
    Texas citrus mite, 187–188
    twospotted spider mite, 186–187
    Yuma spider mite, 188–189
mites (non-Tetranychidae)
    broad mite, 190–191
    citrus bud mite, 2, 192
    citrus rust mite, 193
    flat mite, 80, 193–194
miticides, 183
    See also pesticides
mold
    fruit decay *(Penicillium* spp.), 61, 239f
    sooty, 61–62, 108–110
moles, 216f
monitoring guidelines, overviews, 16–19, 21t
    See also specific pests, e.g., diseases; mites entries; vertebrates
Mononchus spp. (predaceous nematodes), 206
mulberry whitefly *(Tetraleurodes mori)*, 159f
mulch, weed control, 228, 230
multicolored Asian lady beetle *(Harmonia axyridis)*, 94, 166
mycorrhizal fungi, 10–11

native gray ant *(Formica aerata)*, 176, 178f, 180
Nattrassia mangiferae (Henderson tree and branch wilt), 52
natural enemies. *See* biological control methods
nematodes, 201–208
    predaceous *(Mononchus* spp.), 206
Nemorilla pyste (tachinid fly), 139
Neohydatothrips, 118, 121, 125–126
Neotoma fuscipes (riparian woodrat), 220
nesting whitefly *(Paraleyrodes minei)*, 159f
nitrogen, 74f, 75
noise devices, vertebrate management, 213
Norway rat *(Rattus norvegicus)*, 220
nuclear polyhedrosis virus (NPV), 149
nutrient disorders, 26–27, 74–77
nutsedges *(Cyperus* spp.), 227f, 237–238

Odocoileus hemionus (deer), 221
odorous tree ant (*Tapinoma sessile*), 177f
Oedaleonotus enigma (valley grasshopper), 171–172
oil spotting (oleocellosis), 65
oil sprays, 31, 72, 183
oleocellosis, 65
Oligota oviformis (rove beetle), 185, 187, 189, 197
Olla v-nigrum (ashy gray lady beetle), 94, 166
omnivorous leafroller (*Platynota stultana*), 135f, 138, 141–142
onion thrips, 118
Ophion sp. (ichneumonid wasp), 137, 138f
orange production, overview, 2–4
orange tortrix (*Argyrotaenia franciscana*), 131, 135f, 138–139
orangeworms. See caterpillars
orb weavers, 95
Orgyia vetusta (western tussock moth), 131f, 145–146
oriental fruit fly (*Bactrocera dorsalis*), 198–199
Orius spp. See minute pirate bugs (*Orius* spp.)
ozone damage, 72, 207

Pacific gopher snake (*Pituophis catenifer catenifer*), 212f
Panonychus citri (citrus red mite), 34f, 181f, 182–186, 194, 195f
Papilio spp. (orangedogs), 132f, 147–148
Parabemisia myricae (bayberry whitefly), 89t, 158f, 159–161
Paraleyrodes minei (nesting whitefly), 159f
parasitic flies, insect control
    caterpillars, 139, 140, 142, 144f, 149
    scale, 114, 115
parasitic wasps, caterpillar control
    overview, 131f
    amorbia, 143–144
    beet armyworm, 147f
    cutworms, 137, 138f
    leafrollers, 140, 142
    loopers, 149
    orangedogs, 148
    orange tortrix, 139, 140, 142, 143–144
    tussock moth, 146

parasitic wasps, insect control
    aphids, 161f, 163, 165f
    fruit flies, 199
    Fuller rose beetle, 175
    katydids, 171
    mealybugs, 89t, 157, 159
    miners, 151, 153f, 154
    psyllids, 166–167
    sharpshooters, 168
    thrips, 127–128
    whiteflies, 89t, 158f, 160
parasitic wasps, scale control, 89t
    armored species, 100–104, 107
    soft species, 109, 110, 111, 112–113, 176
pavement ant (*Tetramorium caespitum*), 179f
Penicillium spp., 58, 61, 239–240
perennial weeds, life cycles, 226–227
Peromyscus spp. (deer mice), 219
pesticides
    application guidelines, 31–33
    cautions, 29–30, 33–37, 43, 66, 72–73
    harvest guidelines, 239–241, 242
    types, 30–31
    See also herbicides; insecticides
peteca disorder, 64, 65f
pheromone traps, insect monitoring
    overview, 87, 88f
    caterpillars, 136f, 138
    miners, 151–152, 154
    scale, 97f, 104–105
photosynthesis, 9–10
Phyllocnistis citrella (citrus leafminer), 150, 152–154
Phyllocoptruta oleivora (citrus rust mite), 193
Phytophthora spp.
    fruit diseases, 39, 57–58
    nematode presence, 203
    root rot, 23t, 43–44
    trunk diseases, 24f, 42f, 50–51
phytotoxicity, 35, 43, 66, 72–73
picking guidelines, 241
Pierce's disease, 79, 167, 168
pincer traps, 215f
pink hibiscus mealybug (*Maconellicoccus hirsutus*), 154–158
pink scavenger caterpillar (*Pyroderces rileyi*), 134f, 146–147
pirate bugs. See minute pirate bugs (*Orius* spp.)
Pituophis catenifer catenifer (Pacific gopher snake), 212f

Planococcus citri (citrus mealybug), 89t, 154–158
Platynota stultana (omnivorous leafroller), 135f, 138, 141–142
Pnigalio sp., 153f, 154
pocket gophers (*Thomomys* spp.), 209f, 214f, 215, 216f
Polyphagotarsonemus latus (broad mite), 183, 190–191
postemergence herbicides, 230, 232
potassium, 76
potato aphid (*Macrosiphum euphorbiae*), 162f, 164f
potato leafhopper (*Empoasca fabae*), 169
Pratylenchus sp. (root lesion nematode), 201f
precipitation, monitoring, 17–19, 20f
    See also diseases
predatory flies, insect control
    aphids, 94, 165f, 166
    mites, 196–197
preemergent herbicides, 73, 230, 232
prevention of pests, overview, 13–15, 41
    See also specific pests, e.g., diseases; insect management
production areas, citrus, 1–3
pruning guidelines, 27–28, 42
    See also diseases
Pseudococcus spp. (mealybugs), 154–158
Pseudomonas syringae (bacterial blast), 60, 66–67
psocids, 110
psorosis (*Citrus psorosis virus*), 54
psyllid, Asian citrus (*Diaphorina citri*), 15f, 166–167
puff disorder, 65
puncturevine, 230
purple nutsedge, 237, 238
purple scale (*Lepidosaphes beckii*), 106–107
pyramid ant (*Dorymyrmex insanus*), 179f
Pyroderces rileyi (pink scavenger caterpillar), 134f

rabbits, 209, 218
rain gauges, 18f
Rattus spp., 220
recordkeeping, 17, 41, 228–229, 242
red imported fire ant (*Solenopsis invicta*), 176, 181
regional characteristics, citrus production, 1–4

resistance, pesticide, 33–34, 232–233
resurgence problems, pest, 34
Rhyzobius lophanthae (Australian lady beetle), 100, 104, 107, 111
rind disorder, 64
rind stipple, 64
ring-trap fungi, 206–207
riparian woodrat (*Neotoma fuscipes*), 220
robber flies, 172
rodent pests, 209, 214f, 215–218
Rodolia cardinalis (vedalia lady beetle), 93–94, 113f, 114–116
roof rat (*Rattus rattus*), 220–221
root asphyxiation, 69
root damage
    nematodes, 23t, 203–204
    pocket gophers, 214f
root diseases, 42–48
root lesion nematode (*Pratylenchus* sp.), 201f
root rots, 42–48, 49f
rootstock/scion selection, 2, 22, 23t, 41, 206
root weevil, Diaprepes (*Diaprepes abbreviatus*), 172–173
rose beetle, Fuller (*Pantomorus cervinus*), 174–175
Rosellinia root rot (*Rosellinia necatrix*), 43, 48, 49f
rove beetle (*Oligota oviformis*), 185, 187, 189, 197
Rumina decollata (decollate snail), 200

Sacramento Valley, citrus production overview, 1–2
safety guidelines, pesticide application, 31f, 35–36
Saissetia oleae (black scale), 89t, 112–113, 175–176
salinity levels, 74f
sampling techniques
    insects, 88, 124–125, 136f, 140, 169
    mites, 185
    nematodes, 204–205
    nutrient levels, 74
    Phytophthora species, 44
sanitation practices
    overview, 22–23
    disease management, 42
    harvest guidelines, 241–243
    insect management, 183f
    nematode management, 206
    vertebrate management, 212
    weed management, 228

San Joaquin kit fox, 214
San Joaquin Valley, citrus production overview, 1–2
Satsuma mandarin, 11
scale insects
    overview, 96–97
    armored species, 29f, 97–107
    soft species, 107–116, 176
scarring of fruit, pest identification overview, 15, 16t
scavenger caterpillars, 146–147
Schistocerca nitens (vagrant grasshopper), 171f
scion/rootstock selection, 2, 22, 23t, 41, 206
Scirtothrips citri (citrus thrips), 34f, 116–125, 126f, 129f
Sciurus spp. (gray squirrels), 217–218
Scolothrips sexmaculatus. See sixspotted thrips
Scudderia furcata (forktailed katydid), 170–171
seasonal cycles, citrus trees, 5–9
secondary pest outbreaks, 34–35
sedges, identifying, 234, 237–238
seedless cultivars, 8t, 11
Septoria spot (*Septoria citri*), 59–60, 239
sevenspotted lady beetle (*Coccinella septempunctata*), 93
shake and beat sampling, insect monitoring, 88, 136f, 169
sharpshooter, glassy-winged (*Homalodisca vitripennis*), 79, 167–169
sheath nematode (*Hemicycliophora arenaria*), 201, 203, 204
shell bark, 55
shooting, vertebrate management, 213
Simazine damage, 73f
site selection, 41, 42, 49, 243
sixspotted mite (*Eotetranychus sexmaculatus*), 189
sixspotted thrips (*Scolothrips sexmaculatus*)
    overview, 118f, 195f, 196
    mite control, 185, 187, 188, 189
    pesticide cautions, 34f
skirt pruning, 14t, 28
small honey ant (*Prenolepis imparis*), 178f
snails, 199–200
snakes, vertebrate control, 212
sodium, 74

soft scale insects
    overview, 96–97
    brown, 109–111, 112f
    citricola, 107–109
    cotton cushiony, 113–116
soil management, overview, 10–11, 24–26, 242–243
soil pH, nutrient deficiencies, 74–77
soil probes, gopher tunnels, 215
Solenopsis spp. (fire ants), 179f, 181
sooty canker, 52
sooty mold, 61–62, 108–110
Sorghum halepense (johnsongrass), 235f, 236, 237f
Southern fire ant (*Solenopsis xyloni*), 179f, 180f, 181
southern interior, citrus production overview, 1–2
Spermophilus beecheyi (California ground squirrel), 32, 209f, 210–214, 216–217
spider mite destroyer (*Stethorus picipes*), 185, 187, 188, 189, 195–196
spider mites. See mites, spider
spiders, beneficial, 95
spirea aphid (*Aphis spiraecola*), 161–166
Spiroplasma citri (stubborn disease), 39, 81–82, 169
split fruit, 65
Spodoptera exigua (beet armyworm), 131, 134f, 136, 149–150
sprayers, pesticide, 32–33
spray injury, 66, 72–73
squirrel bait stations, 214
starlings, European (*Sturnus vulgaris*), 223
steelblue lady beetle (*Halmus chalybeus*), 107
Stethorus picipes (spider mite destroyer), 185, 187, 188, 189, 195–196
sticky traps/tape, insect monitoring
    overview, 87, 88f
    bean thrips, 128, 130
    fruit fly control, 199
    miners, 151–152, 154
    scale, 104–106
    sharpshooters, 169
    See also pheromone traps
strip weed control strategy, 227
stubborn disease (*Spiroplasma citri*), 39, 81–82, 169
Sturnus vulgaris (European starlings), 223
sunburn, 56, 63, 72
Sus scrofa (wild pig), 223

swallowtails, 147–148
Sylvilagus spp. (rabbits), 218
syrphid fly, aphid control, 94, 165*f*, 166

tachinid flies, caterpillar control
    amorbia, 144*f*
    cabbage looper, 149
    leafrollers, 140, 142
    orange tortrix, 139
Tamarixia radiata (parasitic wasp), 166–167
Tedders traps, 87–88, 174
Telenomus californicus (parasitic wasp), 146
temperatures
    ant activity, 180
    citrus preferences, 11
    freeze/frost damage, 27, 57, 63, 70–71
    monitoring methods, 17, 18–19, 20*f*
    squirrel activity, 216–217
temperatures, insect development
    caterpillars, 138
    miners, 152
    scale, 98, 101*f*, 104–105, 107, 115
    thrips, 120, 127
temperatures, pest development
    degree-days, 18–19
    diseases, 58
    mites, 184, 186, 187, 195
    nematodes, 203
tephritids, 198–199
Tetraleurodes mori (mulberry whitefly), 159*f*
Tetranychus urticae (twospotted spider mite), 186–187
Texas citrus mite (*Eutetranychus banksi*), 187–188
thermometers, 18*f*
    See also temperatures
thief ant (*Solenopsis molesta*), 179*f*
Thomomys spp. (pocket gophers), 209*f*, 214*f*, 215, 216*f*
Thripobius semiluteus, 127–128
thrips
    overview, 116–119, 129*f*
    bean, 128–130
    citrus, 119–125, 126*f*
    greenhouse, 126–128
    Neohydatothrips, 125–126
    pesticide cautions, 34*f*
    See also sixspotted thrips (*Scolothrips sexmaculatus*)
Tortricid larvae, 131
total weed control strategy, 228
tower sprayers, pesticides, 32–33
Toxoptera spp. (citrus aphids), 161–166
transpiration, 10

trap for monitoring, 87–88
    See also pheromone traps; sticky traps/tape
trapping vertebrates, 213, 215, 218*f*
tree guards, 15*f*, 172, 176, 181, 211*t*, 213, 218*f*
trees, citrus
    development stages, 5–9
    growth requirements, 9–11
tree squirrels, 217–218
Trialeurodes spp. (whiteflies), 159*f*
Trichogramma spp. (parasitic wasps), 131*f*, 137, 140, 142, 143–144, 149
    overview, 131*f*
    amorbia control, 143–144
    cabbage looper control, 149
    cutworm control, 137
    leafroller control, 140, 142
Trichoplusia ni (cabbage looper), 133*f*, 136, 149
Tristeza (*Citrus tristeza virus*), 23*t*, 40, 82–83
Trogoderma sternale (predatory beetle), 146
trunk diseases, 48–55
trunk disorders, 55–57
tussock moth, western (*Orgyia vetusta*), 145–146
twicestabbed lady beetle (*Chilocorus orbus*), 107, 109, 111, 112
twig and leaf disorders, 69–77
twig dieback, 70
twospotted spider mite (*Tetranychus urticae*), 186–187
Tylenchulus semipenetrans (citrus nematode), 23*t*, 201–208

vagrant grasshopper (*Schistocerca nitens*), 171*f*
valley grasshopper (*Oedaleonotus enigma*), 171–172
variegated cutworm, 133*f*, 136, 137
vedalia lady beetle (*Rodolia cardinalis*), 93–94, 113*f*, 114–116
vein enation, 68
velvety tree ant, 177*f*
vertebrates, management guidelines
    overview, 32, 209–214, 243
    coyote, 221–222
    deer, 221
    ground squirrel, 210–214, 217*f*
    rabbits, 218
    rodents, 214*f*, 215–221
    starlings, 223
    wild pig, 223
vine mealybug, 155*f*

visual search method, overview, 87
    See also insects
voles (*Microtus* spp.), 209*f*, 219

wasp predators. *See* parasitic wasps *entries*
water management. *See* irrigation guidelines
water quality, pesticide hazards, 36, 207, 233
weather conditions
    monitoring, 17–19, 20*f*, 125
    regional variations, 2
    See also abiotic disorders; diseases; temperatures *entries*
weeds, management guidelines
    overview, 23, 225–228
    control methods, 32, 226*f*, 229–233
    gopher control, 215
    identifying, 233–238
    monitoring, 228–229
weevils, 173–175, 230
western flower thrips, 120–121, 122*f*, 129*f*
western gray squirrel (*Sciurus griseus*), 217
western tussock moth (*Orgyia vetusta*), 131*f*, 145–146
whirligig mite (*Anystis agilis*), 123, 124*f*, 194
whiteflies, 89*t*, 158*f*, 159–161, 176
wildlife, pesticide hazards, 36
wild pig (*Sus scrofa*), 223
wind injury, 62–63, 71
wolf spiders, 95
wood decays, 53–54
woody gall, 49*f*, 68
woolly whitefly (*Aleurothrixus floccosus*), 89*t*, 158*f*, 159–161, 176
World Wide Web sites, 15, 235*t*, 245–246

Xanthomonas axonopodis (citrus bacterial canker), 77–78
Xylella fastidiosa (citrus variegated chlorosis), 79, 167, 168

yellow nutsedge, 227*f*, 237–238
yellow scale (*Aonidiella citrina*)
    overview, 96–97
    biology of, 97–98
    damage characteristics, 100
    identifying, 98–99
    monitoring guidelines, 97*f*, 104–105
    natural enemies, 89*t*, 100–104
Yuma spider mite (*Eotetranychus yumensis*), 188–189

Zelus spp., 93*f*
zinc, 66*f*, 74, 75